# Environment and Plant Metabolism

*flexibility and acclimation*

ENVIRONMENTAL PLANT BIOLOGY series

Editor: W.J. Davies
*Institute of Environmental and Biological Sciences, Division of Biological Sciences, University of Lancaster, Lancaster LA1 4YQ, UK*

---

Abscisic Acid: physiology and biochemistry

Carbon Partitioning: within and between organisms

Pests and Pathogens: plant responses to foliar attack

Water Deficits: plant responses from cell to community

Photoinhibition of Photosynthesis: from molecular mechanisms to the field

Environment and Plant Metabolism: flexibility and acclimation

*Forthcoming titles include:*

Embryogenesis: the generation of a plant

N. SMIRNOFF
*Department of Biological Sciences, University of Exeter, Exeter EX4 4PS, UK*

# Environment and Plant Metabolism

## *flexibility and acclimation*

First published 1995



A CIP catalogue record for this book is available from the British Library.

ISBN 1 872748 93 7

**BIOS Scientific Publishers Ltd**
**St Thomas House, Becket Street, Oxford OX1 1SJ, UK**
**Tel. +44 (0)1865 726286. Fax +44 (0)1865 246823**

DISTRIBUTORS

*Australia and New Zealand*
DA Information Services
648 Whitehorse Road, Mitcham
Victoria 3132

*India*
Viva Books Private Limited
4346/4C Ansari Road
New Delhi 110002

*Singapore and South East Asia*
Toppan Company (S) PTE Ltd
38 Liu Fang Road, Jurong
Singapore 2262

*USA and Canada*
Books International Inc
PO Box 605, Herndon, VA 22070

Typeset by Herb Bowes Graphics, Oxford, UK
Printed and bound in Great Britain by
Biddles Ltd, Guildford and King's Lynn

# Contents

# Contributors

**Bornman, J.F.** Department of Plant Physiology, University of Lund, Box 7007, S-220 07 Lund, Sweden

**Bressan, R.A.** Department of Horticulture, Purdue University, West Lafayette, IN 47907-1165, USA

**Burke, J.J.** USDA-ARS Cropping Systems Research Laboratory, Lubbock, TX 79401, USA

**Burnet, M.** Institut de Recherche en Biologie Végétale, 4101 Rue Sherbrooke Est, Montréal, Québec, Canada H1X 2B2

**Carter, P.J.** Departments of Biochemistry and Botany, University of Glasgow, Glasgow G12 8QQ, UK

**Csonka L.N.** Department of Biology, Purdue University, West Lafayette, IN 47907-1165, USA

**Ferrario, S.** Laboratoire du Metabolisme, INRA, route de St. Cyr, 78026 Versailles Cedex, France

**Fewson, C.A.** Department of Biochemistry, University of Glasgow, Glasgow G12 8QQ, UK

**Foyer, C.H.** Laboratoire du Metabolisme, INRA, route de St. Cyr, 78026 Versailles Cedex, France

**García-Ríos, M.G.** Department of Biology, Purdue University, West Lafayette, IN 47907-1165, USA

**Hanson, A.D.** Institut de Recherche en Biologie Végétale, 4101 Rue Sherbrooke Est, Montréal, Québec, Canada H1X 2B2. *Present address*: Horticultural Sciences Department, University of Florida, PO Box 110690, Gainsville, FL 32611-0690, USA

**Lawlor, D.W.** AFRC Institute of Arable Crops Research, Rothamsted Experimental Station, Harpenden, Herts AL5 2LQ, UK

**Leegood, R.C.** Robert Hill Institute and Department of Animal and Plant Sciences, University of Sheffield, Sheffield S10 2UQ, UK

**Nelson, J.P.S.** Department of Biochemistry, University of Glasgow, Glasgow G12 8QQ, UK

**Nimmo, H.G.** Department of Biochemistry, University of Glasgow, Glasgow G12 8QQ, UK

**Nimmo, G.A.** Departments of Biochemistry and Botany, University of Glasgow, Glasgow G12 8QQ, UK

**Paino D'Urzo, M.** Department of Horticulture, Purdue University, West Lafayette, IN 47907-1165, USA

**Plaxton, W.C.** Departments of Biology and Biochemistry, Queen's University, Kingston, Ontario, Canada K7L 3N6

**Popp, M.** Institute of Applied Botany, Westfalische Wilhelsm-Universität, Hindenburgplatz 55, 48143 Münster, Germany

**Ratcliffe, R.G.** Department of Plant Sciences, University of Oxford, South Parks Road, Oxford OX1 3RB, UK

**Rathinasabapathi, B.** Institut de Recherche en Biologie Végétale, 4101 Rue Sherbrooke Est, Montréal, Québec, Canada H1X 2B2

**Rhodes, D.** Department of Horticulture, Purdue University, West Lafayette, IN 47907-1165, USA

**Rivoal, J.** Institut de Recherche en Biologie Végétale, 4101 Rue Sherbrooke Est, Montréal, Québec, Canada H1X 2B2

**Samaras, Y.** Department of Horticulture, Purdue University, West Lafayette, IN 47907-1165, USA

**Smirnoff, N.** Department of Biological Sciences, University of Exeter, Hatherly Laboratories, Prince of Wales Road, Exeter EX4 4PS, UK

**Sundby-Emanuelsson, C.** Department of Plant Physiology, University of Lund, Box 7007, S-220 07 Lund, Sweden

**Theodorou, M.E.** Departments of Biology and Biochemistry, Queen's University, Kingston, Ontario, Canada K7L 3N6

**Valadier, M.H.** Laboratoire du Metabolisme, INRA, route de St. Cyr, 78026 Versailles Cedex, France

**Wilkins, M.B.** Department of Botany, University of Glasgow, Glasgow G12 8QQ, UK

# Abbreviations

| | |
|---|---|
| *A* | $CO_2$ assimilation rate |
| ABA | abscisic acid |
| ADH | alcohol dehydrogenase |
| a.m.u. | atomic mass units |
| AP | ascorbate peroxidase |
| BADH | betaine aldehyde dehydrogenase |
| $C_a$ | external $CO_2$ concentration |
| $C_i$ | intercellular $CO_2$ concentration |
| CAM | Crassulacean acid metabolism |
| CaMV | cauliflower mosaic virus |
| CF | chloroplast coupling factor |
| CMO | choline monooxygenase |
| CST | choline sulphotransferase |
| DABCO | 1,4-diazobicyclooctane |
| DEAE | diethylaminoethyl |
| DHA | dehydroascorbate |
| DHAR | dehydroascorbate reductase |
| DMPO | dimethylproline-*N*-oxide |
| DMSO | dimethylsulphoxide |
| DMSP | 3-dimethylsulphoniopropionate |
| DTT | dithiothreitol |
| EDTA | ethylenediamine tetraacetic acid |
| EPR | electron paramagnetic resonance |
| FBPase | fructose-1,6-bisphosphatase |
| *F*m | maximum fluorescence |
| *F*o | initial fluorescence |
| Fru-2,6-$P_2$ | fructose-2,6-bisphosphate |
| *F*s | steady-state fluorescence |
| *F*v | variable fluorescence |
| $g_s$ | stomatal conductance |
| G3PDH | glyceraldehyde-3-phosphate dehydrogenase |
| GABA | γ-aminobutyric acid |
| GC–MS | gas chromatography–mass spectroscopy |
| GDC | glutamate decarboxylase |
| GDH | glutamate dehydrogenase |
| GK | γ-glutamyl kinase |
| Gln | glutamine |
| GOGAT | glutamine synthase |
| GPR | γ-glutamyl phosphate reductase |
| GR | glutathione reductase |
| GS | glutamine synthetase |
| GSA | glutamic-γ-semialdehyde |

| | |
|---|---|
| GSH | γ-L-glutamyl-L-cysteinyl-glycine (glutathione) |
| GSSG | oxidized glutathione |
| HFBI | heptafluorobutyryl isobutyl |
| HL | high light |
| HPR | hydroxypyruvate reductase |
| LDH | lactate dehydrogenase |
| LL | low light |
| MDH | malate dehydrogenase |
| MDHA | monodehydroascorbate |
| MDHAR | monodehydroascorbate reductase |
| MPF | M-phase-promoting factor |
| MTP | methylpropionate |
| NMR | nuclear magnetic resonance |
| NR | nitrate reductase |
| NRD | non-radiative dissipation |
| OMMI | D-1-*O*-methyl-*muco*-inositol |
| OMSI | *O*-methyl-*scyllo*-inositol |
| P2C | $\Delta^1$-pyrroline-2-carboxylate |
| P5C | $\Delta^1$-pyrroline-5-carboxylate |
| P5CR | $\Delta^1$-pyrroline-5-carboxylate reductase |
| P5CS | $\Delta^1$-pyrroline-5-carboxylate synthetase |
| PAGE | polyacrylamide gel electrophoresis |
| PAM | pulse-amplified modulation |
| PAR | photosynthetically active radiation |
| PDC | pyruvate decarboxylase |
| PEP | phosphoenolpyruvate |
| PEPC | phosphoenolpyruvate carboxylase |
| PFD | photon flux density |
| PFK | ATP:fructose-6-phosphate 1-phosphotransferase |
| PFP | pyrophosphate:fructose-6-phosphate 1-phosphotransferase |
| PGA | 3-phosphoglycerate |
| Pi | inorganic phosphate |
| PK | pyruvate kinase |
| PPi | inorganic pyrophosphate |
| PRK | ribulose-5-phosphate kinase |
| PSI | photosystem I |
| PSII | photosystem II |
| $Q_A$ | primary quinone electron acceptor in photosystem II reaction centre |
| $Q_B$ | secondary quinone electron acceptor in photosystem II reaction centre |
| QAC | quaternary ammonium compound |
| $q_E$ | energy-related quenching process |
| $q_{NP}$ | non-photochemical quenching process |
| $q_P$ | photochemical quenching process |
| Rubisco | ribulose-1,5-bisphosphate carboxylase-oxygenase |
| RuBP | ribulose bisphosphate |
| RWC | relative water content |
| SAM | *S*-adenosylmethionine |
| SDS | sodium dodecylsulphate |
| SMM | *S*-methylmethionine |

| | |
|---|---|
| SOD | superoxide dismutase |
| SPS | sucrose phosphate synthase |
| $t_{1/2}$ | half-time/half-life |
| TCA | tricarboxylic acid |
| TKW | thermal kinetic window |
| TP | triose phosphate |
| TSC | tertiary sulphonium compound |
| UL | ultraweak luminescence |
| UV | ultraviolet |
| $UV_{BE}$ | biologically active ultraviolet radiation |

# Preface

This book focuses on how the flexibility of metabolism allows plants to deal with their constantly fluctuating environments. Because this is a potentially vast subject area, consideration is limited to pathways closely linked to primary metabolism and to the abiotic environmental factors which have major effects on plant growth (i.e. temperature, water, light, UV radiation and mineral nutrient and oxygen supply). Much research over the last decade has focused on the molecular and cell biology of plant response to extreme environments, and this has revealed the ability of plants to respond rapidly by modification of gene expression and the formation of proteins which might maintain structural integrity and function under extreme conditions. Rather less attention has been paid to primary metabolism. The pathways of primary metabolism are concerned mainly with assimilation of resources (carbon, nitrogen and sulphur) and energy production. These resources, as well as the physical factors which could limit metabolism, such as temperature, light and water and oxygen supply, can fluctuate rapidly. Plants must co-ordinate their metabolic activity with these environmental factors. The topics have been chosen to illustrate what is known about how metabolic pathways interact with the environment. Chapter 2 considers the interaction between carbon and nitrogen assimilation. Chapter 3 considers the control of malate synthesis in Crassulacean acid metabolism (CAM) plants and guard cells in relation to light and temperature. Chapters 3–5 deal with the effect of temperature on metabolic rates and enzyme properties. Chapters 6 and 7 deal with the responses of respiration to shortages of phosphate and oxygen, both of which potentially limit energy production. Chapters 8–11 focus on responses to water deficit, including a consideration of the mechanisms which limit photosynthesis during water deficit (Chapter 8). Chapters 9–11 then review the pathways and control of compatible solute (osmolyte) accumulation in droughted plants. Finally, the role of antioxidant defences in plant response to the environment (Chapter 12) and the effect of UV-B radiation on (Chapter 13) photosynthesis are considered. Some recurrent themes are the role of enzyme phosphorylation/dephosphorylation in co-ordinating metabolism with environmental fluctuations (Chapters 2 and 3), the prospect of metabolic engineering (the use of transgenic plants and mutants to alter metabolism) to alter responses to the environment (Chapters 1, 2, 3, 4, 5, 7, 10, 11 and 12) and the activation of alternative metabolic pathways by changes in environmental conditions (Chapters 6, 7, 9, 10 and 11).

The reader will notice that some major topics are not covered because they have been the subject of recent books or reviews. These include large areas of photosynthesis (e.g. acclimation to different light climates, photoinhibition and responses to elevated carbon dioxide concentration). Also, there is no detailed coverage of $C_4$ photosynthesis and CAM in relation to the environment, although it is through investigation of these two forms of photosynthesis that we have the best understanding of how variations in metabolic pathways might aid plant growth in particular (hot and dry) environments. The contents are also influenced by the usual constraints of the bias of the Editor and the availability of authors. It is hoped that the book will be of interest to research workers in plant biochemistry, physiology and ecophysiology, and those concerned with crop improvement. It should also be useful for advanced undergraduates who have a background knowledge of plant biochemistry.

The book is based on sessions organised by the Plant Metabolism Group of the Society for Experimental Biology at the Annual Meeting held at the University of Wales (Swansea, UK) in April 1994. I am most grateful to all the contributors for their enthusiastic response to the project. I am grateful to Steve Rawsthorne, convenor of the Plant Metabolism Group, for support and encouragement and to Bill Davies for suggesting publication in the Environmental Plant Biology Series.

N. Smirnoff (*Exeter*)

# 1

# Metabolic flexibility in relation to the environment

N. Smirnoff

## 1.1 Introduction

Because plants are rooted in one place, they have a limited capacity to avoid unfavourable changes in their environment such as extremes of temperature, water shortage, insufficient or excessive light and shortage of mineral nutrients. However, growth and developmental patterns can be altered to deal with such problems. For example, increased allocation of biomass to roots occurs in dry and mineral nutrient-deficient conditions, while leaf area and thickness are altered according to light conditions. Plant stature, leaf angle, size, hairiness and waxiness influence light absorption and heat balance and thus affect leaf temperature and transpiration rate. Roots growing into waterlogged soil can produce large intercellular gas channels (aerenchyma) which improve aeration of the root and aid aerobic respiration (Crawford, 1992). Such changes involve alterations to development and source–sink relationships (Pollock *et al.*, 1992) and show that plants have a great deal of plasticity which is influenced by the environment. Many of the developmental responses to the environment are mediated by plant hormones and this could be one of their major roles (Trewavas, 1986). There can be little doubt that all these features are of importance, and perhaps of overriding importance, in the survival of plants in extreme environments. The intrinsically slow growth rate of plants native to extreme environments (Grime, 1979) is partly an inevitable consequence of such features. Fast-growing crop plants, on the other hand, are less well able to survive extremes. Nevertheless, there is interest in the possibility that certain biochemical and metabolic features of plants could contribute to their growth and survival under suboptimal conditions. These biochemical traits can, with the present state of knowledge, be manipulated more easily by recombinant DNA techniques than more complex processes such as development. A better understanding of how metabolism responds to the environment and the extent to which changes in the flux through pathways and the production of particular end-products improve performance is needed before the best use can be made of the increasing number of gene clones which are becoming available.

## 1.2 Variability of primary metabolism

Biochemical diversity in plants is astonishingly high when secondary metabolites (i.e. those not included in the definition of primary metabolism below) are considered. These, for the most part, appear to have important functions in plant defence against herbivores and pathogens, as attractants for pollinators and seed dispersers (Harborne, 1982) and perhaps, in the case of flavonoids, as UV screens (Stapleton and Walbot, 1994; Chapter 13). The subject matter of this book is mostly confined to primary metabolism. This is the network of metabolic pathways concerned with energy production and utilization, the assimilation of mineral nutrients such as nitrogen and sulphur into organic form, the formation of organic osmolytes (molecules used for generating turgor) and formation of precursors for the synthesis of macromolecules. These pathways are considered to be common to all plants. Despite the basic similarity between plants, and indeed other types of organisms, there is a certain degree of variability in the types of pathway, their relative activity and in the major end-products of carbon and nitrogen assimilation in different species and under different conditions. It has often been suggested that such variation is related to the performance of plants in particular types of environments, or that activation of different pathways provides flexibility in relation to the environment (Crawford, 1971). These themes have been explored, with emphasis on animals, in the book by Hochachka and Somero (1984) and this is well worth consulting to compare the biochemical and metabolic features which plants and animals use to deal with temperature variability, water deficit and oxygen shortage.

When considering metabolism in relation to the environment, two different types of question can be raised and aspects of both of these are considered in this book.

(i) Does variation in metabolic pathways and their end-products influence growth and survival in particular kinds of environment? Some examples of this kind of variation are listed in Table 1.1. The role of phylogenetic constraints in variation should not be ignored. For example, molecules like fructans, polyols, quaternary ammonium compounds (betaines) and tertiary sulphonium compounds show strong associations with particular genera or families. Diversity could exist because the same function can be fulfilled by different compounds. For example, starch and fructan may be equally effective as carbon storage compounds, while sucrose and polyols such as mannitol and sorbitol could be equally effective as carbon transport compounds in the phloem.

(ii) How do energy production, assimilation of resources and osmolyte accumulation co-ordinate with fluctuations in physical conditions and resource availability? This question is dealt with in Section 1.3.

***Table 1.1.*** *Biochemical and metabolic variability and flexibility in higher plants and suggested roles in adaptation and acclimation to the environment. The examples selected here have been particularly investigated from the perspective of plant–environment interactions*

| | |
|---|---|
| *Photosynthesis* | |
| $C_4$ photosynthesis | Reduced photorespiration leading to a high temperature optimum for photosynthesis. High water use efficiency[a]. |
| $C_3/C_4$ intermediates | Similar consequences to the above. Also low photorespiration form of aquatics aids $CO_2$ concentration[a,c]. |
| Crassulacean acid metabolism (CAM) | Nocturnal $CO_2$ assimilation increases water use efficiency. Photoprotection afforded by high intercellular $CO_2$ concentration during the day when stomata closed. Aquatic CAM: $CO_2$ concentration from water (Chapter 3)[b,c]. |
| $C_3$/CAM intermediates and inducible CAM | CAM expressed in response to drought/salinity. Similar consequences to above but can use higher capacity $C_3$ photosynthesis under favourable conditions[b]. |
| Bicarbonate uptake by aquatics | Concentrating $CO_2$ from water[c]. |
| Xanthophyll cycle pigments | Pool size increased by high light. Proportion of zeaxanthin increased by high light. Dissipation of excess excitation energy (Chapters 8 and 12)[d]. |
| *Respiration* | |
| Alternative glycolytic enzymes | Bypass phosphate (Pi) and adenylate-requiring steps to allow continued glycolysis during Pi deficiency. Induced by Pi deficiency (Chapter 6). |
| Alternative oxidase | Non-phosphorylating terminal oxidase in mitochondria. Induction or increased activity under environmental control. Acts as energy overflow or adenylate bypass? Thermogenesis in Aroids to attract pollinators (Chapter 6)[e]. |
| Fermentation | Allows energy production during anoxia. Ethanol or lactate as alternative end-products. Pathways activated by hypoxia and anoxia (Chapter 7). |

***Table 1.1.*** *Continued.*

| | |
|---|---|
| 'Dark' $CO_2$ fixation | Catalysed by PEP carboxylase and activated during nitrogen assimilation, cation uptake and pH regulation (Chapter 2). |
| *Carbon storage and translocation compounds* | |
| Fructans | Accumulated by certain species. Suggested role in cold tolerance[f]. |
| Polyols (acyclic and cyclic) | Accumulated by certain species. Suggested role as osmolytes (Chapter 11). |
| Sucrose and oligosaccharides | Desiccation-resistant cells (seeds, pollen, bryophytes) contain high sucrose concentration, some raffinose and stachyose and very low reducing sugar. Protects cytoplasm and prevents phase separation in desiccated membranes[g]. |
| *Nitrogen and sulphur assimilation, translocation and storage* | |
| Nitrate or ammonium preference | Plants from acidic soils may be less sensitive to ammonium toxicity and have a low capacity for nitrate assimilation because of low nitrate reductase activity[h]. |
| Leaf vs. root nitrate assimilation | Leaf assimilation may be less energetically costly because it is linked to photosynthesis[i]. |
| Ureides vs. amides as products of nitrogen fixation in legumes | Ureides (allantoin/allantoic acid) more efficient than amides (asparagine/glutamine) because of higher C/N ratio, but may be confined to tropical species because of low solubility[j]. |
| Proline accumulation | Induced by drought and cold in a wide range of species. May act as a compatible osmolyte (Chapter 9). |
| Accumulation of quaternary ammonium compounds | Induced by drought and salinity. Restricted to certain families and particularly associated with halophytes. May act as compatible osmolytes (Chapter 10). |
| Tertiary sulphonium compounds | As above (Chapter 10). |
| Glutathione | Cysteine-containing tripeptide acts as antioxidant and sulphur store. Homologues with difference amino acid composition occur in some species (Chapter 12). |

References for topics not covered in the book are not meant to be exhaustive.
[a]Pearcy and Ehleringer (1984), Schuster and Monson (1990); [b]Griffiths (1988); [c]Bowes (1987); [d] Demmig-Adams and Adams (1992); [e] McIntosh (1994); [f] Pollock (1986), Hendry (1987); [g]Koster and Leopold (1988), Bianchi *et al.* (1991); [h]Raven *et al.* (1992); [i]Stewart *et al.* (1992); [j]Sprent (1980).

## 1.3 Integration of metabolism with the environment

The environment around plants fluctuates regularly and predictably over daily and seasonal cycles. Highly unpredictable fluctuations are imposed on this regular pattern because of changes in the weather and the availability of nutrients. Many of the environmental factors which fluctuate are associated intimately with metabolic processes. Variation in light, which supplies energy for photosynthesis, has immediate effects on metabolism (Pearcy, 1990), while water deficit decreases stomatal conductance and therefore limits carbon dioxide supply and changes the balance between photosynthesis and photorespiration (Chapter 8). The source (nitrate vs. ammonium) and concentration of nitrogen influences the location (root or shoot) and rate of nitrogen assimilation into amino acids and therefore requires a dynamic balance between photosynthesis, carbon partitioning and nitrogen assimilation (Chapter 2). Phosphate (Pi) supply, which is critical in energy transfer, can influence respiratory pathways and deficiency induces pathways which are less dependent on adenylates and inorganic phosphate (Chapter 6; Table 1.1). Severe phosphate limitation in leaves causes starch accumulation in chloroplasts by limiting the export of triose phosphate from the chloroplasts via the envelope Pi translocator (Sonnewald *et al.*, 1994). Oxygen supply fluctuates in root tissues as a result of waterlogging after heavy rain (Crawford, 1992). This leads to inhibition of aerobic respiration under hypoxia and the induction of fermentation pathways as a means of energy production and NAD regeneration. Activation of ethanolic or lactate fermentation depends on the regulatory properties of the enzymes involved (Chapter 7) as well as on longer term induction of enzymes involved in sucrose mobilization, glycolysis and fermentation (Kennedy *et al.*, 1992). Further to these variations in resource supply, temperature fluctuation influences metabolic rate and has differential effects on enzymes and pathways. Examples of this effect are seen in carbohydrate metabolism, photosynthesis and photorespiration (Chapter 4). It should therefore be obvious to the reader that while all these pathways must work in balance and form an integrated metabolic network under steady-state conditions (as might be found in a controlled laboratory environment), they must also remain co-ordinated in natural environments where all the above factors fluctuate rapidly. Investigation of responses to fluctuating conditions should therefore be an integral part of developing a full understanding of the control of plant metabolism. Many of the control mechanisms of photosynthesis, for example the activation state of enzymes such as ribulose-1,5-bisphosphate carboxylase-oxygenase (Rubisco), fructose-1,6-bisphosphatase and sucrose phosphate synthase, which are influenced by changes in the concentration of effector molecules and by environmentally induced covalent modifications, reflect the necessity of operating with a continually variable supply of light and carbon dioxide. These 'fine' control mechanisms deal with short-term fluctuations. Longer term changes in average conditions lead to adjustment of the relative amounts and concentrations of different components of the photosynthetic apparatus as a result of synthesis and turnover. Similar adjustments to short- and long-term changes in environment are seen in other pathways. Detailed consideration of photosynthesis in relation to the environment is beyond the scope of this book and more information can be found in Yamamoto and Smith (1993) and Baker and Bowyer (1994).

Metabolic processes, absorption of photosynthetically active light and UV radiation produce free radicals and reactive forms of oxygen (e.g. superoxide, hydrogen peroxide, hydroxyl radicals and singlet oxygen). These cause damage, and all organisms have antioxidant systems to deal with them. Active oxygen formation is increased by exposure to high light intensity, water deficit, UV radiation and emergence from anoxia and hypoxia. Antioxidant systems are engaged in these situations and help to maintain metabolic and structural integrity. The antioxidant and photoprotective systems are now being investigated intensively (Chapters 8, 12 and 13) and are an important aspect of the interaction of metabolism with the environment.

## 1.4 Cell biology and metabolism in relation to water deficit and temperature

Responses to temperature and water deficit are treated extensively in the book. Discounting pathogens and nutrient deficiency, these two factors are major limits to crop productivity on a global scale. This section provides a very brief overview of plant response to these environmental factors and attempts to give an integrated view of the relationship between growth, cell biology and the metabolic aspects covered in other chapters of this book.

### 1.4.1 *Water deficit*

The response of plants to water deficit and salinity has been investigated intensively, particularly with respect to improving crop yield under water-limited conditions, and a recent review of the physiological aspects is given in Smith and Griffiths (1993). Despite the interest in the physiological characteristics contributing to drought resistance, it has been argued that crop varieties selected to yield high in favourable conditions also yield relatively high in dry conditions and that the only traits which have been important in breeding for drought resistance are phenological and morphological (Richards, 1993). This view has not detracted from the large research effort which is being devoted to investigations of biochemical and molecular responses to water deficit. The present position is reviewed below.

Taking a simple view, the reduction of growth caused by water deficit can be ascribed to two causes. Firstly, inhibition of growth and leaf expansion results from low turgor and water availability. Secondly, limitation of photosynthesis by stomatal closure prevents biomass accumulation (Chapter 8). This would suggest an essentially passive response to water deficit. However, water deficit also has direct and indirect effects on metabolism which allows these responses to be modified. Cell expansion can be maintained despite loss of turgor in some tissues, for example in cells just behind the apical meristem of maize roots, while in other tissues cell expansion may be inhibited despite maintenance of high turgor (Sharp *et al.*, 1993). The hormone abscisic acid (ABA) has been implicated in these responses. Its synthesis and accumulation are triggered by water deficit (Davies and Jones, 1991). It might be involved in the alteration of cell wall extensibility which occurs during water deficit. Root cell walls may become more extensible, while leaf cell walls become less extensible which prevents leaf expansion. Low water potential alters the composition of cell wall proteins (Bozarth *et al.*, 1987) and perhaps of enzymes involved in cell wall loosening during expansion. These differential responses could alter growth and source–sink relationships, leading to changed growth patterns during drought (e.g. relatively more root growth).

Since a review by Sachs and Ho (1986) suggested that few if any water deficit-induced proteins had been found in plants, the situation has been reversed and large numbers of such proteins have been discovered, many of unknown function (Bray, 1993). Many, but not all, of these proteins are induced by exogenous ABA, and are not formed in ABA-deficient mutants. The genes have a consensus sequence in their promoter regions which confers ABA responsiveness. Two classes of particular interest are the dehydrins, which are included in the LEA and Em groups of proteins which accumulate in embryos during seed desiccation (Bray, 1993; Close *et al.*, 1993), and the aquaporins (Chrispeels and Maurel, 1994). The first group are accumulated in seeds during maturation desiccation and are also accumulated in vegetative tissues during drought. Their accumulation is probably mediated by ABA. Their function is not clear although, being small and hydrophilic, they may interact with other proteins and stabilize them at low cytoplasmic water content. Alternatively, they could sequester ions which would otherwise be damaging when water content is low (Close *et al.*, 1993). The aquaporins appear to act as membrane water channels and a drought-induced increase in their abundance might increase the hydraulic conductivity of cells and therefore influence the ease and direction of water flow (Chrispeels and Maurel, 1994). These responses can be seen as cell biological: stabilization of cell structure and function. They involve the induction of ABA synthesis (via alteration of gene expression) and the effect of ABA, or of water deficit directly, on gene expression.

Metabolism is influenced by water deficit. Some of these effects are secondary, being a consequence of growth reduction and stomatal closure. For example, lowered sink demand could cause feedback inhibition of photosynthesis. Similarly, the partitioning of assimilate between starch and sucrose and nitrate assimilation are influenced by stomatal closure. These responses are mediated by alteration of the phosphorylation state of sucrose phosphate synthase and nitrate reductase (Chapters 2 and 8). There are also responses which are probably more directly mediated by plant water status. The most obvious of these are the accumulation of osmolytes such as proline (Chapter 9), betaines, tertiary sulphonium compounds (Chapter 10) and polyols (Chapter 11). Proline accumulation is an almost universal response to water deficit, occurring in a wide range of organisms from bacteria to animals, while the ability to accumulate the other groups of compounds is more restricted to certain species. The value of osmolyte accumulation for higher plants has not been demonstrated conclusively but roles as compatible (i.e. not harmful to metabolism) solutes in the cytoplasm to allow osmotic adjustment and turgor maintenance have been suggested (Chapters 9 and 11). Additional roles as stabilizers of macromolecules and membranes during desiccation and temperature extremes have also been put forward (Chapters 9 and 11). In bacteria, the uptake of exogenous proline and betaine is stimulated by water deficit, and accumulation of these compounds improves growth at low water potential. Thus, in bacteria, the osmoprotective effect of these solutes is very clear (Csonka and Hanson, 1991). Proline and betaine accumulation involve increased rates of synthesis which result from increased capacities of some of the biosynthetic enzymes and increased levels of their mRNAs (Chapters 9–11). As understanding of these pathways increases, the possibility of 'metabolic engineering' to alter accumulation will enable their function to be tested (see Section 1.5).

The intensively studied inducible Crassulacean acid metabolism (CAM) plant *Mesembryanthemum crystallinum* is an excellent example of altered metabolism during water deficit and salinity (Vernon *et al.*, 1993). In this annual facultative halophyte, the leaves formed later in development exhibit CAM. The switch to CAM is greatly accelerated by exposure to salinity and involves up-regulation of CAM and glycolysis-related enzymes such as phosphoenolpyruvate (PEP) carboxylase (Chapter 3), malic enzyme, phosphorylase and amylase. This increases the rate of nocturnal malic acid accumulation and day time decarboxylation. *M. crystallinum* also accumulates both proline and the cyclitol pinitol during salinity (Chapter 11). The latter involves the induction of a methyltransferase enzyme. Accumulation of these osmolytes, unlike CAM induction, is independent of developmental stage (Vernon *et al.*, 1993).

Water deficit has further implications for leaves because inhibition of photosynthesis will cause leaves to be exposed to much more light energy than can be used for photosynthesis. This could potentially cause damage in the form of photoinhibition and photooxidation (Baker and Bowyer, 1994; Yamamoto and Smith, 1993). Protective systems which dissipate excess energy (e.g. zeaxanthin) and remove active oxygen are activated during drought and high light intensity (Chapters 8, 12 and 13). There is some evidence that drought increases the expression of antioxidant enzymes (Chapter 12). Photorespiration may also have a role in dissipating excess energy when the stomata are closed (Chapters 8 and 12).

The processes mentioned above which respond directly to water deficit (e.g. ABA, betaine and proline synthesis) must be triggered by some aspect of plant water status. At present very little is known about the processes which lead to the perception and transduction of water deficit signals or indeed even if higher plant cells do respond directly to water status in this way. Firstly, the relevant signal or measure of water status (e.g. cell volume, turgor, membrane tension or concentration of a key signal molecule) has not been established. There is evidence that some of the signal transduction systems which have been identified in plant cells might be involved in the response to osmotic shock treatments. Cytosolic free calcium increases in maize root protoplasts exposed to salt (Lynch *et al.*, 1989). Turnover of phosphoinositides increases when carrot cell cultures are exposed to osmotica (Cho *et al.*, 1993). Intracellular pH and proton ATPase activity also change and could be involved in osmoregulation (Spickett *et al.*, 1992). Increased proton pumping could increase rates of ion and sugar uptake and aid the lowering of osmotic potential. The occurrence of stretch-sensitive receptors and ion channels and their role in perceiving and responding to changes in turgor or cell volume needs to be explored (Ding and Pickard, 1993). Experiments on the rapid events associated with the presently known signal transduction systems are of necessity carried out by applying rapid osmotic shock. Since higher plants are usually exposed to slower changes in water status (minutes rather than seconds) the results should be interpreted with caution. Bacteria and algae, on the other hand, are often exposed to rapid changes in salinity and it is perhaps not surprising that they accumulate osmolytes much more rapidly (within minutes rather than hours) and that the mechanisms for sensing water status and activating accumulation of osmolytes are much better understood. In the alga *Poterioochromonas malhamensis*, isofloridoside phosphate synthase is activated by proteolytic cleavage and this leads to accumulation of the osmolyte isofloridoside. The specific protease is activated by Ca-calmodulin after osmotic shock induces an increase in cytosolic calcium (Kauss, 1987). In the halophilic alga *Dunaliella*, increased salinity causes rapid glycerol accumulation. This may be mediated by differential shrinkage of cytosol and chloroplast which alters concentrations of regulatory metabolites in these compartments (Chapter 11). Phosphoinositide and phosphatidylcholine turnover in the plasma membrane of *Dunaliella* is also provoked by changes in salinity (Einspahr *et al.*, 1988, 1989). These processes need to be understood if metabolic responses to water deficit such as solute accumulation are to be manipulated.

### 1.4.2 *Temperature*

Temperature, in the present context, interacts with the plant in two ways. Firstly, it alters reaction rates and influences the kinetic properties of enzymes (Chapters 3–5). Secondly, temperature extremes cause damage. Some tropical species are damaged by temperatures as high as 5–10°C while others can survive to −196°C (Patterson and Graham, 1987). Differences in tolerance to high temperatures are not so extreme, but nevertheless exist. High temperature tolerance for photosynthesis and growth can vary by up to 20°C. The highest temperature tolerances in higher plants are found in partially buried desert succulents (e.g. *Lithops* and *Haworthia* species) which can withstand 60°C for an hour, an exposure lethal to most plants (Nobel, 1989). At the extremes, the problem for survival is to maintain structural and functional integrity of macromolecules and cells. Increased resistance to temperature extremes can occur within hours if plants are hardened by exposure to temperatures markedly higher or lower than the usual growth temperature. The cell and molecular biology of response and acclimation to temperature extremes has been the subject of much recent research (Howarth and Ougham, 1993). Both high and low temperatures induce the synthesis of proteins which are absent or much less abundant at normal temperature. The function of many of these proteins is not known. Some may be involved in the maintenance of structural stability, while others may be enzymes involved in altered metabolism and changes in developmental patterns. Heat-shock proteins accumulate at high temperatures. Many of these are molecular chaperones involved in re-folding heat-denatured protein. They probably represent over-expression of a normal cellular homeostatic mechanism induced by protein unfolding (Howarth and Ougham, 1993). Heat-shock proteins can also be induced by drought in some instances (Almoguera *et al.*, 1993), which would suggest that protein damage could occur during water deficit. At low temperature, acclimation to low non-freezing temperature results in accumulation of a range of cold-regulated proteins and an associated increase in freezing resistance. Some of the cold-induced proteins are similar to drought-induced proteins (dehydrins) and may function in protecting against the cytoplasmic dehydration which occurs when the apoplast freezes. The parallel with drought can be taken further since ABA can sometimes induce freezing tolerance (Orr *et al.*, 1986) and drought pre-treatments can increase freezing tolerance as well as induce similar proteins (Guy *et al.*, 1992). Antifreeze proteins similar to those found in cold water fish may also be induced at low temperature (Kurkela and Franck, 1990). Solutes such as fructans (in certain species), sucrose and proline accumulate when plants are exposed to low temperatures. In the case of sucrose and fructans it can be argued that this is simply a passive accumulation which occurs because growth is more sensitive than photosynthesis to low temperature. In the case of the 'low temperature sweetening', which occurs in some storage organs, the response may be caused by differential temperature sensitivity of carbohydrate metabolizing enzymes (ap Rees *et al.*, 1988). Fructans have been proposed as being particularly associated with low temperature growth, soluble fructans providing a ready source of carbohydrate in cells expanding at low temperature (Hendry, 1987). This hypothesis requires further substantiation (Pollock, 1986). Proline sometimes accumulates in the cold and proline-over-accumulating mutants have increased freezing tolerance (Tantau and Dörffling, 1991). Neither the cause of accumulation of these solutes nor their function is clear. They do not accumulate to high enough levels to lower the freezing point by more than a few degrees, however they could prepare cells for tolerating freezing by acting as compatible osmolytes in the dehydrated cytoplasm.

The other target for extreme temperatures are membranes whose fluidity and therefore function is temperature dependent. This is a complex area and the reader should consult Patterson and Graham (1987) and Quinn (1988) for further information. Correlations between lipid composition and membrane fluidity, freezing tolerance and tolerance of photosynthesis to temperature extremes have been found (Patterson and Graham, 1987). The functioning of photosystem II is particularly sensitive to the fluidity of the thylakoid membrane (Raison *et al.*, 1982). Chilling sensitivity may be caused by membrane dysfunction at low non-freezing temperature. In general, it appears that function at low temperature is better if membranes have lipids with low phase transition temperatures (e.g. the polar lipids contain fatty acids which are unsaturated and of shorter chain length), while function at high temperature is better if the degree of unsaturation is less. Other components such as sterols also influence fluidity. Correlations between lipid phase transition temperatures and temperature tolerance are often found, while acclimation to temperature extremes is associated with altered lipid composition (Patterson and Graham, 1987; Quinn, 1988). Very little is known about how the synthesis, degree of saturation and composition of membrane lipids are regulated by temperature. The prospect of altering lipid metabolism in transgenic plants to change temperature sensitivity is being considered (Section 1.5).

Like drought (Section 1.4.1), low temperature decreases photosynthesis rate and aggravates photoinhibition and photooxidation. Photoprotective mechanisms and antioxidant defences are induced at low temperature (Chapter 12).

## 1.5 Metabolic engineering

Two approaches, to which the term 'metabolic engineering' can be applied, are being used to alter metabolic pathways. The first approach involves the selection of mutants with altered enzyme levels or with enzymes which have altered regulatory properties. The second approach is to produce transgenic plants with altered enzyme levels. The latter approach is becoming widespread as more and more genes encoding enzymes of primary metabolism are cloned. Genes can be over-expressed when attached to a strong constitutive promotor and they can be targeted to particular tissues and organelles by including appropriate regulatory sequences. Alternatively, the introduction of antisense constructs down-regulates expression. This allows production of plants with a range of activities from very high to extremely low. Such plants can be used to study metabolic control (ap Rees and Hill, 1994) as well as the consequences of altering major pathways such as photosynthesis, carbohydrate metabolism and nitrogen assimilation on growth and development. This section will briefly consider examples in which this approach is being, or could be, applied to altering plant response to the environment.

The various osmolytes produced by plants during drought may have an adaptive role, although this has been questioned (see Section 1.4.4. and Chapters 9–11). Their role can be tested by altering their accumulation in mutants and transgenic plants. In the case of proline (Chapter 9) and betaine (Chapter 10) the biosynthetic genes have been cloned and expression in transgenic plants is awaited. Likewise, betaine-deficient and proline-overproducing mutants have been isolated (Chapters 9 and 10). The proline overproducers appear to be more tolerant to freezing and salinity but more detailed characterization is needed. Accumulation of the polyol mannitol, which acts as an osmolyte in some algae and mangroves (Chapter 11), has been achieved in transgenic tobacco. Tobacco does not normally produce mannitol, but plants transformed with a bacterial gene which encodes mannitol 1-phosphate dehydrogenase accumulate mannitol and appear to have higher tolerance to salinity (Tarcynski *et al.*, 1992, 1993). Much more detailed growth analysis is required but it is interesting that diversion of a significant proportion of photosynthate to mannitol has no serious effect on growth (at least in a controlled environment). Tobacco plants have also been transformed with an inositol methyltransferase gene from *M. crystallinum* and accumulate the cyclic polyol ononitol (Vernon *et al.*, 1994). Ononitol is the precursor of pinitol which accumulates in *M. crystallinum* and a number of other species during drought (Chapter 11). In the future, it may prove possible to transform non-polyol-accumulating plants with polyol biosynthesis genes which are under the control of drought- or ABA-regulated promoters so that accumulation occurs only during drought.

Antioxidant defence is important for all aerobic organisms because of the production of reactive forms of oxygen, such as superoxide, singlet oxygen, hydrogen peroxide, as well as organic free radicals. Various situations, for example, high light intensity, UV radiation, drought and pollutants, may increase oxidative damage and could be harmful in this way (Chapters 11 and 13). It is possible that increased activity of antioxidant enzymes could protect plants. Transgenic plants have been produced which over-express superoxide dismutase, glutathione reductase and ascorbate peroxidase. In many instances these are more resistant to oxidative damage caused by pollutants, herbicides or high light (Foyer *et al.*, 1994; Chapter 12). Other potentially protective proteins are the drought-, heat- and cold-induced proteins such as dehydrins and heat-shock proteins (Bartels and Nelson, 1994; Section 1.4). Many of these have been cloned and the effect of altering their expression can be explored. Enzymes from warm- and cold-adapted species often have different thermal responses, for example in the temperature response of substrate affinity ($K_m$). With this in mind, another possibility, raised by Burke (Chapter 5), is to transfer enzymes with different thermal characteristics between species and to investigate the consequences of this for the temperature responses of metabolic pathways and the thermal tolerance of plants. Thermal tolerance is also changed in mutants and transgenic plants with altered lipid metabolism. When the composition of membrane lipids is altered in mutants and in transgenic *Arabidopsis* and tobacco, the tolerance of plants to low temperature is altered. This confirms the conclusion from earlier investigations (Section 1.4.2) that lipid composition influences temperature response. Details can be found in the review by Gibson *et al.* (1994).

It seems likely that altering the expression of key photosynthetic enzymes (Rubisco, PEP carboxylase), carbohydrate-metabolizing enzymes (sucrose phosphate synthase, invertase, chloroplast phosphoglucomutase) and nitrogen-assimilating enzymes (nitrate reductase, glutamine synthetase, glutamate synthase) could have subtle and unexpected consequences for plant growth and interaction with environmental factors. This area is just beginning to be explored and is discussed by Stitt and Schulze (1994), Sonnewald *et al.* (1994), Hoff *et al.* (1994), Lea and Forde (1994), Foyer (Chapter 2) and Leegood (Chapter 4).

Metabolic engineering holds promise for altering plant response to the environment and perhaps for improving crop yield. Will a 'superplant', which has been engineered with drought-inducible enzymes to produce betaines or polyols, with increased expression of antioxidant enzymes to protect against excess light and pollutants, with altered carbon-partioning patterns, with altered membrane lipid composition, with enzymes with altered temperature sensitivity and with more UV-screening compounds, do better than an unaltered plant? At present we do not know the consequences of altering so many components on the function of the whole and there is limited experience of the performance of transgenic plants in the field under naturally fluctuating conditions. We should consider the possibility that, during evolution, metabolic networks and their responses to the environment have approached near enough to optimum that further tinkering will have little effect or could even be deleterious. Also, in our enthusiasm for altering metabolic traits, we must not lose sight of the importance of morphological, phenological and developmental characteristics of plants for dealing with extreme and fluctuating environments (Section 1.1). Nevertheless the progress so far is promising. Further advances and practical applications will depend on a deeper knowledge of plant metabolism and its relation to the environment.

## References


Almoguera, C., Coca, M.A. and Jordano, J. (1993) Tissue-specific expression of sunflower heat shock proteins in response to water stress. *Plant J.* **4**, 947–958.

ap Rees, T., Burrell, M.M., Entwistle, T.G., Hammond, J.B.W., Kirk, D. and Kruger, N.J. (1988) Effects of low temperature on the respiratory metabolism of carbohydrates by plants. In: *Plants and Temperature.* Symposia of the Society for Experimental Biology XXXXII (eds S.P. Long and F.I. Woodward). Company of Biologists, Cambridge, pp. 377–393.

ap Rees, T. and Hill, S.A. (1994) Metabolic control analysis of plant metabolism. *Plant, Cell Environ.* **17**, 587–599.

Baker, N.R. and Bowyer, J.R. (eds) (1994) *Photoinhibition of Photosynthesis. From Molecular Mechanisms to the Field.* BIOS Scientific Publishers, Oxford.

Bartels, D. and Nelson, D. (1994) Approaches to improve stress tolerance using molecular genetics. *Plant, Cell Environ.* **17**, 659–667.

Bianchi, G., Gamba, A., Murelli, C., Salamini, F. and Bartels, D. (1991) Novel carbohydrate metabolism in the resurrection plant *Craterostigma plantagineum. Plant J.* **1**, 355–359.

Bowes, G. (1987) Aquatic plant photosynthesis: strategies that enhance carbon gain. In: *Plant Life in Aquatic and Amphibious Habitats* (ed. R.M.M. Crawford). Blackwell Scientific Publications, Oxford, pp. 79–98.

Bozarth, C.S., Mullet, J.E. and Boyer, J.S. (1987) Cell wall proteins at low water potential. *Plant Physiol.* **85**, 261–267.

Bray, E.A. (1993) Molecular responses to water deficit. *Plant Physiol.* **103**, 1035–1040.

Cho, M.H., Shears, S.B. and Boss, W.F. (1993) Changes in phosphatidylinositol metabolism in response to hyperosmotic stress in *Daucus carota* L. cells grown in suspension culture. *Plant Physiol.* **103**, 637–647.

Chrispeels, M.J. and Maurel, C. (1994) Aquaporins: the molecular basis of facilitated water movement through living plant cells? *Plant Physiol.* **105**, 9–13.

Close, T.J., Fenton, R.D., Yang, A., Asghar, H., DeMason, D.A., Crone, D.E., Meyer, M.C. and Moonan, F. (1993) Dehydrin: the protein. In: *Plant Responses to Cellular Dehydration During Environmental Stress. Current Topics in Plant Physiology,* Volume 10 (eds T.J. Close and E.A. Bray). American Society of Plant Physiologists, Rockville, MD, pp. 104–118.

Crawford, R.M.M. (1971) Some metabolic aspects of ecology. *Trans. Bot. Soc. Edinb.* **41**, 309–322.

Crawford, R.M.M. (1992) Oxygen availability as an ecological limit to plant distribution. *Adv. Ecol. Res.* **23**, 93–185.

Csonka, L. and Hanson, A.D. (1991) Prokaryotic osmoregulation: genetics and physiology. *Annu. Rev. Microbiol.* **45**, 569–606.

Davies, W.J. and Jones,. H.G. (eds) (1991) *Abscisic Acid. Physiology and Biochemistry.* BIOS Scientific Publishers, Oxford.

Demmig-Adams, B. and Adams, W.W. (1992) Photoprotection and other responses to high light stress. *Annu. Rev. Plant Physiol. Mol. Biol.* **43**, 599–626.

Ding, J.P. and Pickard, B.G. (1993) Mechanosensory calcium-selective cation channels in epidermal cells. *Plant J.* **3**, 83–110.

Einspahr, K.J., Maeda, M. and Thompson, G.A. (1988) Concurrent changes in *Dunaliella salina* ultrastructure and membrane phospholipid metabolism after hyperosmotic shock. *J. Cell Biol.* **107**, 529–538.

Einspahr, K.J.,Peeler, T.C. and Thompson, G.A. (1989) Phosphatidylinositol 4,5-bisphosphate phospholipase C and phosphomonoesterase in *Dunaliella salina* membranes. *Plant Physiol.* **90**, 1115–1120.

Foyer, C.H., Descourvières, P. and Kunert, K.J. (1994) Protection against oxygen radicals: an important defence mechanism studied in transgenic plants. *Plant, Cell Environ.* **17**, 507–523.

Gibson, S., Falcone, D.L., Browse, J. and Somerville, C. (1994) Use of transgenic plants and mutants to study the regulation and function of lipid composition. *Plant, Cell Environ.* **17**, 627–637.

Griffiths, H. (1988) Crassulacean acid metabolism: a reappraisal of physiological plasticity in form and function. *Adv. Bot. Res.* **15**, 43–92.

Grime, J.P. (1979) *Plant Strategies and Vegetation Processes.* Wiley, Chichester.

Guy, C., Haskell, D., Neven, L., Klein, P. and Smelser, C. (1992) Hydration-state-responsive proteins link cold and drought stress in spinach. *Planta* **188**, 265–270.

Harborne, J.B. (1982) *Introduction to Ecological Biochemistry.* Academic Press, London.

Hendry, G. (1987) The ecological significance of fructan in a contemporary flora. *New Phytol.* **106 (suppl.)**, 201–216.

Hochachka, P.W. and Somero, G.N. (1984) *Biochemical Adaptation.* Princeton University Press, Princeton, New Jersey.

Hoff, T., Truong, H.-N. and Caboche, M. (1994) The use of mutants and transgenic plants to study nitrate assimilation. *Plant, Cell Environ.* **17**, 489–506.

Howarth, C.J. and Ougham, H. J. (1993) Gene expression under temperature stress. *New*

*Phytol.* **125**, 1–26.
Kauss, H. (1987) Volume regulation in *Poterioochromonas*. Involvement of calmodulin in the $Ca^{2+}$-stimulated activation of isofloridoside synthase. *Plant Physiol.* **71**, 169–172.
Kennedy, R.A., Rumpho, M.E. and Fox, T.C. (1992) Anaerobic metabolism in plants. *Plant Physiol.* **100**, 1–6.
Koster, K.L. and Leopold, A.C. (1988) Sugars and desiccation tolerance in seeds. *Plant Physiol.* **88**, 829–832.
Kurkela, S. and Franck, M. (1990) Cloning and characteristics of a cold and ABA-inducible *Arabidopsis* gene. *Plant Mol. Biol.* **15**, 137–144.
Lea, P.J. and Forde, B.G. (1994) The use of mutants and transgenic plants to study amino acid metabolism. *Plant, Cell Environ.* **17**, 541–556.
Lynch, J., Polito, V.S. and Läuchli, A. (1989) Salinity stress increases cytoplasmic Ca activity in maize root protoplasts. *Plant Physiol.* **90**, 1271–1274.
McIntosh, L. (1994) Molecular biology of the alternative oxidase. *Plant Physiol.* **105**, 781–786.
Nobel, P.S. (1989) Shoot temperatures and thermal tolerances for succulent species of *Haworthia* and *Lithops*. *Plant, Cell Environ.* **12**, 643–651.
Orr, W., Keller, W.A. and Singh, J. (1986) Induction of freezing tolerance in an embryogenic culture of *Brassica napus* by abscisic acid at room temperature. *J. Plant Physiol.* **126**, 23.
Patterson, B.D. and Graham, D. (1987) Temperature and metabolism. In: *The Biochemistry of Plants. A Comprehensive Treatise. Volume 12, Physiology of Metabolism* (ed. D.D. Davies). Academic Press, New York, pp.153–199.
Pearcy, R.W. (1990) Sunflecks and photosynthesis in plant canopies. *Annu. Rev. Plant Physiol. Mol. Biol.* **41**, 421–453.
Pearcy, R.W. and Ehleringer, J. (1984) Comparative ecophysiology of C3 and C4 plants. *Plant, Cell Environ.* **7**, 1–13.
Pollock, C.J. (1986) Fructans and the metabolism of sucrose in vascular plants. *New Phytol.* **104**, 1–24.
Pollock, C.J., Farrar, J.F. and Gordon, A.J. (eds) (1992) *Carbon Partitioning Within and Between Organisms*. BIOS Scientific Publishers, Oxford.
Quinn, P.J. (1988) Effects of temperature on cell membranes. In: *Plants and Temperature. Symposia of the Society for Experimental Biology XXXXII* (eds S.P. Long and F.I. Woodward). Company of Biologists, Cambridge, pp. 237–258.
Raison, J.K., Pike, P.S. and Berry, J.A. (1982) Growth temperature-induced alteration in the thermotropic properties of *Nerium oleander* membrane lipids. *Plant Physiol.* **70**, 215–218.
Raven, R.A., Wollenweber, B. and Handley, L.L. (1992) A comparison of ammonium and nitrate as nitrogen sources for photolithotrophs. *New Phytol.* **121**, 19–32.
Richards, R.A. (1993) Breeding crops with improved stress resistance. In: *Plant Responses to Cellular Dehydration During Environmental Stress. Current Topics in Plant Physiology,* Volume 10 (eds T.J. Close and E.A. Bray). American Society of Plant Physiologists, Rockville, MD, pp. 211–223.
Sachs, M.M. and Ho, T.-H.D. (1986) Alteration of gene expression during environmental stress in plants. *Annu. Rev. Plant Physiol.* **37**, 363–376.
Schuster, W.S. and Monson, R.K. (1990) An examination of the advantages of C3–C4 intermediate photosynthesis in warm environments. *Plant, Cell Environ.* **13**, 903–912.
Sharp, R.E., Voetberg, G.S., Saab, I. and Bernstein, N. (1993) Role of abscisic acid in the regulation of cell expansion in roots at low water potentials. In: *Plant Responses to Cellular*

*Dehydration During Environmental Stress. Current Topics in Plant Physiology,* Volume 10 (eds T.J. Close and E.A. Bray). American Society of Plant Physiologists, Rockville, MD, pp. 57–66.

Smith, J.A.C. and Griffiths, H. (eds) (1993) *Water Deficits. Plant Responses from Cell to Community.* BIOS Scientific Publishers, Oxford.

Sonnewald, U., Lerchl, J., Zrenner, R. and Frommer, W. (1994) Manipulation of sink–source relations in transgenic plants. *Plant, Cell Environ.* **17**, 649–658.

Spickett, C.M., Smirnoff, N. and Ratcliffe, R.G. (1992) Metabolic response of maize roots to hyperosmotic shock. An *in vivo* $^{31}$P nuclear magnetic resonance study. *Plant Physiol.* **99**, 856–863.

Sprent, J.I. (1980) Root nodule anatomy, type of export product and evolutionary origin in some Leguminosae. *Plant, Cell Environ.* **3**, 35–43.

Stapleton, A.E. and Walbot, V. (1994) Flavonoids can protect maize DNA from the induction of ultraviolet radiation damage. *Plant Physiol.* **105**, 881–889.

Stitt, M. and Schulze, D. (1994) Does Rubisco control the rate of photosynthesis and plant growth? An exercise in molecular ecophysiology. *Plant, Cell Environ.* **17**, 465–487.

Stewart, G.R., Joly, C.A. and Smirnoff, N. (1992) Partitioning of nitrogen assimilation between the roots and shoots of cerrado and forest trees of contrasting plant communities of South East Brasil. *Oecologia* **91**, 511–517.

Tantau, H. and Dörffling, K. (1991) *In vitro*-selection of hydroxyproline-resistant cell lines of wheat (*Triticum aestivum*): accumulation of proline, decrease in osmotic potential, and increase in frost tolerance. *Physiol. Plant.* **82**, 243–248.

Tarcynski, M.C., Jensen, R.G. and Bohnert, H.J. (1992) Expression of a bacterial *mtl*D gene in transgenic tobacco leads to production and accumulation of mannitol. *Proc. Natl Acad. Sci. USA* **89**, 2600–2604.

Tarcynski, M.C., Jensen, R.G. and Bohnert, H.J. (1993) Stress protection of transgenic tobacco by production of the osmolyte mannitol. *Science* **259**, 508–510.

Trewavas, A. (1986) Resource allocation under poor growth conditions. A major role for plant growth substances in developmental plasticity. In: *Plasticity in Plants. Symposia of the Society for Experimental Biology XXXX* (eds D.H. Jennings and A. Trewavas). Company of Biologists, Cambridge, pp. 31–76.

Vernon, D., Ostrem, J.A. and Bohnert, H.J. (1993) Stress perception and response in a facultative halophyte – the regulation of salinity-induced genes in *Mesembryanthemum crystallinum. Plant, Cell Environ.* **16**, 437–444.

Vernon, D., Tarcynski, M.C., Jensen, R.G. and Bohnert, H.J. (1994) Cyclitol production in transgenic tobacco. *Plant J.* **4**, 199–205.

Yamamoto, H.Y. and Smith, S.M. (eds) (1993) *Photosynthetic Responses to the Environment. Current Topics in Plant Physiology,* Volume 8. American Society of Plant Physiologists, Rockville, MD.

2

# Co-regulation of nitrogen and carbon assimilation in leaves

C.H. Foyer, M.H. Valadier and S. Ferrario

## 2.1 Introduction

Recent studies have demonstrated unequivocally that the pathways of carbon and nitrogen metabolism are modulated in parallel in higher plants (Evans and Terashima, 1988; Kaiser and Förster, 1989; Pace *et al.*, 1990). In leaves, photosynthetic carbon assimilation provides the driving force for both of these processes, carbohydrate and amino acid synthesis forming major sinks for the products of photosynthesis. Many endogenous metabolic effectors have been implicated in this co-ordination, for example cytokinins enhance chloroplast development and are also involved in the regulation of the synthesis of key enzymes such as phosphoenolpyruvate carboxylase (PEPC) and nitrate reductase (NR), (Flores and Tobin, 1988; Hoff *et al.*, 1994; Sugiharto *et al.*, 1992). However, the success of a plant within a given environment is probably more dependent on opportunity and on compromise within the carbon/nitrogen interaction in order to optimize the use of available resources. Environmental factors such as the availability of light and of $CO_2$, and nutrient status have a profound impact on the supply of energy, carbon skeletons and other substrates that support carbon and nitrogen assimilation. In higher plants, there is little evidence to suggest that carbon and nitrogen assimilation compete for available resources. Rather, continuous reciprocal co-ordination at the molecular and metabolic levels in response to environmental stimuli prevents competition and allows adaptive responses to supply and demand. In Figure 2.1 we have tried to provide a simple framework for the integration of carbon and nitrogen assimilation in leaves. The three enzymes which appear to be the most important as a result of studies conducted to date are sucrose phosphate synthase (SPS), NR and PEPC. These enzymes respond at both the genetic and metabolic levels to changes in environmental conditions. In this chapter, however, we will limit our discussion largely to the short-term regulation of metabolism that precedes the more long-term adaptation at the level of gene expression. When plants are supplied with

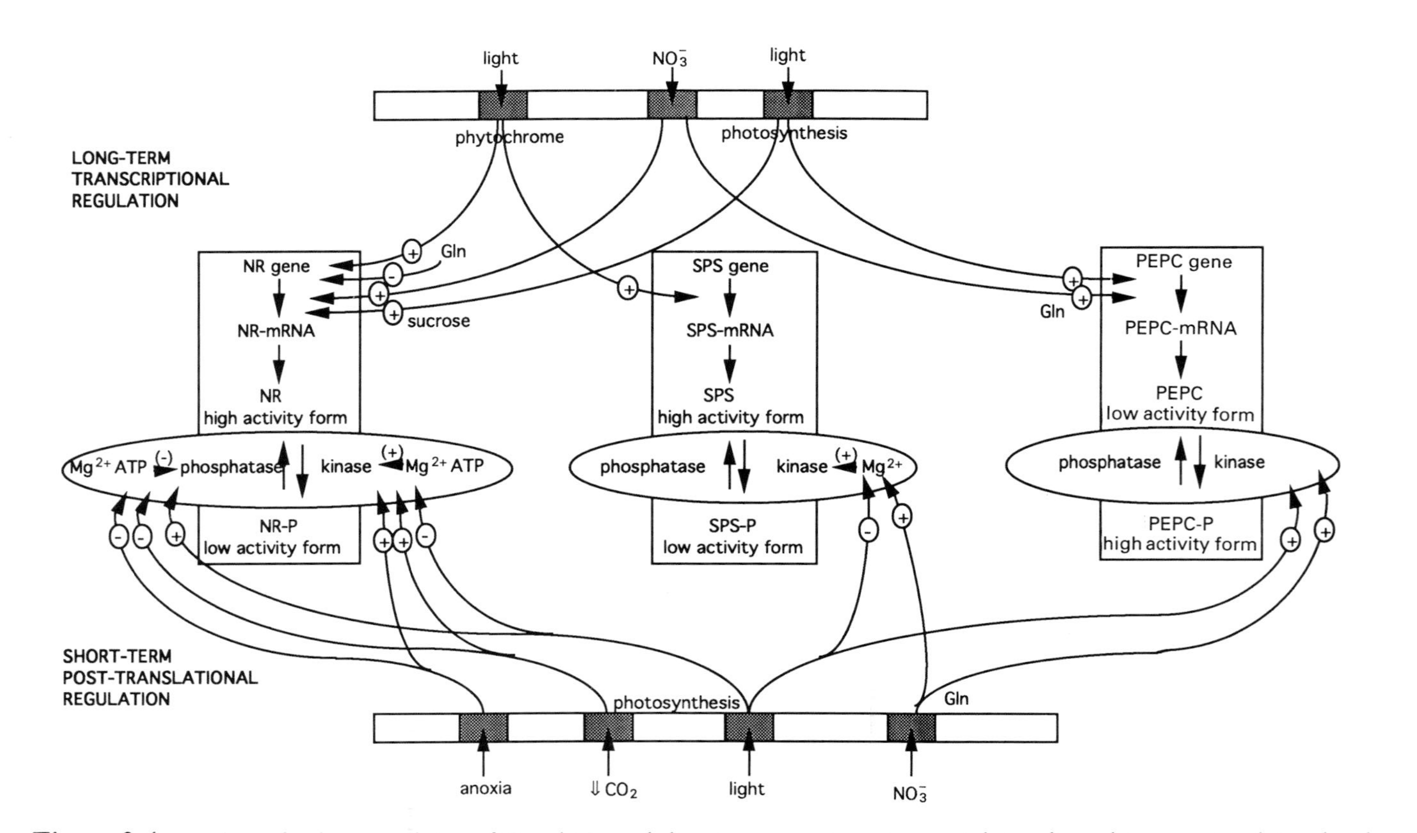


***Figure 2.1.*** *A scheme for the co-regulation of C and N metabolism in response to environmental stimuli involving post-translational and transcriptional regulation of nitrate reductase (NR), sucrose phosphate synthase (SPS) and phosphoenolpyruvate carboxylase (PEPC).*

nitrogen there is an increase in the synthesis of amino acids and a decrease in the synthesis of sucrose. This involves activation of NR and PEPC and inhibition of SPS. Interestingly, the activities of all these enzymes are regulated by protein phosphorylation and it has been proposed that modulation of the respective protein kinases and phosphatases directs carbon flow (Champigny and Foyer, 1992). The influence of the environment on the relationship between photosynthesis and nitrogen assimilation has been studied for many years by modifying supply and demand, for example by causing N deprivation, or by removing $CO_2$ and thus inhibiting photosynthesis. This classical physiological and biochemical approach alone left many unresolved questions concerning the nature of the co-regulation of C and N metabolism and its ultimate effects on biomass accumulation. Transgenic plants offer us a means to achieve a more complete understanding of the roles of individual enzymes in the C/N relationship and are enabling us to assess the degree to which the modification of the activities of single enzymes ultimately could affect total biomass accumulation by the plant.

## 2.2 Co-ordination of N/C assimilation

Plants respond quickly to changes in the supply of $CO_2$ or $NO_3^-$. A decrease in the ambient $CO_2$ levels causes a rapid decrease in the rate of N assimilation by modulation of NR, both in the shoots and in the roots (Kaiser and Förster, 1989; Pace *et al.*, 1990). Similarly, when N is supplied to leaves, there is a limitation of net sucrose synthesis and an enhancement of PEPC activity. This enzyme plays a key role in regulating the flow of carbon through the anaplerotic pathway. It is perhaps the most responsive of all enzymes to N availability (Sugiharto and Sugiyama, 1992). NR, SPS and PEPC all respond to N availability and to light–dark transitions. We will examine the mechanisms of regulation of each of these enzymes in turn and consider how this regulation may be used to achieve co-ordinate modulation of C and N metabolism.

NR is generally considered to be a limiting factor for growth, development and protein production in plants (Solomonson and Barber, 1990). This enzyme, which catalyses the initial step of nitrate assimilation, the reduction of nitrate to nitrite, is known to be influenced by environmental factors such as light and nitrate availability. We are far from a complete understanding of the mechanisms of regulation of NR *in vivo*. However, a complex interaction between the regulation of gene transcription and post-translational modification of the protein, by protein phosphorylation, in response to environmental factors such as light and $NO_3^-$ or $CO_2$ availability, has been demonstrated.

Studies using inhibitors and *in vitro* translation of mRNA isolated from plants grown in the presence or absence of $NO_3^-$, provided the evidence for the control of NR gene expression by $NO_3^-$ at the transcriptional level (Cheng *et al.*, 1986). Further studies clearly showed that the transcription of the NR gene is induced by $NO_3^-$ and light (Cheng *et al.*, 1986; Galangau *et al.*, 1988). This phenomenon is very rapid in roots and in leaves. NR mRNA could be detected within 40 min of supplying $NO_3^-$ to roots (Melzer *et al.*, 1989). The light effect is mediated, at least in part, by sucrose (Cheng *et al.*, 1992; Vincentz *et al.*, 1993). There is a circadian rhythm of NR gene expression in tobacco. An increase in the level of NR mRNA occurs during the night and a decrease during the day (Deng et al., 1990). Glutamine (Gln), which acts as a repressor, is involved in the circadian expression of the NR gene (Deng *et al.*, 1991). Gln accumulation is correlated with low NR mRNA accumulation. Levels of the NR protein reach a peak some hours after that in NR mRNA (Galangau *et al.*, 1988). The lag between mRNA and protein accumulation seems to be species dependent. Moreover, there does not appear to be a simple correlation between NR protein and NR activity, presumably because post-translational modifications of NR activity are superimposed on the transcriptional variations.

The modulation of NR activity by protein kinases is rapid (half-time 2.5 min) and is induced in leaves by environmental factors such as light–dark transitions (Huber *et al.*, 1992b) or a restriction of the $CO_2$ supply (Kaiser and Brendle-Behnish, 1991) and in roots by anoxia (Glaab and Kaiser, 1993). The phosphorylation of the protein has been shown to decrease NR activity *in vitro* under 'selective assay' conditions (Kaiser *et al.*, 1993). NR has a phosphorylated form which shows a lower activity when assayed *in vitro* in the presence of $Mg^{2+}$ (5 mM). $Ca^{2+}$ can sometimes also cause inhibition but regulation is considered to be mediated by $Mg^{2+}$ *in vivo* (Kaiser and Brendle-Behnish, 1991). The phosphorylated form of NR is predominant in leaves in the dark (Huber *et al.*, 1992b) and also in the light when the leaves are deprived of $CO_2$ (Kaiser and Brendle-Behnish, 1991). Phosphorylation is reversible with a half-time of 20 min (Riens and Heldt, 1992) and occurs on seryl residues (Huber *et al.*, 1992a). Partial purification of two putative proteins, involved in the NR phosphorylation, has been achieved (Spill and Kaiser, 1994). Dephosphorylation of the NR protein *in vivo* in the light can be prevented by okadaic acid. This indicates the involvement of a protein phosphatase in the reactivation process of NR (Huber *et al.*, 1992a).

The role of $Mg^{2+}$ and its importance in the regulation of NR activity are not understood completely. NR inactivation has been observed *in vitro* by incubation with ATP in the presence of $Mg^{2+}$ and reactivation was subsequently achieved in the presence of an excess of AMP or ethylenediamine tetraacetic acid (EDTA) (Kaiser and Spill, 1991; MacKintosh, 1992). Therefore, $Mg^{2+}$ may act both on the protein phosphatase (as an inhibitor) and on the protein kinase (as an activator). AMP could also be involved in the inhibition of the protein phosphatase (Spill and Kaiser, 1994). Post-translational modifications of NR have a physiological role since they allow rapid adaptation of nitrate reduction to environmental conditions. The way in which the cell perceives these modifications remains unresolved. The light effect cannot be dissociated from photosynthesis, since glucose and sucrose mimic the light effect on NR gene expression (Vincentz *et al.*, 1993). Gln is also involved in this regulation. A model for signal transduction from the recognition of $NO_3^-$ by a 'sensor protein' to the first events in the nucleus has been proposed by Redinbaugh and Campbell (1991). In reality, the situation may be much more complicated, since single signals such as light induce both short- and long-term regulation involving regulation of NR transcription and dephosphorylation of the enzyme protein. The relative importance of endogenous factors such as $Mg^{2+}$ and AMP is questionable. Their *in vivo* concentrations are not known for certain. Free $Mg^{2+}$ is required for inactivation of NR in the range of 1–5 mM. Such quantities have been measured in isolated chloroplasts, but NR is a cytosolic enzyme and the cytosolic concentration of $Mg^{2+}$ is unknown. Indeed, whether the cytosolic $Mg^{2+}$ concentration varies in response to environmental stimuli is a matter for speculation (Kaiser and Brendle-Behnish, 1991).

The sucrose synthesis pathway is considered to be regulated in mature leaves largely by the modulation of two key enzymes: fructose-1,6-bisphosphatase (FBPase) and SPS. Photosynthetic carbon fixation produces triose phosphate (TP) which is exported from the chloroplast to the cytosol via the phosphate translocator (Riesmeier *et al.*, 1993). In the cytosol, TP is converted to fructose-1,6-bisphosphate. Fructose-1,6-bisphosphate is hydrolysed by FBPase resulting in formation of hexose phosphates. Glucose-1-phosphate is used to form UDP-glucose. SPS then catalyses sucrose phosphate synthesis from UDP-glucose and fructose-6-phosphate. Sucrose phosphate is hydrolysed, using sucrose phosphate phosphatase, to sucrose (Stitt *et al.*, 1987). Sucrose is then translocated from the leaf to the sink organs. Sucrose synthesis is co-ordinated with the rate of photosynthesis and is also adjusted to meet the demands of the sink organs. Environmental factors are known to influence SPS activity. Of these, light and the availability of $CO_2$ are well known regulators (Battistelli *et al.*, 1991). Nitrogen nutrition also plays a role in regulating the activity of the enzyme (Champigny and Foyer, 1992; Kerr *et al.*, 1984). SPS activity is often considered to be subject to both ‘coarse’ and ‘fine’ control (Stitt *et al.*, 1987). ‘Fine control’ consists of allosteric modulation achieved by metabolic effectors that modify the kinetic properties of the enzyme. Glucose-6-phosphate is an activator increasing the affinity of SPS for its substrate, fructose-6-phosphate. Inorganic phosphate (Pi) a is non-competitive inhibitor with respect to fructose-6-phosphate and competitive with respect to the second substrate, UDP-glucose (Amir and Preiss, 1982; Doehlert and Huber, 1983).

Fructose-2,6-bisphosphate regulates the cytosolic FBPase and is thus involved in the control of the initial part of the pathway of sucrose synthesis (Gerhardt *et al.*, 1987; Stitt and Quick, 1989). ‘Fine control’ of the pathway of sucrose synthesis may be important *in vivo* in achieving the required balance between photosynthetic rate, the rate of sucrose export and starch accumulation within the chloroplasts. ‘Coarse control’ results in changes in the amount of extractable SPS activity. This is due to protein turnover and to covalent modification of the protein. During leaf development, SPS activity increases in parallel with the amount of total protein (Walker and Huber, 1989). Diurnal fluctuations of SPS activity are observed in many species. An endogenous rhythm in SPS activity has been characterized in soybean leaves (Kerr *et al.*, 1985). Huber *et al.* (1989a) divided plants into three groups according to the regulation of foliar SPS by light and its associated metabolism. The 12 species studied by Huber *et al.* (1989a) could be distinguished as follows: (a) plants where light activation of foliar SPS involved an increase in the maximum velocity of catalysis; (b) species where light activation caused little or no change in maximum velocity but modified the kinetic properties of the enzyme; and (c) plant types where SPS activity was unchanged following light–dark transitions. The mechanism of the light activation of SPS is suggested to involve covalent modification by protein phosphorylation. This has been observed *in vivo* and *in vitro* (Huber *et al.*, 1989b; Walker and Huber, 1989). The phosphorylated form of SPS is characterized by an increased sensitivity to Pi inhibition (Stitt *et al.*, 1988). The 'activation state' of the enzyme is defined as the ratio of the activity measured with limiting substrates plus Pi, to that with saturating substrates. The 'activation state' of SPS appears to be correlated with the phosphorylation state of the protein (Huber and Huber, 1990). Furthermore, multi-site phosphorylation of the SPS protein has been observed. Surprisingly, a constitutive 'stimulatory site' has been identified in addition to the 'inhibitory' sites. Both sites of phosphorylation involve seryl residues (Huber and Huber, 1990). The protein kinase specific for SPS has been partially characterized and seems to be activated by $Mg^{2+}$ as is the protein phosphatase.

The phosphorylation/dephosphorylation modulation of SPS occurs during the light–dark transitions, and is the mechanism used to modify SPS activity in order to co-ordinate nitrogen and carbon metabolism. Short-term changes in the activation states of SPS and PEPC have been observed when high $NO_3^-$ (40 mM) levels are supplied to leaves grown on low levels of N (Foyer *et al.*, 1994a; Van Quy *et al.*, 1991; Van Quy and Champigny, 1992). Co-regulation of C and N metabolism by the stimulation of protein phosphorylation leads to the inhibition of SPS activity and activation of PEPC. PEPC activation increases malate synthesis and diverts the flow of the C skeletons through the anaplerotic pathway towards amino acid synthesis (Champigny and Foyer, 1992). PEPC has been shown to be regulated in both $C_3$ and $C_4$ plants by phosphorylation/dephosphorylation changes in response to light–dark transitions and to the N availability (Jiao and Chollet, 1990; see also Chapter 3). Gln and glutamate control the activity of PEPC by (i) induction and repression of PEPC RNA synthesis, respectively (ii) activation and inhibition of the PEPC protein kinase activity, respectively and (iii) direct modification of the enzyme (Manh *et al.*, 1993). A hypothesis for the co-ordination of C and N metabolism, where light and $NO_3^-$ act as environmental signals enhancing the activity of cytosolic protein kinases regulating SPS and PEPC, was presented by Champigny and Foyer (1992). The N metabolites which are the signal and mediate this regulation are likely to be Gln and glutamate which act as positive and negative effectors of protein kinases, respectively. Although the phosphorylation/dephosphorylation changes indicate some similarity in the regulation of SPS, PEPC and NR, the enzymes modulating this interconversion are most probably quite distinct (Huber *et al.*, 1992b).

## 2.3 The use of transgenic plants to study the regulation of carbon and nitrogen assimilation

Advances in molecular biology have provided the means to decrease, modify or increase the expression of specific proteins in plant cells. The ability to generate transgenic plants with different levels of NR or SPS has been particularly useful in allowing us to clarify the roles and importance of these enzymes in plant metabolism and biomass production.

### 2.3.1 *Modulation of NR activity*

*Nicotiana plumbaginifolia* mutant E23, which completely lacks NR activity, has been a useful tool for the production of transgenic plants expressing various levels of the enzyme. Vaucheret *et al.* (1990) described the expression of a native tobacco NR structural gene, *nia-2*, in the *N. plumbaginifolia* mutant (Figure 2.2). The transgenic plants produced in this way showed only poor expression of the NR gene because of the absence of activation sequences in the promoter.

***Figure 2.2.*** *Transgenic* Nicotiana plumbaginifolia *plants expressing different levels of foliar NR activity. The foliar NR activities of the transformed lines E23 (307), E23 (304) and E23 (100) were approximately 4.5%, 5.5% and 7.0% of the values obtained in the wild-type, respectively. In contrast, the foliar NR activity of the high NR expressor (30.C$^+$) can be up to five times that of the wild-type, depending on growth conditions.*

Plants over-expressing NR activity were produced by transformation of the NR-deficient mutant E23 with the *nia-2* cDNA under the control of the constitutive 35S promoter from cauliflower mosaic virus (CaMV; Vincentz and Caboche, 1991). These plants have an increased foliar NR activity measured *in vitro*, since the expression of the NR gene is no longer regulated by $NO_3^-$ levels (Figure 2.2). No adverse effects of the latter transformation were observed (Foyer *et al.*, 1994b; Vincentz and Caboche, 1991). These leaves showed a marked decrease of foliar $NO_3^-$ content and an increase in the Gln pool

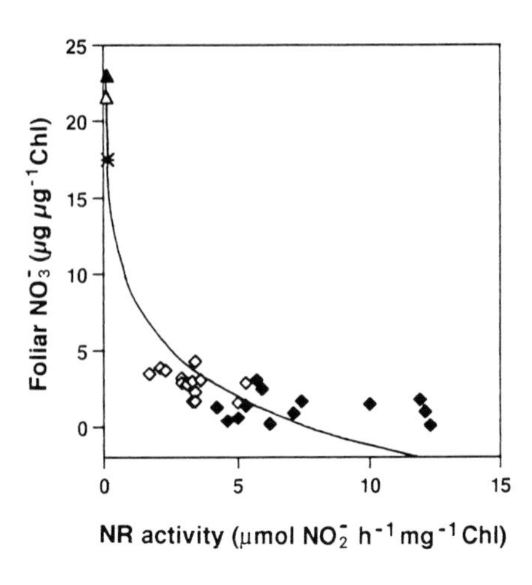


***Figure 2.3.*** *The relationship between foliar NR activity and nitrate content in* Nicotiana plumbaginifolia *grown on 12 mM nitrogen. Individual plants from the transgenic line of high NR expressor 30.C$^+$ (◆), wild-type controls (◊) and average values for the populations of transgenic low NR expressing lines 307 (▲), 304 (△) and 100 (*) are given.*

(Foyer *et al.*, 1993; Quilleré *et al.*, 1994). Indeed, there is an inverse relationship between foliar NR activity and the $NO_3$ content of the leaf (Figure 2.3).

The consequences of long-term N deprivation are generally severe. Reduced N assimilation leads to decreases in photosynthesis, biomass accumulation and growth (Henry and Raper, 1991; Khamis *et al.*, 1990; Rideout *et al.*, 1994). However, such studies are complicated by the absence of $NO_3^-$, which is an ion that makes a large contribution to the overall ionic balance of plant cells. Transgenic plants modified in NR activity allow an examination of the effects of decreases in the capacity to assimilate N while maintaining the $NO_3^-$ supply to the plants. In addition, we can study the effects of $NO_3^-$ deprivation in a plant system where the NR activity is not determined by the $NO_3^-$ content, that is the expression of NR is not regulated by $NO_3^-$ (Figure 2.2). The leaves of low NR expressors contain considerable amounts of $NO_3^-$ (Figure 2.4) This large store of $NO_3^-$ has repercussions for the ionic and osmotic balance within these leaves which are noticeably more succulent than the wild-type.

The leaves of the high NR expressors were not very different from the wild-type controls with similar or very slightly increased rates of photosynthesis and respiration, comparable SPS and PEPC activities, and similar levels of protein, chlorophyll, starch and sucrose (Foyer *et al.*, 1993, 1994b). It is clear that the increases in NR activity were insufficient to have much impact on photosynthesis, growth or biomass production. This suggests that, in the natural state, NR is present at levels above those required for maximal operation of the N assimilation pathway and production of photosynthetic components. Consequently, any further increase in NR activity has only minor effects on the partitioning of carbon between amino acid biosynthesis and sucrose synthesis and is without significant effect on photosynthesis. Our results suggest that overproduction of this single enzyme has only a marginal effect on plant metabolism. In marked contrast, large changes in C and N partitioning were observed in the plants expressing very low levels of NR activity where ability to assimilate N clearly restricts growth and photosynthesis (Foyer *et al.*, 1993, 1994a).

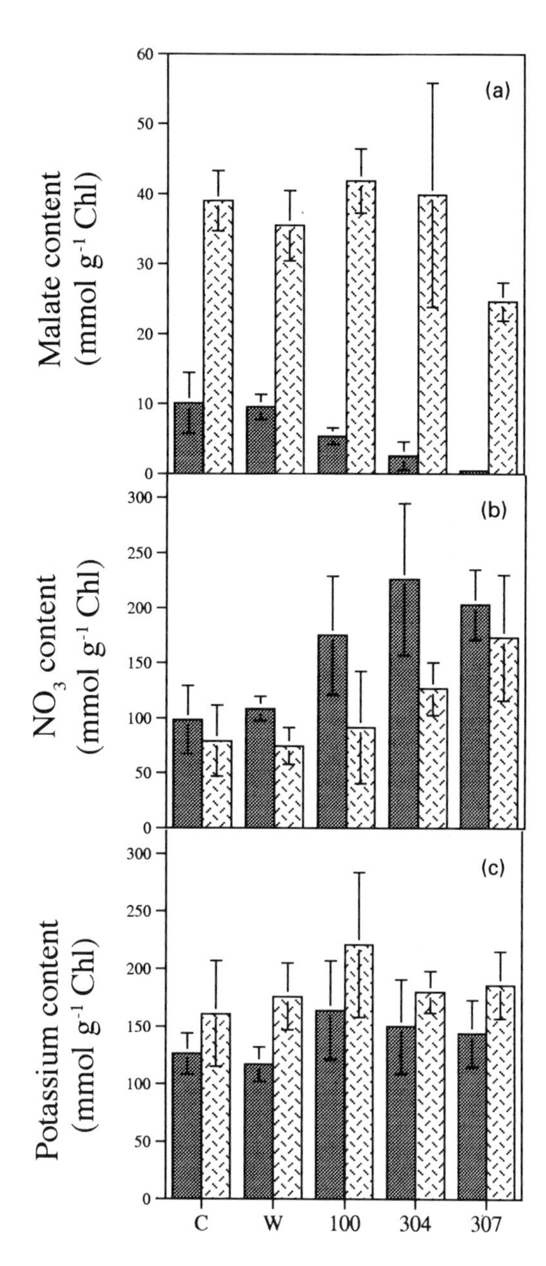


***Figure 2.4.*** *The foliar contents of malate (a), $NO_3^-$ (b) and potassium (c) in wild-type (W), the high NR expressing line 30.$C^+$ (C) and the low NR expressing lines 100, 304 and 307 of* Nicotiana plumbaginifolia *at two levels of irradiance, 170 μmol photons $m^{-2}$ $s^{-1}$ (grey bars) and 450 μmol photons $m^{-2}$ $s^{-1}$ (speckled bars).*

## 2.3.2 *Modulation of SPS activity*

Transgenic tomato plants expressing both the native SPS and also the SPS gene from maize were produced by Worrell *et al.* (1991) and Galtier *et al.* (1994). The active enzyme was expressed from the cloned maize cDNA under the control of either the promoter for the gene encoding the small subunit of ribulose-1,5-bisphosphate carboxylase-oxygenase from tobacco or the promoter from CaMV. In the former, expression was largely limited to photosynthetic cells, while in the latter expression was constitutive. In the transgenic plants, the levels of foliar SPS were about three times those of the wild-type controls in both the light and the dark (Galtier *et al.*, 1993, 1994). The diurnal modulation of the native tomato SPS resulted in a marked decrease in SPS activity in the untransformed plants in the dark. In the transformed plants, however, the differences in enzyme activity between the light and the dark were much smaller. In maize leaves, SPS is subject to light–dark modulation. When the maize SPS is expressed in tomato, however, this type of regulation is largely lost. (Galtier *et al.*, 1993, 1994; Worrell *et al.*, 1991). Carbon partitioning between starch and sucrose in the leaves was greatly modified. Sucrose formation was favoured and sucrose accumulated in the leaves (Galtier *et al.*, 1993). Starch accumulation in the leaves was much decreased. There was a strong positive correlation between the foliar ratio of sucrose to starch and SPS activity (Galtier *et al.*, 1993). SPS is, thus, a major controlling factor determining assimilate partitioning in leaves.

The photosynthetic capacity was increased in all plants expressing high SPS activity. The increase in photosynthetic rate in air was significant. The light- and $CO_2$-saturated rate of photosynthesis was increased by about 20% in leaves expressing high SPS activity (Galtier *et al.*, 1993). High SPS activity thus has the potential to increase ambient photosynthesis even when light and $CO_2$ are limiting $CO_2$ assimilation.

## 2.4 Possibilities for improved biomass production

The overall strategy of using over-expression of single enzymes as a tool to modify biomass accumulation is equivocal. We are aware that many factors contribute to the ultimate plant productivity. Nevertheless, our initial results are encouraging. Biomass production, expressed as net carbon gain, is governed essentially by the relative rates of input (photosynthesis) and output (respiration). In many species, a significant portion (nearly half) of the carbon fixed during the photoperiod is used in dark respiration. Indeed, there is a negative correlation between the rate of dark respiration and the net carbon gain (Robson, 1982; Winzeler *et al.*, 1988).

The high NR expressor and wild-type untransformed controls of *N. plumbaginifolia* (see Section 2.3.1) were grown with complete nutrient solution containing 12 mM $NO_3^-$ and then transferred to 0.2 or 0.05 mM $NO_3^-$ for 21 days. The decrease in N availability resulted in a decrease in total biomass in both the high NR expressors and wild-type plants (Figure 2.5) The net $CO_2$ exchange rate was also reduced in both types of plants as a result of N deprivation (Figure 2.6). Under N-replete conditions, foliar $NO_3^-$ was lower in the transgenic plants than in the untransformed controls, while NR activity was higher (Figure 2.6). N deprivation induced marked decreases in the foliar $NO_3^-$ content and NR activity of both types of plants (Figure 2.6). The differences between the transgenic and control plants disappeared in terms of both $NO_3^-$ accumulation and foliar NR activity (Figure 2.6). N deficiency must limit the

***Figure 2.5.*** *The high NR expressor line (30.C⁺ front line of plants) and the wild-type controls (rear line of plants) of* Nicotiana plumbaginifolia *grown with either 12, 0.2 or 0.05 mM* $NO_3^-$.

synthesis of protein in general. Whether this is sufficient to explain the relative increase in the loss of NR activity in the high NR expressors relative to controls is not known. Measurements of NR gene expression and NR protein levels are required in order to elucidate this further.

High NR activity may not really be useful *in situ* in the absence of the substate ($NO_3^-$). In the case of N deprivation, virtually no $NO_3^-$ was present in the leaves. The $NO_3^-$ required for NR activity depends largely on $NO_3^-$ uptake by the roots and its subsequent transport to the leaf. In addition, $NO_3^-$ can be remobilized from the vacuole. The measured values indicate that the vacuolar pool was depleted (foliar $NO_3^-$ fell below measurable levels). These results suggest that the uptake and translocation of $NO_3^-$ from roots to leaves may be a major limiting step of the $NO_3^-$ assimilation in both types of plants. The inducibility of the system of $NO_3^-$ absorption by the roots must be similar in the two plant types and would be affected equally by the N deprivation. If similar levels of $NO_3^-$ were taken up by the wild-type and transgenic plants but the transgenic plants reduced more of the $NO_3^-$, then total protein N and amino acid N would increase. Although amino acid N increases, the total protein N level does not change (Quilleré *et al.*, 1994). The transgenic plants have similar root systems on a fresh or dry weight basis (Quilleré *et al.*, 1994). Thus, the quantity of root is not affected by over-expression of NR activity, but the amount of $NO_3^-$ absorption, per unit surface area, is decreased in this situation. It is possible that $NO_3^-$ absorption is decreased as a result of the increases in the Gln pool size that are found in the high NR expressors. Gln is known to decrease $NO_3^-$ absorption by the roots. This study clearly shows that $NO_3^-$ uptake is a major limiting step for plant growth; constitutive NR expression alone is not sufficient to compensate for a deficit in $NO_3^-$ supply.

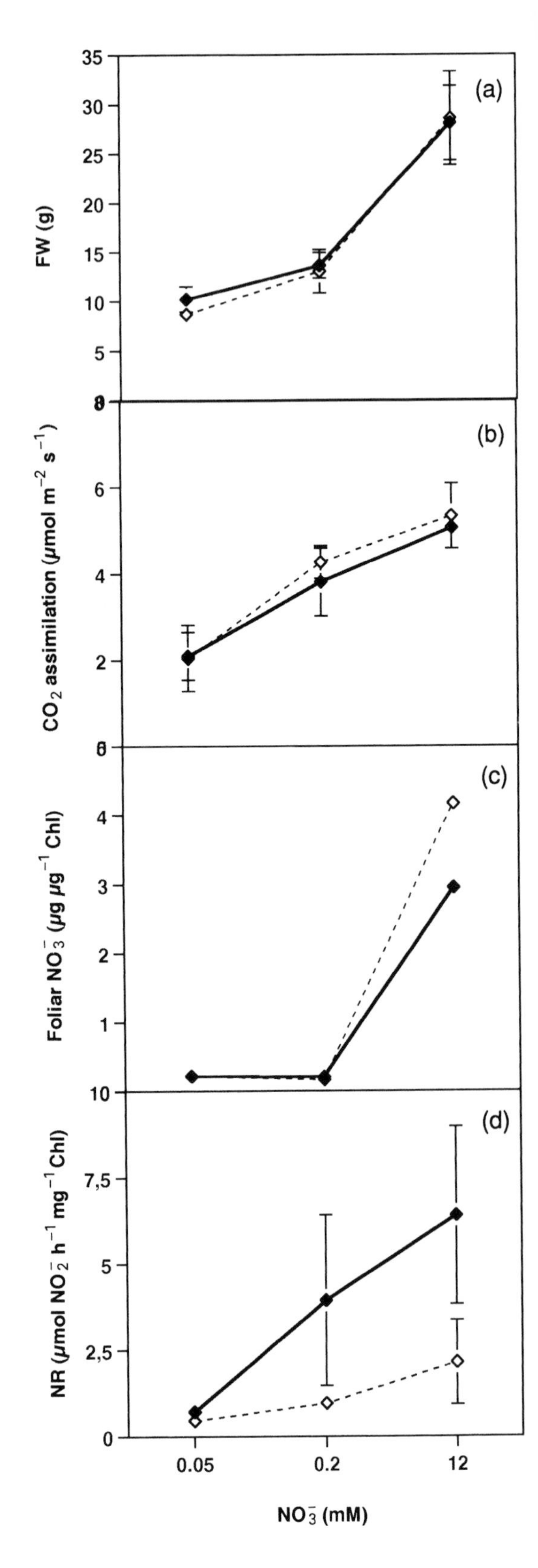


***Figure 2.6.*** *Total biomass accumulation (fresh weight, a), $CO_2$ assimilation rate (b), foliar $NO_3^-$ content (c) and nitrate reductase activity (d) in transgenic high NR expressors (bold line) and wild-type plants (dotted line) of* Nicotiana plumbaginifolia *grown at different levels of nitrogen.*

In contrast to the results obtained with over-expression of NR (Figure 2.6), the constitutive expression of SPS caused a marked increase in biomass accumulation (Figure 2.7). When SPS was increased in photosynthetic cells alone there was no increase in biomass accumulation (Galtier *et al.*, 1994) and these plants had very similar characteristics to the wild-type controls. The increase in photosynthesis required to sustain the measured increase in biomass in the plants with the constitutive over-expression of SPS was only about 10% above that of the wild-type. This level of increase was observed in all types of SPS over-expressor, but it was only translated into an increase in biomass when SPS was over-expressed in all tissue types (Galtier *et al.*, 1994).

***Figure 2.7.*** *Wild-type tomato plants (left) and transgenic tomato plants (right) expressing the maize gene for SPS under the control of the 35S promoter from cauliflower mosaic virus, in addition to the native forms of the enzyme.*

## 2.5 Perspectives and conclusions

The existence of strict reciprocal control between the pathways of C and N metabolism is rationalized by the necessity to achieve the appropriate balance between supply and demand in an environment that is constantly changing and often hazardous. The use of transgenic plants to study the C/N interaction has provided some information on the roles of individual enymes, such as NR and SPS, in plant growth. Interestingly, the over-expression and loss of regulation of NR, which would normally be considered to be rate-limiting, has little effect. Other rate-limiting steps or limiting environmental factors take over, for example the availability of $NO_3^-$ appears to be more important than NR activity in this regard. $NO_3^-$ uptake and transport could be a crucial limiting step for growth rather than $NO_3^-$ assimilation *per se*. When $NO_3^-$ supply to the roots is saturating, constitutive NR expression does not confer any increase in biomass compared to the wild-type control plants. These studies indicate that much further experimentation is needed into the relative importance of absorption of nutrients, their assimilation and the partitioning of carbon in plants in order to engineer greater plant productivity.

The relative importance of the post-translational regulation of the enzymes SPS, NR and PEPC *in vivo* remains to be established, particularly the relevance of the phosphorylation/dephosphorylation modulation of such enzymes in response to N. This could be studied in the case of NR by comparing the phosphorylation state of the enzyme in the transgenic NR over-expressors to the wild-type controls. The influence of environmental factors on phosphorylation/dephosphorylation changes is being studied in our laboratory. In the future, molecular genetic techniques must be used in order to elucidate the role of the individual seryl residues on proteins that are phosphorylated and their importance in the co-regulation of SPS and PEPC activities *in situ*, in response to environmental stimuli. In the future we also need to address the quantitative aspects of the C/N control network, to describe the limits of its robustness and to assess the relationships between regulation, adaptation and acclimatization to the environment.

## References


Amir, J. and Preiss, J. (1982) Kinetic characterisation of spinach leaf sucrose-phosphate synthase. *Plant Physiol.* **69**, 1027–1030.

Battistelli, A., Adcock, M.D. and Leegood, R.C. (1991) The relationship between the activation state of sucrose-phosphate synthase and the rate of $CO_2$ assimilation in spinach leaves. *Planta* **183**, 620–622.

Champigny, M.L. and Foyer, C.H. (1992) Nitrate activation of cytosolic protein kinases diverts photosynthetic carbon from sucrose to amino acid biosynthesis. Basis for a new concept. *Plant Physiol.* **100**, 7–12.

Cheng, C.L., Dewdney, J., Kleinhofs, A. and Goodman, H.M. (1986) Cloning and nitrate induction of nitrate reductase mRNA. *Proc. Natl Acad. Sci. USA* **83**, 6825–6828.

Cheng, C.L., Acedo, G.N., Cristinsin, M. and Conkling, M.A. (1992) Sucrose mimics the light induction of *Arabidopsis* nitrate reductase gene transcription. *Proc. Natl Acad. Sci. USA* **89**, 1861–1864.

Deng, M.D., Moureaux, T., Leydecker, M.T. and Caboche, M. (1990) Nitrate reductase expression is under the control of a circadian rhythm and is light inducible in *Nicotiana tabacum* leaves. *Planta* **18**, 257–261.

Deng, M.D., Moureaux, T., Cherel, I., Boutin, J.P. and Caboche, M. (1991) Effects of nitrogen metabolites on the regulation and circadian expression of tobacco nitrate reductase. *Plant Physiol. Biochem.* **29**, 239–247.

Doehlert, D.C. and Huber, S.C. (1983) Regulation of spinach leaf sucrose-phosphate synthase by glucose 6-phosphate, inorganic phosphate, and pH. *Plant Physiol.* **73**, 989–994.

Evans, J.R. and Terashima, I. (1988) Photosynthetic characteristics of spinach leaves grown with different nitrogen treatments. *Plant Cell Physiol.* **29**, 157–165.

Flores, S. and Tobin, E.M. (1988) Cytokinin modulation of LHCP mRNA levels: the involvement of post-transcriptional regulation. *Plant Mol. Biol.* **11**, 409–415.

Foyer, C.H., Lefebvre, C., Provot, M., Vincentz, M. and Vaucheret, H. (1993) Modulation of nitrogen and carbon metabolism in transformed *Nicotiana plumbaginifolia* mutant E23 lines expressing either increased or decreased nitrate reductase activity. In: *Aspects of Applied Biology, Volume 34, Physiology of Varieties* (eds E. White, P.S. Kettlewell, M.A. Parry and R.P. Ellis). Association of Applied Biologists, Wellesbourne, Warwick, pp. 137–145.

Foyer, C.H., Noctor, G., Lelandais, M., Lescure, J.-C., Valadier, M.H., Boutin, J.-P. and Horton, P. (1994a) Short term effects of nitrate, nitrite and ammonium assimilation on chlorophyll *a* fluorescence, thylakoid protein phosphorylation, $CO_2$ assimilation and amino acid biosynthesis in maize. *Planta* **192**, 211–220.

Foyer, C.H., Lescure, J.-C., Lefebvre, C., Morot-Gaudry, J.-F., Vincentz, M. and Vaucheret, H. (1994b) Adaptations of photosynthetic electron transport, carbon assimilation and carbon partitioning in transgenic *Nicotiana plumbaginifolia* plants to changes in nitrate reductase activity. *Plant Physiol.* **104**, 171–178.

Galangau, F., Daniel-Vedele, F., Moureaux, T., Dorbe, M.F., Leydecker, M.T. and Caboche, M. (1988) Expression of leaf nitrate reductase genes from tomato and tobacco in relation to light–dark regimes and nitrate supply. *Plant Physiol.* **88**, 383–388.

Galtier, N., Foyer, C.H., Huber, J., Voelker, T.A. and Huber, S.C. (1993) Effects of elevated sucrose-phosphate synthase activity on photosynthesis, assimilate partitioning and growth in tomato (*Lycopersicum esculentum* var. UC28B). *Plant Physiol.* **101**, 535–543.

Galtier, N., Alred, R., Quick, P., Voelker, T.A. and Foyer, C.H. (1994) Constitutive expression of sucrose phosphate synthase leads to an increase in biomass production. Manuscript in preparation.

Gerhardt, R., Stitt, M. and Heldt, H.W. (1987) Subcellular metabolite levels in spinach leaves. Regulation of sucrose synthesis during diurnal alterations in photosynthetic partitioning. *Plant Physiol.* **83**, 399–407.

Glaab, J. and Kaiser, W.M. (1993) Rapid modulation of nitrate reductase in pea roots. *Planta* **191**, 173–179.

Henry, L.T. and Raper, C.D. Jr (1991) Soluble carbohydrate allocation to roots, photosynthetic rate of leaves, and nitrogen assimilation as affected by nitrogen stress and irradiance. *Bot. Gaz.* **152**, 23–39.

Hoff, T., Truong, H.N. and Caboche, M. (1994). The use of mutants and transgenic plants to study nitrate assimilation. *Plant, Cell Environ.* **17**, 489–506.

Huber, S.C. and Huber, J.L.A. (1990) Regulation of spinach leaf sucrose-phosphate synthase by multisite phosphorylation. *Curr. Top. Plant Biochem. Physiol.* **9**, 329–343.

Huber, S.C., Nielsen, T.H., Huber, J.L.A. and Pharr, D.M. (1989a) Variation among species in light activation of sucrose-phosphate synthase. *Plant Cell Physiol.* **30**, 277–285.

Huber, J.L., Huber, S.C. and Nielsen, T.H. (1989b) Protein phosphorylation as a mechanism for regulation of spinach leaf sucrose-phosphate synthase activity. *Arch. Biochem. Biophys.* **270**, 681–690.

Huber, J.L., Huber, S.C., Campbell, W.H. and Redinbaugh, M.G. (1992a) Reversible light/dark modulation of spinach leaf nitrate reductase activity involves protein phosphorylation. *Arch. Biochem. Biophys.* **296**, 58–65.

Huber, S.C., Huber, J.L., Campbell, W.H. and Redinbaugh, M.G. (1992b) Comparative studies of the light modulation of nitrate reductase and sucrose-phosphate synthase activities in spinach leaves. *Plant Physiol.* **100**, 706–712.

Jiao, J.A. and Chollet, R. (1990) Regulatory phosphorylation of serine 15 in maize phosphoenolpyruvate carboxylase by a C4 leaf protein-serine kinase. *Arch. Biochem. Biophys.* **283**, 300–305.

Kaiser, W.M. and Brendle-Behnisch, E. (1991) Rapid modulation of spinach leaf nitrate reductase by photosynthesis. I. Modulation *in vivo* by $CO_2$ availability. *Plant Physiol.* **96**, 363–367.

Kaiser, W.M. and Förster, J. (1989) Low $CO_2$ prevents nitrate reduction in leaves. *Plant Physiol.* **91**, 970–974.

Kaiser, W.M. and Spill, D. (1991) Rapid modulation of spinach leaf nitrate reductase by photosynthesis. II. *In vitro* modulation by ATP and AMP. *Plant Physiol.* **96**, 368–375.

Kaiser, W.M., Spill, D. and Glaab, J. (1993) Rapid modulation of nitrate reductase in leaves and roots: indirect evidence for the involvement of protein phosphorylation/dephosphorylation. *Physiol. Plant.* **89**, 557–562.

Kerr, P.S., Huber, S.C. and Israel, D.W. (1984) Effects of N source on soybean leaf sucrose phosphate synthase, starch formation, and whole plant growth. *Plant Physiol.* **75**, 483–488.

Kerr, P.S., Rufty, T.W. and Huber, S.C. (1985) Endogenous rhythms in photosynthesis, sucrose phosphate synthase activity, and stomatal resistance in leaves of soybean *Glycine max* (L.) Merr. *Plant Physiol.* **77**, 275–280.

Khamis, S., Lamaze, T., Lemoine, Y. and Foyer, C.H. (1990) Adaptation of the photosynthetic apparatus in maize leaves as a result of nitrogen limitation. *Plant Physiol.* **94**, 1436–1443.

MacKintosh, C. (1992) Regulation of spinach-leaf nitrate reductase by reversible phosphorylation. *Biochim. Biophys. Acta* **1137**, 121–126.

Manh, C.T., Boutin, J.P., Provot, M. and Champigny M.L (1993) Metabolite effectors for short-term nitrogen-dependent enhancement of phosphoenolpyruvate carboxylase activation and decrease of net sucrose synthesis in wheat leaves. *Plant Physiol.* **89**, 460–466.

Melzer, J.M., Kleinhofs, A. and Warner, R.L. (1989) Nitrate reductase regulation: Effects of nitrate and light on nitrate reductase mRNA accumulation. *Mol. Gen. Genet.* **217**, 341–346.

Pace, G.H., Volk, R.J. and Jackson, W.A. (1990) Nitrate reduction in response to $CO_2$–limited photosynthesis. Relationship to carbohydrate supply and nitrate reductase activity in maize seedlings. *Plant Physiol.* **92**, 286–292.

Quilleré, I., Dufosse, C., Roux, Y., Foyer, C.H., Caboche, M. and Morot-Gaudry, J.F. (1994) The effects of deregulation of NR gene expression on growth and nitrogen metabolism of *Nicotiana plumbaginifolia* plant. *J. Exp. Bot.* in press.

Redinbaugh, M.G. and Campbell, W.H. (1991) Higher plant responses to environmental nitrate. *Physiol. Plant.* **82**, 640–650.

Reins, B. and Heldt, H.W. (1992) Decrease in nitrate reductase activity in spinach leaves during a light–dark transition. *Plant Physiol.* **98**, 573–577.

Rideout, J.W., Chaillou, S., Raper, C.D. Jr and Morot-Gaudry, J.F. (1994) Ammonium and nitrate uptake by soybean during recovery from nitrogen deprivation. *J. Exp. Bot.* **45**, 23–33.

Riesmeier, J.W., Flugge, U.I., Schulz, B., Heineke, D., Heldt, H.W., Willmitzer, L. and Frommer, W.B. (1993) Antisense repression of the chloroplast triose phosphate translocator affects carbon partitioning in transgenic potato plants. *Proc. Natl Acad. Sci. USA* **90**, 6160–6164.

Robson, M.J. (1982) The growth and economy of selection lines of *Lolium peressue* cv. 523 with differing rates of respiration. 2. Growth as young plants from seed. *Ann. Bot.* **49**, 331–339.

Solomonson, P.L. and Barber, M.J. (1990) Assimilatory nitrate reductase: functional properties and regulation. *Annu. Rev. Plant Physiol. Mol. Biol.* **41**, 225–253.

Spill, D. and Kaiser, W.M. (1994) Partial purification of two proteins (100 kDa and 67 kDa) cooperating in the ATP-dependent inactivation of spinach leaf nitrate reductase. *Planta* **192**, 183–188.

Stitt, M. and Quick, W.P. (1989) Photosynthetic carbon partitioning: its regulation and possibilities for manipulation. *Physiol. Plant.* **77**, 633–641.

Stitt, M., Huber, S. and Kerr, P. (1987) Control of photosynthetic sucrose formation. In: *The Biochemistry of Plants, Volume 10, Photosynthesis.* (eds M.D. Hatch and N.K. Boardman). Academic Press, New York, pp. 327–409.

Stitt, M., Wilke, I., Fiel, R. and Heldt, H.W. (1988) Coarse control of sucrose-phosphate synthase in leaves: alterations of the kinetic properties in response to the rate of photosynthesis and the accumulation of sucrose. *Planta* **174**, 217–230.

Sugiharto, B. and Sugiyama, T. (1992) Effects of nitrate and ammonium on gene expression of phosphoenolpyruvate carboxylase and nitrogen metabolism in maize leaf tissue during recovery from nitrogen stress. *Plant Physiol.* **98**, 1403–1408.

Sugiharto, B., Burnell, J.N. and Sugiyama, T. (1992) Cytokinin is required to induce the nitrogen-dependent accumulation of mRNA's for phosphoenolpyruvate carboxylase and carbonic anhydrase in detached maize leaves. *Plant Physiol.* **100**, 153–156.

Van Quy, L. and Champigny, M.L. (1992) $NO_3^-$ enhances the kinase activity for phosphorylation of phosphoenolpyruvate carboxylase and sucrose phosphate synthase proteins in wheat leaves. Evidence from the effects of mannose and okadic acid. *Plant Physiol.* **99**, 344–347.

Van Quy, L., Foyer, C.H. and Champigny, M.L. (1991) Effect of light and $NO_3^-$ on wheat leaf phosphoenolpyruvate carboxylase activity. Evidence for covalent modulation of the $C_3$ enzyme. *Plant Physiol.* **97**, 1476–1482.

Vaucheret, H., Chabaud, M., Kronenburger, J. and Caboche, M. (1990) Functional complementation of tobacco and *Nicotiana plumbaginifolia* nitrate reductase deficient mutants by transformation with the wild type alleles of the tobacco structural genes. *Mol. Gen. Genet.* **220**, 468–474.

Vincentz, M. and Caboche, M. (1991) Constitutive expression of nitrate reductase allows normal growth and development of *Nicotiana plumbaginifolia* plants. *EMBO J.* **10**, 1027–1035.

Vincentz, M., Moureaux, T., Leydecker, M.T., Vaucheret, H. and Caboche, M. (1993) Regulation of nitrate and nitrite reductase expression in *Nicotiana plumbaginifolia* leaves by nitrogen and carbon metabolites. *Plant J.* **3**, 315–324.

Walker, J.L. and Huber, S.C. (1989) Regulation of sucrose-phosphate synthase in spinach leaves by protein level and covalent modification. *Planta* **177**, 116–120.

Winzeler, M., McCullough, D.E. and Hunt, L.A. (1988) Genotypic differences in dark respiration of mature leaves in winter wheat (*Triticum aestivium* L.) *Can. J. Plant Sci.* **68**, 669–675.

Worrell, A.C., Bruneau, J.M., Summerfelt, K., Boersig M. and Voelker, T.A. (1991) Expression of a maize sucrose-phosphate synthase in tomato alters leaf carbohydrate partitioning. *Plant Cell* **3**, 1121–1130.

3

# Regulation of malate synthesis in CAM plants and guard cells; effects of light and temperature on the phosphorylation of phosphoenolpyruvate carboxylase

H.G. Nimmo, P.J. Carter, C.A. Fewson, J.P.S. Nelson, G.A. Nimmo and M.B. Wilkins

## 3.1 Introduction

Phosphoenolpyruvate carboxylase (PEPC) (EC 4.1.1.31) catalyses the fixation of $CO_2$ (as $HCO_3^-$) into oxaloacetate. It plays an anaplerotic role in non-photosynthetic tissues and in the leaves of $C_3$ plants. This role is discussed in Chapter 2 in relation to co-ordination between photosynthesis and nitrate assimilation. In the leaves of $C_4$ plants it is located in the mesophyll tissue and catalyses the first committed step in the $C_4$ pathway of $CO_2$ assimilation (Andreo *et al.*, 1987; Jiao and Chollet, 1991). In this chapter, however, we concentrate on the regulation of PEPC in two tissue types which exhibit periodic accumulation and loss of malate. In Crassulacean acid metabolism (CAM), PEPC is responsible for the nocturnal fixation of atmospheric $CO_2$; the oxaloacetate produced is reduced to malate which is stored as malic acid in the vacuole. During the following light period, the malate is released from the vacuole and decarboxylated, and the resulting $CO_2$ is fixed in the Calvin cycle (Osmond and Holtum, 1981). CAM occurs in many succulents and tropical epiphytes and the nocturnal stomatal opening associated with dark $CO_2$ fixation increases the water use efficiency of photosynthesis (Ting, 1985). During severe drought, when stomata are closed night and day, CAM may prevent photoinhibition by recycling respired $CO_2$ (Griffiths, 1988). Consideration of the metabolic pathways involved in CAM suggests that mechanisms must exist to permit flux through PEPC at night and reduce or eliminate it during the day. In the guard cells of many species, accumulation of malate is at least partly responsible for the increase in anion content that is required to balance the uptake of $K^+$ ions during stomatal opening, while the malate content of guard cells declines during stomatal closure (Allaway, 1973; Raschke, 1979; Raschke *et al.*, 1988). Hence, it is likely that flux through guard cell PEPC is tightly regulated in co-ordination with stomatal movements. In this chapter, we describe the role of phosphorylation in the regulation of PEPC in leaves of the CAM plant *Bryophyllum (Kalanchoë) fedtschenkoi* and in guard cells of *Commelina communis*, and the effects of light and temperature on the phosphorylation system.

## 3.2 Phosphorylation of CAM PEPC

Like the PEPCs from other types of higher plant (Andreo *et al.*, 1987), CAM PEPC is an allosteric enzyme subject to activation by glucose-6-phosphate and inhibition by L-malate. These characteristics gave rise to the view that fluctuations in the cytosolic concentration of malate may play a role in regulating flux through PEPC. Some years ago, however, it became clear from the work of several groups that CAM PEPC must be controlled at an additional level, in that the enzyme is more sensitive to inhibition by malate (and therefore less active under physiological conditions) during the day than at night (Buchanan-Bollig and Smith, 1984; Kluge *et al.*, 1981; Nimmo *et al.*, 1984; Winter, 1982). Detailed investigation of the mechanism underlying this effect led to the discovery that *B. fedtschenkoi* PEPC is regulated by reversible phosphorylation (Nimmo *et al.*, 1984, 1986).

Nimmo *et al.* (1984) studied changes in the activity of PEPC throughout the diurnal cycle by examining the properties of the enzyme in freshly prepared and desalted extracts of leaf tissue from *Bryophyllum* plants adapted to a short day (8 h) photoperiod. This work revealed four important points. First, PEPC was some 10-fold more sensitive to inhibition by malate during the light period (apparent $K_i$ for malate 0.3 mM) than in the middle of the dark period (apparent $K_i$ 3.0 mM). Secondly, the specific activity of PEPC measured under $V_{max}$ conditions did not change significantly throughout the diurnal cycle. Thirdly, the conversions of the 'night' to the 'day' form of PEPC and vice versa coincided with the cessation and onset of malate accumulation, respectively, suggesting that these interconversions are very important in regulating the flux through PEPC. Finally, both conversions occurred during the dark period, about 1–2 h before the lights came on in the morning and about 4–6 h after the lights went off at night. Since the kinetic differences between the two forms of the enzyme were observed even in desalted extracts, it was suggested that the conversions between the two kinetically distinct forms of PEPC represent some form of covalent modification of the enzyme. Moreover, it appeared that the interconversions between the forms are controlled by a circadian rhythm rather than by light–dark transitions.

The molecular mechanism responsible for these conversions was then investigated, first by immunoprecipitation of PEPC from detached leaves pre-labelled by $^{32}$Pi. This showed that the 'night' form of the enzyme contained $^{32}$P whereas the 'day' form did not (Nimmo *et al.*, 1984). On purification of the two forms of the enzyme, it was found that the 'night' form was phosphorylated on serine residues (Nimmo *et al.*, 1986). In addition, removal of the phosphate groups by treatment with alkaline phosphatase increased the malate sensitivity of the purified 'night' form of PEPC to that of the 'day' form. These data indicated that the conversion of the 'day' to the 'night' form of *Bryophyllum* PEPC was caused by protein phosphorylation. Similar conclusions have been reported for the PEPCs from other CAM species (e.g. Baur *et al.*, 1992; Brulfert *et al.*, 1986; Kluge *et al.*, 1988).

It is well known that detached *Bryophyllum* leaves kept in constant environmental conditions exhibit persistent rhythms of $CO_2$ metabolism (for a review, see Wilkins, 1992). For example, leaves in continuous darkness and $CO_2$-free air at 15°C maintain a rhythm of $CO_2$ output that is attributable directly to changes in flux through PEPC (Warren and Wilkins, 1961); $CO_2$ that would otherwise be released is re-fixed periodically into malate. Nimmo *et al.* (1987) investigated the phosphorylation of PEPC in such leaves. Kinetic studies and immunoprecipitation from $^{32}$Pi-labelled leaves showed that PEPC exhibited a persistent circadian rhythm of interconversions between a malate-sensitive, dephosphorylated form and a less sensitive, phosphorylated form. Importantly, the changes in the $K_i$ of the PEPC for malate were exactly in phase with the rhythm of $CO_2$ output (Nimmo *et al.*, 1987). Interconversions between the two forms of PEPC were also observed in leaves kept in continuous light. Overall, these results significantly strengthened the conclusions that phosphorylation of *Bryophyllum* PEPC (a) regulates flux through the enzyme; and (b) is controlled by an endogenous circadian rhythm rather than by changes in illumination.

## 3.3 Properties and regulation of PEPC kinase and phosphatase

Reversible phosphorylation of proteins is achieved through the activities of one or more protein kinases and protein phosphatases. The phosphorylation state of a protein *in vivo* represents the steady-state balance between the activities of the kinase(s) and the phosphatase(s), and can be regulated by changes in the activities of either or both of the kinase(s) and phosphatase(s). In order to understand the molecular basis of the diurnal and circadian changes in the phosphorylation state of PEPC, the properties of the enzymes responsible for its phosphorylation and dephosphorylation were investigated.

Recently, a significant insight into plant protein phosphatases has been obtained by comparing them with their mammalian counterparts. Mammalian tissues contain four main types of protein phosphatase catalytic subunit (termed

1, 2A, 2B and 2C) that dephosphorylate protein phosphoserine or phosphothreonine residues. These enzymes can be distinguished by their substrate specificities, sensitivities to certain inhibitors and requirements for activity (Cohen, 1989). It is now clear from biochemical studies that higher plants contain activities that are very similar to the mammalian phosphatases 1, 2A and 2C (Carter *et al.*, 1990; MacKintosh and Cohen, 1989; MacKintosh *et al.*, 1991).

Carter *et al.* (1990) showed that the phosphorylated form of *Bryophyllum* PEPC could be dephosphorylated by the purified catalytic subunit of protein phosphatase 2A from rabbit skeletal muscle, but not by the type 1 phosphatase. Dephosphorylation of PEPC completely reversed the effects of phosphorylation on its sensitivity to malate. Carter *et al.* (1990) also partially purified the type 2A catalytic subunit from *Bryophyllum* and found that it, too, could dephosphorylate PEPC. However, the activity of this phosphatase did not change significantly during the normal diurnal cycle in *Bryophyllum* (Carter *et al.*, 1991), suggesting that the phosphorylation state of PEPC might be regulated largely through changes in the activity of PEPC kinase.

Carter *et al.* (1991) partially purified PEPC kinase from 'night' *Bryophyllum* leaves and showed that it could phosphorylate the 'day' form of PEPC with a stoichiometry approaching one phosphate group per subunit. This *in vitro* phosphorylation caused essentially the same changes in the malate sensitivity of PEPC as are observed with intact tissue. The activity of PEPC kinase was not affected by $Ca^{2+}$, but the enzyme was inhibited by malate and glucose-6-phosphate (Carter *et al.*, 1991). However, it is not yet clear if these effectors are physiologically significant.

Measurements of PEPC kinase activity in desalted leaf extracts prepared throughout the diurnal cycle in *Bryophyllum* proved very informative. Kinase activity appeared some 4–6 h after the lights went off at night and disappeared some 2 h before they went on in the morning. Leaf extracts that contained PEPC kinase activity also contained PEPC in the phosphorylated form with a high $K_i$ for malate, whereas in those extracts lacking PEPC kinase activity, PEPC was in the dephosphorylated form (Carter *et al.*, 1991). These results, allied to the observation that the activity of protein phosphatase 2A did not change significantly during the diurnal cycle, suggested that the major factor which determines the phosphorylation state of PEPC is the presence or absence of PEPC kinase activity.

The mechanism underlying the nocturnal appearance of PEPC kinase was investigated by allowing leaves to take up inhibitors of protein or RNA synthesis. Treatment of detached leaves with puromycin, cycloheximide, cordycepin or actinomycin D during the day prevented the appearance of PEPC kinase activity during the following night (Carter *et al.*, 1991 and unpublished data). Similar treatments during the middle of the night resulted in a rapid loss (within 1–2 h) of kinase activity. This indicates that *de novo* synthesis of a protein is required for the appearance of PEPC kinase, and that the kinase activity turns over rapidly. The simplest hypothesis is that the component synthesised is PEPC kinase itself. However, it could also be another component that is required to activate the kinase. Irrespective of the nature of the component, its synthesis must be followed by its destruction or inactivation. *Bryophyllum* leaves kept in constant environmental conditions show a persistent rhythm of appearance and disappearance of PEPC kinase activity (Carter *et al.*, 1991), indicating that the protein synthesis and destruction steps exhibit a circadian rhythm. This rhythm in PEPC kinase activity is presumed to play an important role in generating the rhythms of $CO_2$ metabolism (Wilkins, 1992).

## 3.4 Effects of light and temperature on PEPC

It is evident from the previous section that environmental factors could affect $CO_2$ fixation via PEPC in several ways, for example by affecting the steady-state level of phosphorylation of the enzyme or simply by altering its kinetic properties, either its $V_{max}$ or its sensitivity to malate. We have therefore studied the effects of light and temperature on the phosphorylation state of PEPC and the activity of PEPC kinase in intact leaves, and the effects of temperature on the kinetic properties of purified PEPC.

As noted in Section 3.3, *Bryophyllum* leaves exhibit a rhythm of PEPC kinase activity, owing to protein synthesis and destruction, during the normal diurnal cycle or in constant environmental conditions. If leaves containing PEPC kinase activity (i.e. those in the middle of a normal dark period at 15°C) are transferred to 3°C in darkness, PEPC kinase activity remains constant (Carter *et al.*, 1994). However, if such leaves are illuminated, PEPC kinase disappears within 2 h and PEPC becomes dephosphorylated. These results indicate that low temperature can block the protein destruction step but that this effect can be overcome by light. If leaves containing PEPC kinase activity are transferred to 30°C in darkness, PEPC kinase again disappears and PEPC becomes dephosphorylated (Carter *et al.*, 1994). In other words, both light and high temperature cause disappearance of PEPC kinase. To assess the physiological significance of these data, we have examined the fixation of $CO_2$ by detached leaves maintained in constant darkness and normal air at different temperatures. At 15°C, leaves show a single bout of $CO_2$ fixation corresponding to a normal night. This is followed by conversion of PEPC to the dephosphorylated form and a slow, steady rate of $CO_2$ output (Nimmo *et al.*, 1987); similar behaviour is observed over the range 8–25°C. However, below 5°C, PEPC remains in the phosphorylated form and dark $CO_2$ fixation is prolonged for up to 36 h (Carter *et al.*, 1994).

Lowering the temperature from 25 to 3°C not only decreases the catalytic capacity of PEPC but also causes a considerable reduction (about 10-fold) in the sensitivity to malate of both the phosphorylated and dephosphorylated forms of PEPC when assayed *in vitro* (Carter *et al.*, 1994). Further effects of temperature on the properties of PEPC are discussed in Chapter 6. When leaves maintained in constant darkness and normal air at 25°C have reached a steady rate of $CO_2$ output, reducing the temperature to 3°C results in a bout of $CO_2$ fixation. The PEPC remains dephosphorylated and the leaf malate content does not change significantly (Carter *et al.*, 1994), so the most likely explanation of this behav-

of ability to accommodate these shifts in control will result in a limitation by one process and predispose the system to stress. Of course, it is in the nature of stress that once one part of the system has failed to function, then the rest of the network of control will start to collapse. It is, therefore, often difficult unambiguously to identify the point at which a stress is primarily sensed.

In this chapter, I shall outline two situations in which temperature exerts profound differential effects on metabolism. The first concerns temperature-dependent alterations in the relationship between photosynthesis and carbohydrate metabolism, while the second concerns the influence of temperature on photorespiration and the profound consequences that this has had for the evolution and operation of photosynthesis in $C_3$ plants.

## 4.2 Differential effects of temperature illustrated by changes in the relationship between photosynthesis and carbohydrate synthesis

Although research has only really scratched the surface of the myriad of differential effects of temperature on the component processes of photosynthesis, photorespiration and product synthesis, these are further complicated because (i) relationships between these processes change with other environmental conditions, such as light and $CO_2$ concentration and (ii) acclimation occurs within these processes at different rates, so that there is a constantly shifting pattern of the allocation of control.

As an example, we can consider the relationship between the rate of sucrose synthesis and its acclimation to that of photosynthetic $CO_2$ fixation and carbon export. Immediately after transfer to low temperature and limiting light, increases in the capacities of light-activated enzymes, such as Rubisco and the stromal fructose-1,6-bisphosphatase (FBPase), occur extremely rapidly, even though the photosynthetic capacity (measured at 25°C) does not rise immediately (Table 4.1 and Holaday *et al.*, 1992). This can be viewed as a means of temperature compensation (Hazel and Prosser, 1974) by which photosynthetic capacity is maintained. It is important to note that the temperature dependence of photosynthesis changes greatly with changes in environmental conditions, such as light and $CO_2$ concentration (Ludlow and Wilson, 1971). Of course, under low $CO_2$, the temperature response tends to be flat because of increases in photorespiration at higher temperatures. However, under low light in saturating $CO_2$ (Stitt and Grosse, 1988), the response of photosynthesis to temperature is also essentially flat. These changes in photosynthetic response imply the existence of factors which increase the capacity of enzymes, either by light activation, increases in their substrates or appropriate temperature-dependent changes in their kinetic parameters. (See Chapters 3 and 5 for further discussion of temperature effects on enzymes.)

In contrast to enzymes of the Benson–Calvin cycle, sucrose phosphate synthase (SPS) is not activated immediately upon transfer to low temperature (Holaday *et*

***Table 4.1.*** *Changes in photosynthesis, carbohydrates and enzymes in leaves of spinach plants after transfer from growth at 25°C to growth at 10°C for a period of 10 days*

| Parameter | Time after transfer to 10°C (days) | | |
|---|---|---|---|
| | 0 | 2 | 10 |
| Photosynthesis ($\mu$mol $O_2$ $m^{-2}$ $s^{-1}$) | 10.3 | 11.1 | 15.3 |
| Sucrose content ($\mu$mol $mg^{-1}$ Chl) | 16.8 | 35.7 | 28.8 |
| Partitioning ratio (sucrose:starch) | 4.0 | 3.6 | 4.8 |
| Sucrose phosphate synthase | | | |
| Total activity ($\mu$mol $h^{-1}$ $mg^{-1}$ Chl) | 34 | 39 | 57 |
| Activation state (%) | (41%) | (40%) | (32%) |
| Rubisco | | | |
| Total activity ($\mu$mol $h^{-1}$ $mg^{-1}$ Chl) | 478 | 500 | 578 |
| Activation state | (70%) | (99%) | (99%) |

Photosynthesis and carbon partitioning were determined in saturating $CO_2$ in a leaf-disc electrode at 25°C. Data for enzymes are from Holaday *et al.* (1992), the remainder are from W. Martindale and R.C. Leegood (unpublished results).

*al.*, 1992). This is consistent with the now well-documented observation that sucrose synthesis is particularly sensitive to transfer of plants to low temperature and that the rate of sucrose synthesis can limit the rate of photosynthesis. This is shown by a number of phenomena, including reversed $O_2$ sensitivity of photosynthesis (Sage and Sharkey, 1987), a flattening of the response of photosynthesis to $CO_2$ at above ambient $CO_2$ partial pressures and a response to the provision of phosphate fed via the transpiration stream (Labate and Leegood, 1988). This limitation by the rate of sucrose synthesis results in decreased recycling of Pi to the chloroplast and hence a limitation on the rate at which ATP can be generated to sustain regeneration of ribulose bisphosphate (RuBP) in the Benson–Calvin cycle. Exactly why this limitation arises remains to be resolved. There is evidence that it is not the maximum capacity of the enzymes involved (Leegood and Furbank, 1986), but instead may be related to such factors as the intrinsic temperature sensitivity of enzymes such as the cytosolic FBPase or SPS (Stitt and Grosse, 1988), the lack of ability rapidly to activate SPS referred to above (Holaday *et al.*, 1992) or to a temporarily suboptimal phosphate status of the cytosol (Leegood and Furbank, 1986). Testing these various hypotheses ultimately requires the use of appropriate transgenic or mutant plants in which one factor alone can be manipulated. Whatever the cause(s), it is clear that this limitation is overcome in a matter of hours (Labate and Leegood, 1990) and that such a feedback limitation via phosphate status is not a long-term means of regulating the relationship between photosynthesis and carbohydrate synthesis. Nevertheless, despite the fact that the rate of sucrose synthesis is initially inhibited, the sucrose content of the leaf rises as a result of diminished export (Table 4.1 and Farrar, 1988).

In the longer term, changes in the capacities of enzymes, mediated by changes in gene expression, appear to be more important in determining rates of photosynthesis and product synthesis (Stitt, 1991). There is clear evidence for a general upward regulation of the capacities of a range of enzymes after transfer to low temperature. These are required to sustain increased photosynthetic rates (i.e. acclimation) at low temperatures. In particular, there are many reports of increases in Rubisco activity (e.g. Badger *et al.,* 1982; see also references in Leegood and Edwards, 1995) as well as of other enzymes of the Benson–Calvin cycle (Badger *et al.,* 1982; Holaday *et al.,* 1992). In contrast to the short-term behaviour of sucrose synthesis, the amount (but not the activation state) of SPS increases over several days, as does photosynthetic capacity and partitioning of carbon into sucrose (Table 4.1). The imbalance between sucrose synthesis and photosynthesis which existed in the short-term is thus restored. The extent to which export resumes is less clear, but the accumulation of sucrose which occurs, probably in the extravacuolar compartment (Koster and Lynch, 1992) can be viewed as a mechanism which leads to cryoprotection. Sucrose and other soluble sugars can act as effective cryoprotectants of membranes and proteins at least *in vitro* (Guy, 1990; Guy *et al.*, 1992). These changes in carbon partitioning within the leaf can therefore be considered adaptive and it is important to realize that induction of freezing tolerance and responses to low temperature are part of a continuum. There are also obvious parallels with the responses of plants to increases in atmospheric $CO_2$ concentrations (in which carbohydrates accumulate) and with water stress (in which sucrose accumulates and SPS is activated in the short-term (Quick *et al.*, 1989)) which may contribute towards osmoregulation (although longer term water stress leads to $O_2$ insensivity, perhaps as a result of a decrease in SPS activity (Vassey and Sharkey, 1989)). However, mechanisms clearly differ between low temperature and responses to elevated $CO_2$ and water stress. The fact that sugars can accumulate at lower temperatures and, at the same time, are accompanied by acclimatory increases in the capacities of enzymes of the Benson–Calvin cycle and other metabolic pathways (Holaday *et al.*, 1992) suggests that sugars do not affect gene expression in the same way that they have been shown to at higher temperatures. Under these conditions, activities of Rubisco and other enzymes, as well as amounts of protein and mRNAs have been shown to decline in several systems in response to the supply or accumulation of sugars (Krapp *et al.*, 1991,1993; Schäfer *et al.*, 1992; Sheen, 1989), including plants exposed to elevated $CO_2$ concentrations (van Oosten and Besford, 1994).

In contrast to responses to low temperature such as those just described, recent evidence suggests that failure to make such adjustments in photosynthetic capacity may be an important factor leading to the syndrome of chilling sensitivity. In two chilling-sensitive plants, tomato and bean, there is evidence that inadequate light activation of enzymes occurs very early in the sequence of events following chilling. In tomato, for example, Sassenrath *et al.* (1991) showed that there was an accumulation of fructose-1,6-bisphosphate and sedoheptulose-1,7-bisphosphate after a 1-h exposure to high light at low temperatures. There was also a depletion of RuBP, indicating a restriction within the regenerative phase of the Benson–Calvin cycle relative to carboxylation. Sassenrath *et al.* (1991) showed that the light activation of the stromal FBPase was restricted by chilling tomato plants, suggesting that chilling disrupts the normal thioredoxin-dependent activation of the bisphosphatases. In contrast, in bean, chilling did not affect the light activation of NADP-malate dehydrogenase but did restrict the activation of Rubisco and the stromal FBPase (Holaday *et al.*, 1992). This indicates that the electron transport system was not affected and that, at this stage, loss of membrane function was not involved. The fact that the restriction occurred immediately after the chilling treatment suggests that the inability to activate these key enzymes might be one of the first steps in the response to chilling. In the longer term in bean, SPS is also very susceptible to loss of maximum catalytic activity and inactivation at low temperatures (Holaday *et al.*, 1992). However, as mentioned above, it is in the nature of stress that once one part of the system has failed to function (e.g. light activation via the thioredoxin system), then the rest of the network of control will start to collapse. Inability to activate these enzymes in chilling-sensitive plants might possibly be connected to differences in the redox potentials of these enzymes (see, for example, Kramer *et al.*, 1990).

## 4.3 The origins, consequences and control of photorespiration in relation to temperature

### 4.3.1 *The oxygenase activity of Rubisco*

Rubisco catalyses both the carboxylation and the oxygenation of RuBP. Photorespiration originates in the oxygenation reaction catalysed by Rubisco. Oxygenation of RuBP leads to the production of one molecule of glycerate-3-phospate and one of glycolate-2-phosphate.

$$\text{RuBP} + \text{H}_2\text{O} + \text{CO}_2 \xrightarrow{\text{Rubisco}} 2(\text{glycerate-3-phosphate}) + 2\text{H}^+$$
$$\text{RuBP} + \text{O}_2 \xrightarrow{\text{Rubisco}} \text{glycerate-3-phosphate} + \text{glycolate-2-phosphate} + 2\text{H}^+$$

Evolution has led to improvements in the specificity factor ($\Omega$) of Rubisco. $\Omega$ is a measure of the relative specificity for $CO_2$ and $O_2$. $\Omega = V_cK_o/V_oK_c$, where $V_c$, $V_o$ are $V_{max}$ values for carboxylation and oxygenation and $K_c$, $K_o$ are $K_m$ values for $CO_2$ and $O_2$, and it varies from about 10 in the bacteria, to 50 in the cyanobacteria, 60 in the algae and about 80 in higher plants. Intermediate values in the cyanobacteria and algae are offset by the presence of $CO_2$-concentrating mechanisms in these organisms (Jordan and Ogren, 1981). However, oxygenation has clearly not been eliminated during the course of evolution. Oxygenation occurs by direct attack of oxygen on the 2,3-enediol intermediate in the Rubisco reaction (Andrews and Lorimer, 1987) and is viewed as an inevitable consequence of the reaction mechanism of Rubisco.

The $K_m$ ($CO_2$) of Rubisco is about 10 μM, which is about the same as the concentration of $CO_2$ dissolved in water at 20°C, but the $CO_2$ concentration would be rather less in the mesophyll cells of a leaf in air (~ 5 μM). The $K_m$ ($O_2$) is about 535 μM, which is about double the $O_2$ concentration in a leaf at 20°C. Therefore, under ambient conditions, Rubisco catalyses both carboxylation and oxygenation of RuBP. The two substrates are competitive, so that raising the $CO_2$ concentration will inhibit oxygenation and vice versa.

In $C_3$ plants, $CO_2$-saturated photosynthesis has a steep response to temperature but, in low $CO_2$, photosynthesis has an essentially flat response to temperature (see Figure 4.2). This occurs because both the $K_m$ and $V_{max}$ for carboxylation by Rubisco have a similar temperature dependence when $CO_2$ is at or below the $K_m$ ($CO_2$) ($Q_{10}$ values of ~ 2.2) (Farquhar and von Caemmerer, 1982; Hall and Keys, 1983; Jordan and Ogren, 1984) and because the rate of photorespiratory $CO_2$ release increases with increasing temperature (Figure 4.1). Higher temperatures promote oxygenation, and hence photorespiration, in two ways. First, the solubility of $CO_2$ in water declines more rapidly than that of $O_2$ as the temperature is increased. For example, at 10°C, the ratio of the solubilities of $O_2$ to $CO_2$ in water is 20, whereas at 40°C it is 28 (Edwards and Walker, 1983). Second, the kinetic parameters change and the specificity factor (Ω) of

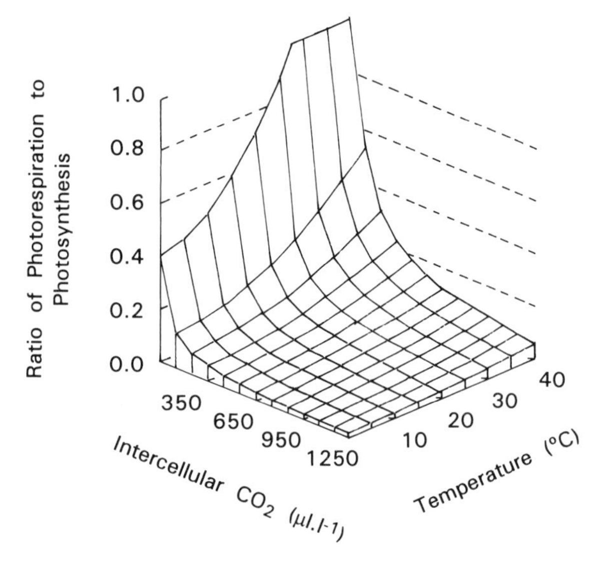


***Figure 4.1.*** *The ratio of photorespiration to photosynthesis in the leaf of a $C_3$ plant as a function of temperature and intercellular $CO_2$ concentration. The intercellular $CO_2$ concentration in air would typically be around 200 μl $l^{-1}$. Redrawn from Ehleringer* et al. *(1991) with permission from Elsevier Trends Journals.*

Rubisco decreases with increasing temperature in the range 5–40°C both *in vitro* (Jordan and Ogren, 1984) and *in vivo* (Brooks and Farquhar, 1985). This is because the reaction of the 2,3-enediol intermediate (Andrews and Lorimer, 1987) with $O_2$ has a higher free energy of activation than the reaction with $CO_2$. The difference in activation energies ($\Delta G_o^{\ddagger} - \Delta G_c^{\ddagger}$) is about 10.8 kJ $mol^{-1}$ in spinach and tobacco. This means that oxygenation is more sensitive to temperature and increases faster than carboxylation as the temperature rises (Chen and Spreitzer, 1992). Jordan and Ogren (1984) considered changes in Rubisco activity with temperature to be attributable about one-third to changes in solubility and about two-thirds to changes in the specificity factor and kinetic parameters. The response of photosynthesis and photorespiration in $C_3$ plants to varying temperature at a given $CO_2$ concentration can be calculated from the kinetic properties of Rubisco (Farquhar, 1988; Leegood and Edwards, 1995). The ratio of oxygenase ($V_o$) to carboxylase ($V_c$) activities (Jordan and Ogren, 1984) is given by:

$$\frac{V_o}{V_c} = \frac{[O_2]}{\Omega[CO_2]}$$

where $\Omega$ is the relative specificity factor. The $V_o/V_c$ ratio can be calculated using known values of $\Omega$ and $[CO_2]$, and $[O_2]$ at a given temperature (Jordan and Ogren, 1984).

Under light saturation, when RuBP is saturating and photosynthesis is limited by Rubisco, then according to the model of Farquhar (Sage and Sharkey, 1987; Farquhar, 1988):

$$A = \frac{V_c[CO_2](1 - 0.5\ V_o / V_c)}{[CO_2] + K_c(1 + [O_2]/K_o)} - R_d$$

$V_c$ and $V_o$ are the velocities of carboxylase and oxygenase respectively at substrate saturation, and $K_c$ and $K_o$ are the Michaelis constants for $CO_2$ and $O_2$. $R_d$, respiration in the dark, is used as an estimate of mitochondrial respiration with a $Q_{10}$ of approximately 2 (Collatz *et al.*, 1992; Edwards and Baker, 1993; Long, 1991).

Alternatively, the temperature dependence of carboxylation and oxygenation can be calculated using the relationship (Brooks and Farquhar, 1985):

$$V_o / V_c = 2\Gamma^* / [CO_2]$$

where $\Gamma^*$ is the measured $CO_2$ compensation point for photosynthesis in the absence of dark respiration, with the following temperature dependence (Brooks and Farquhar, 1985):

$$\Gamma^* = P[42.7 + 1.68(T - 25) + 0.012(T - 25)^2]$$

where T is the temperature (°C) and P is the atmospheric pressure in bars.

Aside from the differential effects of temperature at the level of Rubisco, temperature might also be expected to have differential effects on various steps within the photorespiratory pathway and thus alter the manner in which it is regulated. One example is the non-enzymic decarboxylation of glyoxylate by isolated peroxisomes fed glycolate. This is a result of the higher temperature coefficient of glycolate oxidase activity (which generates $H_2O_2$) relative to that of catalase (which destroys $H_2O_2$). As a result, glycolate is oxidized to $CO_2$ about three times more rapidly at 35°C than at 25°C (Grodzinski and Butt, 1977). Although this pathway is not usually considered to be an important component of photorespiration, it can become so in null mutants of the photorespiratory pathway (Somerville and Ogren, 1981) and may also be important in heterozygous plants with lowered activities of chloroplastic glutamine synthetase (Häusler *et al.*, 1994b).

The glycolate-2-phosphate which is produced by Rubisco when it oxygenates RuBP cannot be utilized within the Calvin cycle. Instead, it is salvaged, albeit inefficiently, in the photorespiratory pathway. This pathway involves three subcellular compartments; the chloroplasts, peroxisomes and mitochondria. The key features of this pathway are the conversion of the two-carbon molecule, glycolate-2-phosphate, to glycine and decarboxylation of two molecules of glycine to serine, $CO_2$ and $NH_3$. The three-carbon molecule, serine, is then converted to glycerate-3-phosphate, which re-enters the Benson–Calvin cycle. $CO_2$ release results in the release of one-quarter of the carbon in glycolate and decreases the efficiency of photosynthesis. The principal factor influencing the rate of photorespiration is temperature. In the present atmosphere, photorespiration in $C_3$ plants results in substantial losses of fixed carbon, particularly at higher temperatures. In air (350 p.p.m. $CO_2$, 21% $O_2$) the ratio of the rate of photorespiration to photosynthesis is around 0.1 at 10°C, rising to about 0.3 at 40°C, but low intercellular concentrations of $CO_2$, as may occur, for example, under water stress, can result in ratios of 0.6 or higher at high temperatures. The photorespiratory pathway also emphasizes the close integration of carbon and nitrogen metabolism in a leaf, because $NH_3$ is also released during photorespiration at a rate up to 10 times faster than the primary assimilation of $NH_3$ from nitrate (Keys *et al.*, 1978). Since $NH_3$ is a more valuable resource than $CO_2$, $NH_3$ must be efficiently refixed by glutamine synthetase (GS) and glutamine synthase (GOGAT) in the chloroplast.

Historically, photorespiration has not always been as important as it is currently. It was only at the end of the Cretaceous (63 million years ago), during the rise of the angiosperms, that atmospheric $CO_2$ concentrations declined close to present concentrations (from around 5–10 times the current concentration), with an intervening rise during the Eocene and Oligocene (30–50 million years ago) (Ehleringer *et al.*, 1991). At 200 p.p.m. $CO_2$ (pre-industrial levels), photorespiration could easily reach 50% of net photosynthesis at elevated temperatures in $C_3$ plants. One possible consequence of this decrease in atmospheric $CO_2$ concentrations may have been the evolution of $CO_2$-concentrating mechanisms in land plants, that is the development of $C_3$–$C_4$ intermediate and $C_4$ modes of photosynthesis. $C_4$ photosynthesis is thought to have evolved not more than 30 million years ago, perhaps as recently as 7 million years ago (the date of the oldest known fossilized $C_4$ material). There is evidence, from $\delta^{13}C$ measurements of mammalian tooth enamel and palaeosol carbonate deposits, for a considerable expansion of $C_4$ ecosystems between 5 and 7 million years ago, perhaps as a response to a detectable decrease in atmospheric $CO_2$ (Cerling *et al.*, 1993), although recent evidence suggests widespread $C_4$ ecosystems which preceded these (Morgan *et al.*, 1994). Morgan *et al.* also suggest that changes in $CO_2$ may not have been the only factor leading to the evolution of $C_4$ photosynthesis. Apart from changes in $CO_2$ concentration, it is quite possible that changes in temperature also played an important role. Comparisons of $C_3$, $C_3$–$C_4$ intermediate and $C_4$ species in the genus *Flaveria* suggest that high temperatures may well be a factor which led to advantages in carbon gain of $C_3$–$C_4$ intermediate modes of photosynthesis over $C_3$ photosynthesis. At least some of these $C_3$–$C_4$ intermediate plants are believed to be evolutionary intermediates between $C_3$ and $C_4$ species (Monson and Jaeger, 1991; Monson and Moore, 1989; Schuster and Monson, 1990).

### 4.3.2 *Mutants of the photorespiratory pathway*

Mutants of barley, pea and *Arabidopsis thaliana* have been isolated which lack particular enzymes of the photorespiratory pathway (Blackwell *et al.*, 1988b; Ogren, 1984). The screening for these mutants has been done by growing mutagenized plants in 2% $CO_2$, which suppresses photorespiration, and then transferring them to air. Plants lacking enzymes of the photorespiratory pathway show visible symptoms and die rapidly. However, rescue of these plants by return to elevated $CO_2$ has allowed identification of a range of mutants. These include different plants lacking glycolate-2-phosphate phosphatase, catalase, glycine decarboxylase, serine hydroxymethyltransferase and serine:glyoxylate aminotransferase as well the enzymes involved in $NH_3$ assimilation (GS and GOGAT) and the chloroplast dicarboxylate transporter. All these mutants indicate that, since growth in high $CO_2$ in such plants is normal, the photorespiratory pathway is not required for synthesis of intermediates for biosynthesis.

A number of recent studies have employed mutants and transgenic plants to study the control of metabolism (e.g. Kruckeberg *et al.*, 1989; Stitt *et al.*, 1991). Flux control analysis involves asking how much the flux changes for a small given change in the amount of an enzyme (Kacser and Burns, 1973). Recently, photorespiratory mutants and wild-type barley have been crossed to generate a series of heterozygous plants with varying activities of GS between 48 and 96% of the wild-type and GOGAT activities down to 70% of the wild-type (Häusler *et al.*, 1994a). Decreased GS led to an increase in leaf $NH_3$, a large decrease in total amino acids and had a small negative impact on leaf protein and Rubisco (Häusler *et al.*, 1994a), showing that even small reductions in GS limit $NH_3$ assimilation and may bring about a loss of nitrogen from the plants. These plants were used to study the control of carbon and nitrogen metabolism and photosynthetic electron transport under different conditions of light, $CO_2$ and $O_2$ concentration and temperature.

### 4.3.3 *Effects of $CO_2$, light and temperature on photosynthesis and photorespiration in photorespiratory mutants*

Häusler *et al.* (1994a,b) have investigated the control of photosynthesis and photorespiratory carbon metabolism in photorespiratory mutants, asking how changes in the activities of these enzymes affect fluxes of $CO_2$ and of electron transport under changing environmental conditions which lead to changes in the relative rates of photosynthesis and photorespiration. The influence of temperature on the rate of $CO_2$ assimilation is shown in Figure 4.2. At an intercellular $CO_2$ concentration ($C_i$) of 250 p.p.m., an increase in temperature between 11 and 22°C resulted in a marked increase in $CO_2$ assimilation in the wild-type and the ferredoxin (Fd)-dependent-GOGAT mutant, respectively. With a further increase in temperature to 30°C, rates of $CO_2$ assimilation were enhanced only in the Fd-GOGAT mutant, but remained virtually unchanged in the wild-type. In the 47% GS mutant, the increase in the rates of $CO_2$ assimilation with an increase in temperature from 11 to 22°C was less steep compared to the wild-

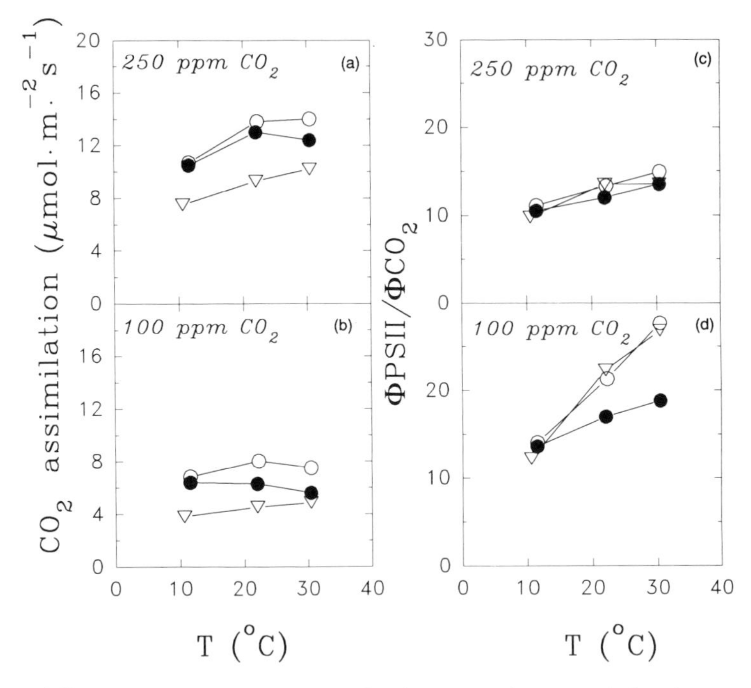


***Figure 4.2.*** *Temperature dependence of carbon assimilation and the electron requirement for $CO_2$ assimilation (ΦPSII/Φ$CO_2$) for wild-type barley (○), a 47% GS mutant (●), and a 63% Fd-GOGAT mutant (▽). The external $CO_2$ concentration was adjusted to give an intercellular $CO_2$ concentration of either 250 p.p.m. (a,c) or 100 p.p.m. (b,d) at each temperature. The PFD was 660 μmol $m^{-2}$ $s^{-1}$. From Häusler* et al. *(1994b).*

type and there was a decrease in $CO_2$ assimilation as the temperature was increased to 30°C. The response of $CO_2$ assimilation to changes in temperature at lower concentrations of $CO_2$ (100 p.p.m.) was less marked than in ambient $CO_2$. The relative changes in $CO_2$ assimilation with changing temperatures in both the wild-type and the Fd-GOGAT mutant were relatively similar in low $CO_2$, whereas $CO_2$ assimilation rates in the 47% GS mutant declined considerably with an increase in the temperature above 11°C.

Although the absolute magnitude of $CO_2$ assimilation was governed partially by Rubisco activities, which exhibited some variation between the plants (Häusler *et al.*, 1994a,b), these results indicate that there were temperature-dependent changes in the $CO_2$ assimilation rates which were associated with the decrease in GS or Fd-GOGAT. The inhibition of $CO_2$ assimilation with an increase in temperature in the GS mutant under conditions of enhanced photorespiratory flux suggests that a decrease in GS feeds back on $CO_2$ assimilation, although the data do not reveal the mechanism by which this might occur.

Apart from measurements of gas exchange, the quantum yield of electron transport by photosystem II (ΦPSII) can be estimated from chlorophyll fluorescence parameters (Genty *et al.*, 1989; Harbinson *et al.*, 1990) and can be used to estimate the rate of linear electron transport. It has been shown that a linear relationship exists between ΦPSII and the apparent quantum yield of $CO_2$ assimilation ($\Phi CO_2$) under non-photorespiratory conditions in $C_3$ plants (Krall and Edwards, 1990). Under photorespiratory conditions, this relationship becomes non-linear (Genty *et al.*, 1989; Harbinson *et al.*, 1990; Krall and Edwards, 1990) and ΦPSII increases in 21% $O_2$, particularly at low $CO_2$ concentrations. This is due to an additional energy requirement for processes which result from the oxygenation of RuBP and loss of $CO_2$, the recycling of carbon to the Benson–Calvin cycle, and the re-assimilation of ammonium via the GS/Fd-GOGAT cycle. The relationship between ΦPSII and $\Phi CO_2$ therefore provides information on the relative flux of electrons required to support both photosynthesis and photorespiration.

In barley leaves photosynthesizing in ambient $CO_2$, the relative electron requirement for $CO_2$ assimilation, expressed as $\Phi PSII/\Phi CO_2$ ratio, increased linearly with an increase in temperature from 11 to 30°C (Figure 4.2). These data are therefore consistent with enhanced photorespiratory fluxes at higher temperatures. There were only small differences in the $\Phi PSII/\Phi CO_2$ ratio between the different plants. In low $CO_2$, there was a marked increase in the electron requirement with temperature in both the wild-type and the Fd-GOGAT mutant. This is consistent with increased photorespiratory fluxes as the temperature was increased in low $CO_2$. However, the electron requirement in the GS mutant was only slightly greater than in ambient $CO_2$ and was lower than in the wild-type and the Fd-GOGAT mutant. This observation suggests that a lower rate of electron transport is required to maintain a given rate of $CO_2$ assimilation in the GS mutant as $C_i$ is lowered (see also Häusler *et al.*, 1994b).

Thus a decrease in GS or Fd-GOGAT activity can influence rates of $CO_2$ assimilation and/or the apparent electron requirement for $CO_2$ assimilation under photorespiratory conditions. Although the 47% GS mutant had rates of $CO_2$ assimilation over a wide range of intercellular $CO_2$ concentrations and photon flux densities (PFDs) which were comparable to the wild-type, the electron requirement per $CO_2$ assimilated was decreased under conditions of enhanced photorespiratory flux (Häusler *et al.*, 1994b). Figure 4.3 summarizes the electron requirement for $CO_2$ assimilation ($\Phi$PSII/$\Phi CO_2$ ratios) determined for the whole range of mutants under different environmental conditions. At moderate PFDs and ambient $CO_2$, there were apparently no differences in the

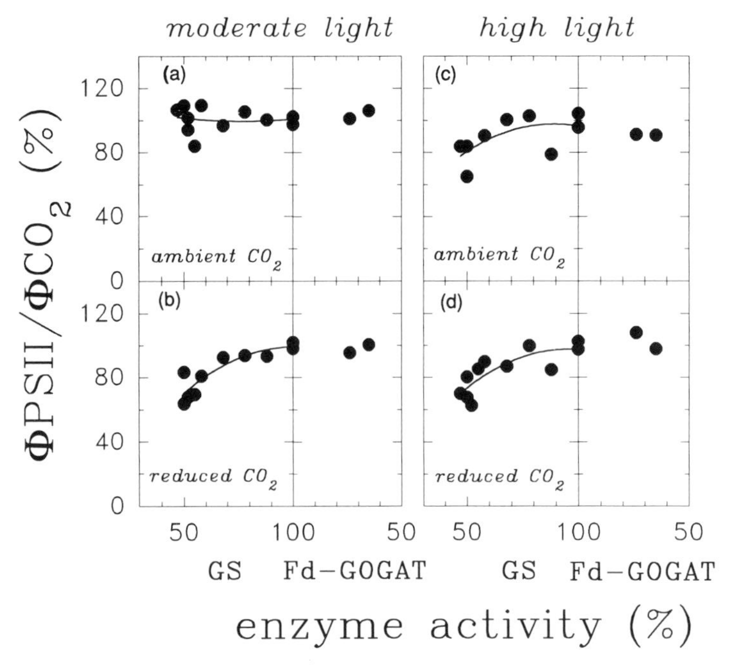


***Figure 4.3.*** *Dependence of the electron requirement for $CO_2$ assimilation ($\Phi$PSII/$\Phi CO_2$ ratio) on the fractional decrease in the activities of glutamine synthetase (GS) and glutamate synthase (Fd-GOGAT) in barley leaves. The intercellular $CO_2$ concentration was approximately 230 p.p.m. (a,c) or 100 p.p.m. (b,d) at an intermediate PFD of 360 μmol $m^{-2}$ $s^{-1}$ (a,b) or at a high PFD of 1090 μmol $m^{-2}$ $s^{-1}$ (c,d). From Häusler* et al. *(1994b).*

*In the wild-type, the $\Phi$PSII/$\Phi CO_2$ ratios at moderate PFDs of 12 and 18 in ambient and low $CO_2$, respectively, are consistent with theoretical electron requirements expected for non-cyclic electron transport (a, b). The $\Phi$PSII/$\Phi CO_2$ ratios at high PFDs were further increased to approximately 16 and 24 in ambient and low $CO_2$, respectively, which is likely to reflect an additional electron flux to $O_2$ (i.e. Mehler reaction) (c, d).*

electron requirement per $CO_2$ assimilated for the whole range of GS mutants. With an increase in the PFD in ambient $CO_2$, the $\Phi PSII/\Phi CO_2$ ratio decreased at a GS activity below 70% and was reduced by 15% in mutants with 50% less GS. This effect was more pronounced in low $CO_2$ at both moderate and high light. At both PFDs, the $\Phi PSII/\Phi CO_2$ ratios decreased by 30–40% with a decrease in GS to 50% of the wild-type.

These data and the temperature dependence of the $\Phi PSII/\Phi CO_2$ show that the electron requirement in the GS mutants was decreased appreciably only under conditions of enhanced photorespiratory flux. There are three possible reasons for a decrease in the $e/CO_2$ ratio with a reduction in GS activity under photorespiratory conditions.

(i) The decrease in electron transport represents that part of electron transport which is normally used to assimilate ammonium via GS and Fd-GOGAT. This is unlikely because, even at the $CO_2$ compensation point, assimilation of all of the ammonium generated in photorespiration would only account for about 15% of the rate of electron transport (Farquhar and von Caemmerer, 1982).

(ii) The second possibility is an inhibition of photorespiratory $CO_2$ release. This could occur either by metabolically induced changes in the specificity factor of Rubisco, which is unlikely (Jordan and Ogren, 1984) or it could occur by an inhibition of glycine decarboxylation. An inhibition of glycine decarboxylation would result in a more favourable ratio of $CO_2$ fixation to $CO_2$ release as well as in a decrease in the electron requirement for the recycling of intermediates back into the Benson–Calvin cycle and for the re-assimilation of ammonium. Measurements of the ammonium contents of these plants under photorespiratory conditions also lend support to the notion that there may be an inhibition of glycine decarboxylation, because ammonium contents rose with a decrease to 66% GS, but then fell as GS was further decreased (Häusler *et al.*, 1994a). Although it is conceivable that ammonium might cause end-product inhibition of the conversion of glycine to serine, it is known that ammonium is an inhibitor of the transamination of glyoxylate to glycine (Havir, 1986).

If the conversion of glycine to serine were inhibited under conditions of enhanced photorespiratory flux, this might also lead to an accumulation of glyoxylate, rather than glycine, because the leaf would be depleted of amino acids and hence of amino donors (Blackwell *et al.*, 1988a; Häusler *et al.*, 1994a; Ta and Joy, 1986). A shortage of glycine might then be another mechanism which could limit glycine decarboxylation. Alternatively, an accumulation of glyoxylate and/or formate may influence the rate of glycine decarboxylation (Shingles *et al.*, 1984). Glyoxylate could then either be partially decarboxylated in the peroxisomes to yield $CO_2$ and formate (Somerville and Ogren, 1981), or it could be converted into a variety of organic acids (e.g. malate, tartrate, oxalate, isocitrate, glycolate, hydroxycitrate and 4-hydroxy-2-oxoglutarate) via several pathways known to exist in leaves (Igamberdiev, 1989). One possibility is formation of malate by malate synthase (Murray *et al.*, 1987). Malate accumulation has been observed in *Arabidopsis* mutants lacking serine transhydroxymethylase (Somerville and Ogren, 1981) and in plants treated with the herbicidal inhibitor of GS, L-methionine sulphoximine (Walker *et al.*, 1984). Barley mutants lacking serine:glyoxylate aminotransferase also convert [$^{14}$C]glyoxylate into malate (Murray *et al.*, 1987). Another possibility is the condensation of glyoxylate with succinate yielding isocitrate, which is catalysed by isocitrate lyase (Zelitch, 1988; Zemlyanukhin *et al.*, 1987). Both malate synthase and isocitrate lyase are readily induced in senescent leaves (Gut and Matile, 1988).

(iii) The third possibility is that an additional carboxylation process occurs which does not depend directly upon the provision of ATP and NADPH by photosynthetic electron transport. One candidate could be phosphoenolpyruvate carboxylase (PEPC). This would also be consistent with the observations of malate accumulation, as discussed above, although at present there is little evidence either to support or to refute the involvement of PEPC (Häusler *et al.*, 1994b).

These observations on the barley mutants with decreased GS therefore suggest that photorespiration is suppressed when plants are transferred from air to lower $CO_2$ concentrations, higher light intensities or higher temperatures. This has the effect of decreasing the $CO_2$ evolved by glycine decarboxylation. This would have the additional advantage that $NH_3$ evolution would also be decreased. In circumstances in which evolution of $NH_3$ exceeds the capacity to refix it via GS, such a diversion could have an important role in sparing the loss the $NH_3$ and thus improving the nitrogen economy of the plant. Even the wild-type plants showed control over the electron requirement for $CO_2$ assimilation, with values of the control coefficient, $C^{Je}_{GS} = 0.1$ at intermediate light intensities and low $CO_2$ and $C^{Je}_{GS} = 0.13$ at high light intensities and low $CO_2$ (where $J_e$ = the electron requirement for $CO_2$ assimilation) (see Figure 4.3 and Häusler *et al.*, 1994b), suggesting that such a mechanism operates in the wild-type as well as in the mutants. Indeed, the amount of ammonia is usually higher in illuminated than in darkened leaves of $C_3$ plants (see, for example, Häusler *et al.*, 1994a), suggesting that the capacity to generate $NH_3$ often slightly exceeds the capacity to refix it. It is also known that plants have both an ammonia compensation point and lose ammonia to the atmosphere (Farquhar *et al.*, 1980; Schjoerring *et al.*, 1993). Thus at high temperatures, under water stress (and hence low intercellular concentrations of $CO_2$), or under conditions of poor nitrogen nutrition, which would all promote high rates of photorespiration, and hence high rates of $NH_3$ release, the temporary diversion of photorespiratory carbon may be of considerable physiological importance even in wild-type plants which are usually considered to possess sufficient GS.

## References

Andrews, T.J. and Lorimer, G.H. (1987) In: *Biochemistry of Plants*, Volume 10 (eds P.K. Stumpf and E.G. Conn). Academic Press, New York, pp. 132–218.

Badger, M.R., Björkman, O. and Armond, P.A. (1982) An analysis of photosynthetic response and adaptation to temperature in higher plants: Temperature acclimation in the desert evergreen, *Nerium oleander* L. *Plant, Cell Environ.* **5**, 85–99.

Blackwell, R.D., Murray, A.J.S., Lea, P.J. and Joy, K.W. (1988a) Photorespiratory amino donors, sucrose synthesis and the induction of $CO_2$ fixation in barley deficient in glutamine synthetase and/glutamate synthase. *J. Exp. Bot.* **39**, 845–858.

Blackwell, R.D., Murray, A.J.S., Lea, P.J., Kendall, A.C., Hall, N.P., Turner, J.C. and Wallsgrove, R.M. (1988b) The value of mutants unable to carry out photorespiration. *Photosynth. Res.* **16**, 155–176.

Brooks, A. and Farquhar, G.D. (1985) Effect of temperature on the $CO_2/O_2$ specificity of ribulose-1,5-bisphosphate carboxylase/oxygenase and the rate of respiration in the light. *Planta* **165**, 397–406.

Cerling, T.E., Wang, Y. and Quade, J. (1993) Expansion of $C_4$ ecosystems as an indicator of global ecological change in the late Miocene. *Nature* **361**, 344–345.

Chen, Z. and Spreitzer, R.J. (1992) How various factors influence the $CO_2/O_2$ specificity of ribulose-1,5 bisphosphate carboxylase/oxygenase. *Photosynth. Res.* **31**, 157–164.

Collatz, G.J., Ribas-Carbo, M. and Berry, J.A. (1992) Coupled photosynthesis–stomatal conductance model for leaves of $C_4$ plants. *Aust. J. Plant Physiol.* **19**, 519–538.

Edwards, G.E. and Baker, N.R. (1993) Can $CO_2$ assimilation in maize leaves be predicted accurately from chlorophyll fluorescence analysis? *Photosynth. Res.* **37**, 89–102.

Edwards, G.E. and Walker, D.A. (1983) *$C_3$, $C_4$: Mechanisms and Cellular and Environmental Regulation of Photosynthesis.* Blackwell Scientific Publications, Oxford.

Ehleringer, J.R., Sage, R.F., Flanagan, L.B. and Pearcy, R.W. (1991) Climate change and the evolution of $C_4$ photosynthesis. *Trends Ecol. Evol.* **6**, 95–99.

Farquhar, G.D. (1988) Models relating subcellular effects of temperature to whole plant responses. In: *Plants and Temperature,* Volume 42 (eds S.P. Long and F.I. Woodward), Symposia for the Society for Experimental Biology. Company of Biologists, Cambridge, pp. 395–409.

Farquhar, G.D. and von Caemmerer, S. (1982) Modelling of photosynthetic response to environmental conditions. In: *Encyclopedia of Plant Physiology New Series, Volume 12B. Physiological Plant Ecology II. Water Relations and Carbon Assimilation* (eds P.L. Lange, P.S. Nobel, C.B. Osmond, and H. Ziegler). Springer Verlag, Berlin, pp. 549–588.

Farquhar, G.D., Firth, P.M., Wetslaar, R. and Weir, B. (1980) On the gaseous exchange of ammonia between leaves and the environment: Determination of the ammonia compensation point. *Plant Physiol.* **66**, 710–714.

Farrar, J.F. (1988) Temperature and the partitioning and translocation of carbon. *SEB Symp.* **42**, 203–235.

Genty, B., Briantais, J.M. and Baker, N.R. (1989) The relationship between the quantum yield of photosynthetic transport and quenching of chlorophyll fluorescence. *Biochim. Biophys. Acta* **990**, 87–92.

Grodzinski, B. and Butt, V.S. (1977) The effect of temperature on glycolate decarboxylation in leaf peroxisomes. *Planta* **133**, 261–266.

Gut, H. and Matile, P. (1988) Apparent induction of key enzymes of the glyoxylic acid cycle in senescent barley leaves. *Planta* **176**, 548–550.

Guy, C.L. (1990) Cold acclimation and freezing tolerance: role of protein metabolism. *Annu. Rev. Plant Physiol. Plant Mol. Biol.* **41**, 187–223.

Guy, C.L., Huber, J.L.A. and Huber, S.C. (1992) Sucrose phosphate synthase and sucrose accumulation at low temperature. *Plant Physiol.* **100**, 502–508.

Hall, N.P. and Keys, A.J. (1983) Temperature dependence of the enzymic carboxylation and oxygenation of ribulose 1,5-bisphosphate in relation to effects of temperature on photosynthesis. *Plant Physiol.* **72**, 945–948.

Harbinson, J., Genty, B. and Baker, N.R. (1990) The relationship between $CO_2$ assimilation and electron transport in leaves. *Photosynth. Res.* **25**, 213–224.

Häusler, R.E., Blackwell, R.D., Lea, P.J. and Leegood, R.C. (1994a) Control of photosynthesis in barley leaves with reduced activities of glutamine synthetase and glutamate synthase. I. Plant characteristics and changes in ammonium, nitrate and amino acids. *Planta* **194**.

Häusler, R.E., Lea, P.J. and Leegood, R.C. (1994b) Control of photosynthesis in barley leaves with reduced activities of glutamine synthetase and glutamate synthase. II. Control of electron transport and $CO_2$ assimilation. *Planta* **194**.

Havir, E.A. (1986) Inactivation of serine/glyoxylate and glutamate:glyoxylate aminotransferases from tobacco leaves by glyoxylate in the presence of ammonium ions. *Plant Physiol.* **80**, 473–478.

Hazel, J.R. and Prosser, C.L. (1974) Molecular mechanisms of temperature compensation in poikilotherms. *Physiol. Rev.* **54**, 620–677.

Holaday. A.S., Martindale, W., Alred, R., Brooks, A.L. and Leegood, R.C. (1992) Changes in activities of enzymes of carbon metabolism in leaves during exposure of plants to low temperature. *Plant Physiol.* **98**, 1105–1114.

Igamberdiev, A.U. (1989) Pathways of glycolate conversion in plants. *Biol. Rundsch.* **27**, 137–144.

Jordan, D.B. and Ogren, W.L. (1981) Species variation in the specificity of ribulose bisphosphate carboxylase oxygenase. *Nature* **291**, 513–515.

Jordan, D.B. and Ogren, W.L. (1984) The $CO_2/O_2$ specificity of ribulose 1,5-bisphosphate carboxylase/oxygenase. Dependence on ribulosebisphosphate concentration, pH and temperature. *Planta* **161**, 308–313.

Kacser, H. and Burns, J.A. (1973) The control of flux. In: *Rate Control of Biological Processes,* Volume 27 (ed. D.D. Davies), Symposia for the Society of Experimental Biology. Cambridge University Press, Cambridge, pp. 65–104.

Keys, A.J., Bird, I.F., Cornelius, M.J., Lea, P.J., Wallsgrove, R.M. and Miflin, B.J. (1978) Photorespiratory nitrogen cycle. *Nature* **275**, 741–743.

Koster, K.L. and Lynch, D.V. (1992) Solute accumulation and compartmentation during the cold acclimation of puma rye. *Plant Physiol.* **98**, 108–113.

Krall, J.P. and Edwards, G.E. (1990) Quantum yields of photosystem II electron transport and carbon dioxide fixation in $C_4$ plants. *Aust. J. Plant Physiol.* **17**, 579–588.

Kramer, D.M., Wise, R.R., Frederick, J.R., Alm, D.M., Hesketh, J.D., Ort, D.R. and Crofts, A.R. (1990) Regulation of coupling factor in field-grown sunflower: A redox model relating coupling factor activity to the activities of other thioredoxin-dependent enzymes. *Photosynth. Res.* **26**, 213–222.

Krapp, A., Quick, W.P. and Stitt, M. (1991) Ribulose-1,5-bisphosphate carboxylase-oxygenase, other photosynthetic enzymes and chlorophyll decrease when glucose is supplied to mature spinach leaves via the transpiration stream. *Planta* **186**, 58–69.

Krapp, A., Hofmann, B, Schäfer, C. and Stitt, M. (1993) Regulation of the expression of *rbcS* and other photosynthetic genes by carbohydrates: a mechanism for the 'sink regulation' of photosynthesis. *Plant J.* **3**, 817–828.

Kruckeberg, A., Neuhaus, H.E., Feil, R., Gottlieb, L. and Stitt, M. (1989) Decreased-activity mutants of phosphoglucose isomerase in the cytosol and chloroplast of *Clarkia xantiana*. I. Impact on mass–action ratios and fluxes to sucrose and starch, and estimation of flux control coefficients and elasticity coefficients. *Biochem. J.* **261**, 932–940.

Labate, C.A. and Leegood, R.C. (1988) Limitation of photosynthesis by changes in temperature. Factors affecting the response of carbon dioxide assimilation to temperature in barley leaves. *Planta* **173**, 519–527.

Labate, C.A. and Leegood, R.C. (1990) Factors influencing the capacity for photosynthetic carbon assimilation in barley leaves at low temperature. *Planta* **182**, 492–500.

Leegood, R.C. and Edwards, G.E. (1995) Carbon metabolism and photorespiration: Temperature dependence in relation to other environmental factors. In: *Advances in Photosynthesis* (ed. N.R. Baker). Kluwer, Dordrecht, in press.

Leegood, R.C. and Furbank, R.T. (1986) Stimulation of photosynthesis by 2% oxygen at low temperatures is restored by phosphate. *Planta* **168**, 84–93.

Long, S.P. (1991) Modification of the response of photosynthetic productivity to rising temperature by atmospheric $CO_2$ concentrations: Has its importance been underestimated? *Plant, Cell Environ.* **14**, 729–739.

Ludlow, M.M. and Wilson, G.L. (1971) Photosynthesis of tropical pasture plants. I. Temperature, carbon dioxide concentration, leaf temperature and leaf–air vapour pressure difference. *Aust. J. Biol. Sci.* **24**, 449–470

Monson, R.K. and Jaeger, C.H. (1991) Photosynthetic characteristics of $C_3$–$C_4$ intermediate *Flaveria floridana* (Asteraceae) in natural habitats: Evidence of advantages to $C_3$–$C_4$ photosynthesis at high leaf temperatures. *Am. J. Bot.* **78**, 795–800.

Monson, R.K. and Moore, B.D. (1989) On the significance of $C_3$–$C_4$ intermediate photosynthesis to the evolution of $C_4$ photosynthesis. *Plant, Cell Environ.* **12**, 689–699.

Morgan, M.E., Kingston, J.D. and Marino, B.D. (1994) Carbon isotopic evidence for the emergence of $C_4$ plants in the Neogene from Pakistan and Kenya. *Nature* **367**, 162–165.

Murray, A.J.S., Blackwell, R.D., Joy, K.W. and Lea, P.J. (1987) Photorespiratory N donors, aminotransferase specificity and photosynthesis in a mutant of barley deficient in serine:glyoxylate aminotransferase activity. *Planta* **172**, 106–113.

Ogren, W.L. (1984) Photorespiration: Pathways, regulation, and modification. *Annu. Rev. Plant Physiol.* **35**, 415–442.

Quick, P., Siegl, G., Neuhaus, E., Feil, R. and Stitt, M. (1989) Short-term water stress leads to a stimulation of sucrose synthesis by activating sucrose-phosphate synthase. *Planta* **177**, 535–546.

Sage R.F. and Sharkey T.D. (1987) The effect of temperature on the occurrence of $O_2$ and $CO_2$-insensitive photosynthesis in field grown plants. *Plant Physiol.* **84**, 658–664.

Sassenrath, G.F., Ort, D.R. and Portis, A.R. Jr (1991) Impaired reductive activation of stromal bisphosphatases in tomato leaves following low-temperature exposure at high light. *Arch. Biochem. Biophys.* **282**, 302–308.

Schäfer, C., Simper, H. and Hofmann, B. (1992) Glucose feeding results in coordinated changes of chlorophyll content, ribulose-1,5-bisphosphate carboxylase-oxygenase activity and photosynthetic potential in photoautotrophic suspension cultured cells of *Chenopodium rubrum*. *Plant, Cell Environ.* **15**, 343–350.

Schjoerring, J.K., Kyllingsbaek, A., Mortensen, J.V. and Byskov-Nielsen, S. (1993) Field investigations of ammonia exchange between barley plants and the atmosphere. I. Concentration profiles and flux densities of ammonia. *Plant, Cell Environ.* **16**, 161–167.

Schuster, W.S. and Monson, R.K. (1990) An examination of the advantages of $C_3$–$C_4$ intermediate photosynthesis in warm environments. *Plant, Cell Environ.* **13**, 903–912.

Sheen, J. (1989) Metabolic represssion of transcription in higher plants. *Plant Cell* **2**, 1027–1038.

Shingles, R., Woodrow, L. and Grodzinski, B. (1984) Effects of glycolate pathway intermediates on glycine decarboxylation and serine synthesis in pea (*Pisum sativum* L.). *Plant Physiol.* **74**, 705–710.

Somerville, C.R. and Ogren, W.L. (1981) Photorespiration-deficient mutants of *Arabidopsis thaliana* lacking mitochondrial serine transhydroxymethylase activity. *Plant Physiol.* **67**, 666–671.

Stitt, M. (1991) Rising $CO_2$ levels and their potential significance for carbon flow in photosynthetic cells. *Plant, Cell Environ.* **14**, 741–762.

Stitt, M. and Grosse, H. (1988) Interactions between sucrose synthesis and $CO_2$ fixation. IV. Temperature-dependent adjustment of the relation between sucrose synthesis and $CO_2$ fixation. *J. Plant Physiol.* **133**, 392–400.

Stitt, M., Quick, W.P., Schurr, U., Schulze, E.D., Rodermel, S.R. and Bogorad, L. (1991) Decreased ribulose-1,5-bisphosphate carboxylase-oxygenase in transgenic tobacco transformed with 'antisense' *rbcS* II. Flux-control coefficients for photosynthesis in varying light, $CO_2$, and air humidity. *Planta* **183**, 555–566.

Ta, T.C. and Joy, K.W. (1986) Metabolism of some amino acids in relation to the photorespiratory nitrogen cycle of pea leaves. *Planta* **169**, 117–122.

van Oosten, J.-J. and Besford, R.T. (1994) Sugar feeding mimics effect of acclimation to high $CO_2$: Rapid down regulation of RuBisCo small subunit transcripts but not of the large subunit transcripts. *J. Plant Physiol.* **143**, 306–312.

Vassey, T.L. and Sharkey, T.D. (1989) Mild water stress of *Phaseolus vulgaris* plants leads to reduced starch synthesis and extractable sucrose phosphate synthase activity. *Plant Physiol.* **89**, 1066–1070.

Walker, K.A., Keys, A.J. and Givan, C.V. (1984) Effect of L-methionine sulphoximine on the products of photosynthesis in wheat (*Triticum aestivum*) leaves. *J. Exp. Bot.* **35**, 1800–1810.

Zelitch, I. (1988) Synthesis of glycolate from pyruvate via isocitrate lyase by tobacco leaves in the light. *Plant Physiol.* **86**, 463–468

Zemlyanukhin, A.A., Igamberdiev, A.U., Eprintsev, A.T. and Los, A.A. (1987) Alternative pathway of glyoxylate metabolism in plants. *Sov. Plant. Physiol.* (Engl. Transl.) **34**, 772–777.

5

# Enzyme adaptation to temperature

J.J. Burke

## 5.1 Introduction

Enzyme adaptations to temperature occur constantly as temperature patterns modulate diurnally, seasonally, or over centuries. These adaptations entail qualitative and/or quantitative metabolic changes that often provide a competitive advantage, impact migration to new environments and effect the survival of the species. Changes in isozymes or allozymes, changes in enzyme concentration, modification by substrate and effectors and metabolic regulation of enzyme function without changing enzyme composition, are all possible strategies for adaptation to changes in temperature. A detailed evaluation of these adaptive mechanisms has been provided by Hochachka and Somero (1984) and will only be discussed briefly. The intent of this chapter is to provide representative examples of (i) genetic diversity in the temperature dependence of the apparent $K_m$s of enzymes, (ii) mechanisms for *in vivo* modification of the thermal dependence of apparent $K_m$ values of existing enzymes, (iii) changes in enzyme quantity to compensate for temperature stress, and (iv) alteration of enzyme quality by synthesis of distinct isoforms. This information provided the foundation upon which the concept of thermal kinetic windows (TKWs) for optimal plant metabolism was developed. Comparison of these adaptation strategies of diverse organisms to changing temperatures provides insight into mechanisms for future modification of the temperature characteristics of plant metabolism.

## 5.2 Relationship between the temperature dependence of the apparent $K_m$ of enzymes and the adaptation of organisms to unique thermal environments

The investigation of homologous enzymes from organisms adapted to different temperatures has revealed conservation of the apparent Michaelis–Menten constants ($K_m$) of substrates and co-factors (Graves and Somero, 1982; Hochachka and Somero, 1984; Place and Powers, 1979, 1984; Somero, 1986; Yancey and Somero, 1978). An early example of this relationship was reported by Somero and Low (1976) in the fish, *Trematomas*, which is found in nearly constant 0°C Antarctic waters. In *Trematomas*, as temperature rises from 5 to 20°C, the affinity of pyruvate kinase for phosphoenolpyruvate (PEP) decreases severalfold, as shown by the increase in the apparent $K_m$ of PEP. Other investigations have also demonstrated a correlation between the ecological niche of the organism and the temperature dependency of enzymes (Hall, 1985; Simon *et al.*, 1983; Teeri and Peet, 1978). A study by Yancey and Somero (1978) investigated temperature–$K_m$ relationships in the skeletal muscle ($M_4$, $A_4$, LDH-5) isozyme of lactate dehydrogenase (LDH). They found that, at physiological temperatures for the species studied, the $K_m$ values remained within the approximate range of 0.15–0.35 mM pyruvate. They suggested that the increase in $K_m$ that accompanies rising temperature for each LDH studied may parallel an increase in the intracellular pyruvate concentrations with temperature, such that a stable ratio of $K_m$ to pyruvate is maintained over the full body temperatures experienced by a species.

The establishment of a relationship between the conservation of the $K_m$ and the physiological temperatures of the organism raised the question of how much difference in the average body temperature of an organism is required before there is a detectable modification in the temperature dependence of the apparent $K_m$ of enzymes. Graves and Somero (1982) investigated the question of what minimal difference in average body temperature is needed to select for temperature-adaptive changes in enzymes. They evaluated Barracuda species (genus *Sphyraena*) of the Eastern Pacific coast from cool temperate zones, slightly warmer waters of the Gulf of California and in warm tropical waters. The habitats differed by no more than 6–8°C in temperature. They found that the temperature dependence of the $K_m$ of pyruvate of the purified $M_4$-LDH homologues shifted between these species, but were virtually identical at the average body temperatures, and that $k_{cat}$ values at body temperature were highly conserved. They conclude that this comparison of congeners from different temperature regimes strongly argues that the interspecific differences noted in the broad comparisons of mammals and differently adapted ectotherms do represent important facets of biochemical adaptation to temperature. Place and Powers (1979, 1984) reached a similar conclusion reporting that the strength of selection for $K_m$ conservation was indicated in studies of populations of killifish (*Fundulus heteroclitus*) which occur along a steep latitudinal temperature gradient on the East Coast of the United States.

The observation of the proposed selection for conservation of $K_m$ at average body temperature is supported by a recent report by Dahlhoff and Somero (1993). They measured the effects of temperature on cytosolic malate dehydrogenases (cMDHs) from the shell muscle of five species of eastern Pacific abalone found at different latitudes and/or tidal heights. They reported that the $K_m$ for NADH was conserved within a narrow range (11–21 μmol $l^{-1}$) at physiological temperatures for all species. The $K_m$ of NADH for cMDHs of the two species living at higher latitudes and/or lower tidal heights (*Haliotos rufesens* (red) and *H. kamtschatkana* (pinto)) were perturbed to a much greater extent at elevated temperatures than for cMDHs of congeners from lower latitudes and/or higher tidal heights (*H. fulgens* (green), *H. corregata* (pink) and *H. cracherodii* (black)). Dahlhoff and Somero (1993) showed that cMDHs from the abalone species living in the warm habitats were adapted to function optimally at higher temperatures than the cMDHs of the two species living in the cooler habitats. They proposed that these results provide further evidence that interspecific variation in protein structure and function may be driven by natural selection based on only small (i.e. several degrees Celsius) differences in average body temperature, and that such selection is an important element of the mechanisms of species formation and the maintenance of biogeographic patterning.

Which aspects of body temperature, that is the range of temperatures experienced or maximal temperatures experienced, appear to relate more closely to the observed changes in the temperature characteristics of the apparent $K_m$? Coppes and Somero (1990) investigated whether the adaptive differences in the effect of temperature on $K_m$ reflect selection based on the range of temperatures experienced by the species, the highest temperature experienced by the species, or both of these factors. For most eurythermal species, $K_m$ of pyruvate varied only about twofold between 10 and 30°C, the temperatures which encompass most of the habitat ranges of these species. Stenothermal species with habitat temperature ranges of 14–18°C, showed sharp increases in $K_m$ at temperatures above 20°C. They hypothesize that the thermal stability of $K_m$ values is more strongly selected by maximal habitat temperature than by the amount of variation in habitat temperature. The importance of the maximal habitat temperature for adaptation is supported by a recent report by Huey and Kingsolver (1993). They reported the analysis of comparative data that illustrated the historical evolution of thermal sensitivity of locomotion in iguanid lizards. Taxa that experience high body temperatures in nature have evolved high optimal temperatures for sprinting. They concluded that critical thermal maxima are co-adapted with optimal temperatures but not with critical thermal minima. Thus some, but not all, aspects of thermal sensitivity are co-adapted.

In a study investigating the molecular forms and kinetic properties of MDH and glutamate oxaloacetate transaminase in Glenlea and Kharkov wheat cultivars, Simon *et al.* (1989) found that for MDH the apparent $K_m$ for oxaloacetate ranged from 0.23 to 1.22 mM and increased as a positive function of assay temperature. They reported few significant differences in $K_m$ values of MDH from plants of the two cultivars acclimated to the same thermoperiod when enzyme preparations were subjected to the same assay temperature. However, they felt that it may be significant that, for the 35°C assays, the $K_m$ values of MDH from Kharkov (winter wheat) plants acclimated to 25–28°C were significantly higher than those of Glenlea (spring wheat) subjected to the same conditions. These findings tend to support the hypothesized importance of adaptation to the maximal temperature experienced during the evolution of the organism.

## 5.3 Modification of the thermal dependence of apparent $K_m$ values of existing enzymes by changes in pH

One important modulator of the temperature dependence of the apparent $K_m$ of many enzymes is pH. Reeves (1972) described an imidazole alphastat hypothesis for vertebrate acid–base regulation related to tissue carbon dioxide content and body temperature in bullfrogs. He reported that the temperature dependence of acid-base regulation serves to maintain the structural and functional integrity of proteins. A study by Yancey and Somero (1978) investigated temperature–$K_m$ relationships in the skeletal muscle ($M_4$, $A_4$, LDH-5) isozyme of LDH. They determined $K_m$ across a range of temperatures at either a constant pH of 7.4 or where pH was allowed to vary in accord with values reported for intracellular fluids of turtle skeletal muscle (Malan *et al.*, 1976). They found that the temperature dependency of $K_m$ varied less at physiological pH compared with values obtained at a constant pH. Heisler (1984) reported that studies on the thermal dependence of metabolic parameters must take into account the thermal dependence of acid–base regulation, and Somero (1986) stated that changes in pH can effectively negate the effect of temperature on protein function. The impact of pH on the temperature characteristics of the apparent $K_m$ of enzymes has been identified in numerous studies (Heisler, 1984; Reeves, 1972; Somero, 1986; Walsh and Somero, 1982; Wilson, 1977; Yancey and Somero, 1978).

A recent report by Blier and Guderley (1993) measured the kinetics of pyruvate decarboxylation by partially purified pyruvate dehydrogenase from rainbow trout red muscle. Measurements were made at two pH conditions (a stable pH and an adjusted pH that mimics the cellular pH adjustments with body temperature) and at temperatures of 8, 15 and 22°C. At constant pH, a change in temperature from 15 to 8°C decreased the apparent $K_m$ of mitochondria for pyruvate. When pH co-varied with temperature, the apparent $K_m$ did not change. They conclude that pH regulation would minimize the functional impact of changes in temperature. In an earlier study, Yacoe (1986) also demonstrated a buffering of thermal effects on the mitochondrial apparent $K_m$ for succinate oxidation through alphastat pH variation.

Trout exercised to exhaustion can develop considerable lactacidosis which can lower intramuscular pH from the typical resting value of pH 7.3 to a post-exercise value of 6.5 (Tang and Boutilier, 1991). Purified fructose-1,6-bisphosphatase (FBPase) from rainbow trout (*Oncorhynchus mykiss*) has been analysed by Ferguson and Storey (1992) at 20°C under conditions reflective of 'rest' and 'exercise/recovery' intramuscular pH *in vivo*. Affinity for fructose-1,6-bisphosphate was increased as was FBPase activity when pH was lowered from 7.0 to 6.5. Inhibition of FBPase by fructose-2,6-bisphosphate was strongly alleviated by a reduction in pH from 7.0 to 6.5. FBPase demonstrated maximal activity at pH 6.5, whereas the optimal pH for fructose-6-phosphate 1-phosphotransferase (PFP) was 7.0 or greater. The data provided by Ferguson and Storey suggests that pH plays an important role in determining net flux through the FBPase/PFK locus *in vivo*.

The introduction of $^{31}$P-nuclear magnetic resonance (NMR) spectroscopy as a non-invasive and non-destructive method for pH measurements in intact cells and tissues (Burt *et al.*, 1979; Gadian *et al.*, 1979; Moon and Richards, 1973) offered an improvement over traditional methods. $^{31}$P-NMR allowed for simultaneous determination of both cytoplasmic and vacuolar pH in higher plant cells (Roberts *et al.*, 1980, 1981). Aducci *et al.* (1982) used $^{31}$P-NMR spectroscopy to evaluate both the cytoplasmic and vacuolar pH in maize root tips. They detected a selective pH decrease of about 0.5 units in the cytoplasmic compartment when the temperature was increased from 4 to 28°C. This effect was completely reversible with temperature. They reported that pH variations between 6.5 and 8.5 of the external buffer solution did not alter the respective position of the two Pi peaks in maize root tips, nor did it induce variations in the temperature dependence of the cytoplasmic pH.

The determination of the temperature dependence of the apparent $K_m$ of NADH-hydroxypyruvate reductase (HPR) has been used extensively in the identification of the TKWs of optimal enzyme function (Burke, 1990; Burke and Oliver, 1993; Ferguson and Burke, 1991). In these studies, NADH-HPR was purified according to the procedure of Titus *et al.* (1983). Titus *et al.* (1983) reported that the purified enzyme exhibited a pH optimum between 6.9 and 7.3 with hydroxypyruvate. They identified the Michaelis constants for HPR within this pH optimum. At pH 7.1, $K_m$ values were 62 $\pm$ 6 and 5.8 $\pm$ 0.7 μM for hydroxypyruvate and NADH, respectively. The pH optimum was 5.8–6.2 with glyoxylate and, at pH 6.0, $K_m$ values were 5700 $\pm$ 600 and 2.9 $\pm$ 0.5 μM for glyoxylate and NADH, respectively. Because of the identified effects of pH on the apparent $K_m$ of enzymes discussed previously in this chapter, Burke (1990) took care to adjust the pH of the assay medium to 7.0 at 25°C, and to allow the pH to vary with temperature (Figure 5.1) in a manner similar to the cytoplasmic pH changes reported by Aducci *et al.* (1982). The pH of the NADH-HPR assay medium used by Burke (1990) declined 0.1 pH unit with every 5°C increase in temperature (Figure 5.1). By allowing the pH to co-vary with temperature in a manner similar to cytoplasmic pH *in vivo*, the changes in apparent $K_m$ with temperature measured *in vitro* should reflect more closely the physiological responses within the cells.

Hochachka and Somero (1984) discuss the significance of the conservation of apparent $K_m$. They stated that one of the most critical aspects of enzyme function is the capacity of the enzyme to maintain its apparent $K_m$ of substrate within the range appropriate for the proper catalytic rate and regulatory sensitivity.

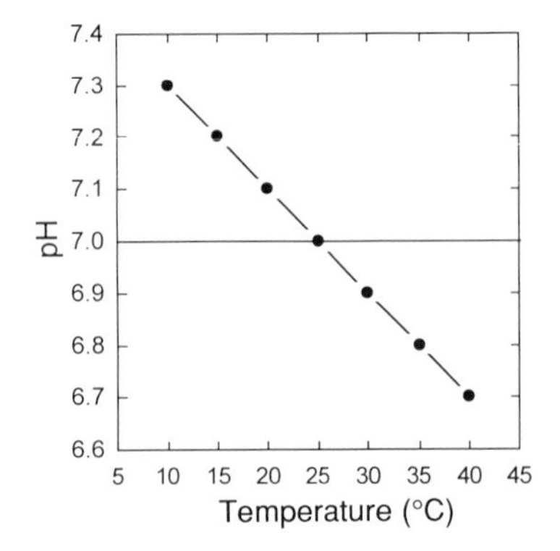


***Figure 5.1.*** *Temperature-dependent changes in the pH of the medium used to determine NADH-hydroxypyruvate reductase activity.*

A 'good' enzyme will not function at or near its potential $V_{max}$, but instead will maintain a reserve capacity which enables it to increase its rate of function in response to regulatory signals (changes in pH or changes in concentrations of metabolite modulators) and to increases in substrate concentration (Hochachka and Somero, 1984). Somero *et al.* (1983) showed that the apparent $K_m$ of pyruvate versus temperature for $M_4$-LDHs of vertebrates having different adaptation temperatures exhibited a conservation of $K_m$ within the physiological temperatures of the organisms. An example of this conservation of apparent $K_m$ in plants for the enzyme NADH-HPR from wheat and cotton using glyoxylate as a substrate, and cucumber using hydroxypyruvate as a substrate, is shown in Figure 5.2. Minimum apparent $K_m$ values of approximately 3–5 μM NADH were observed in the wheat and cotton (Burke *et al.*, 1988), and a minimum apparent $K_m$ value of 8 μM NADH was observed in cucumber (Burke and Oliver, 1993). These apparent $K_m$ values for NADH are similar to the values reported by Titus *et al.* (1983) of 2.9 μM NADH with glyoxylate and 5.8 μM

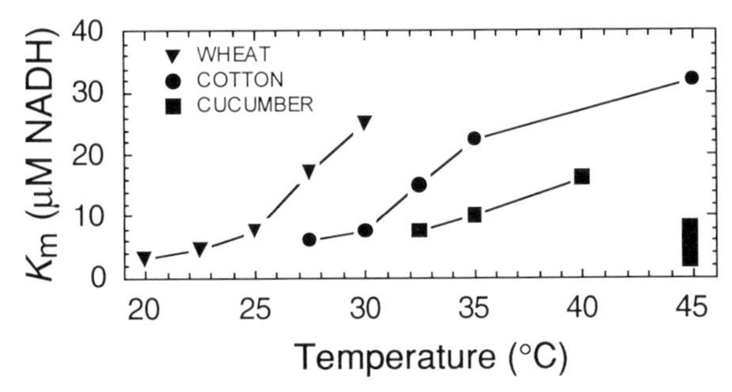


***Figure 5.2*** *Temperature characteristics of the apparent* $K_m$ *for NADH for hydroxypyruvate reductase with glyoxylate as substrate in wheat and cotton, and hydroxypyruvate as substrate in cucumber. The bar to the right illustrates the conservation of* $K_m$*s for these species. Only data for temperatures above the temperature providing the minimum apparent* $K_m$ *are shown.*

NADH with hydroxy-pyruvate for the enzyme from cucumbers and measured at a pH of 7.1. A summary of minimum $K_m$s for several species is shown in Table 5.1. The minimum $K_m$s for NADH for NADH-HPR are highly conserved, ranging from 1.5 to 8 μM, for the eight species. Although similar apparent $K_m$ values for NADH were obtained, the temperatures at which the minimum occurred varied from 20 to 35°C. These data strongly resemble the observation that apparent $K_m$s are conserved at the physiological temperatures of the organism (Dahlhoff and Somero, 1993; Graves and Somero, 1982; Place and Powers, 1979, 1984).

***Table 5.1*** *Comparison of minimum apparent* $K_m$ *values for NADH for NADH-HPR with either glyoxylate or hydroxypyruvate as substrate*

| Species | Variety | Substrate | Minimum apparent $K_m$ for NADH (μM) | Temperature of minimum apparent $K_m$ (°C) |
|---|---|---|---|---|
| Wheat *Triticum aestivum* L. | Kanking | Glyoxylate | 2.5 | 20 |
| Potato *Solanum tuberosum* L. | Norgold M | Hydroxypyruvate | 2.5 | 20 |
| Tomato *Lycopersicon esculentum* L. | Del Oro | Hydroxypyruvate | 1.5 | 22.5 |
| Soybean *Glycine max* L. | Wayne | Hydroxypyruvate | 3 | 25 |
| Petunia *Petunia hybrida* | Red Sail | Hydroxypyruvate | 4 | 25 |
| Cotton *Gossypium hirsutum* L. | Paymaster 145 | Glyoxylate | 6 | 27.5 |
| Cucumber *Cucumis sativum* L. | Ashley | Hydroxypyruvate | 8 | 32.5 |
| Bell Pepper *Capiscum annuum* L. | California Wonder | Hydroxypyruvate | 5 | 35 |

From Burke (1990); Burke and Oliver (1993); Burke et al. (1988); Ferguson and Burke (1991).

It is interesting to compare the reported concentration of NADH within plant microbodies with the apparent $K_m$ values obtained from purified NADH-HPR to see if this enzyme falls within the definition of a 'good' enzyme described by Hochachka and Somero (1984). Mettler and Beevers (1980) evaluated the oxidation of NADH in glyoxysomes by a malate–aspartate shuttle. They reported that, based on estimates of compartment size, the concentrations of pyridine nucleotides were calculated to be 0.2–0.4 mM in the cell cytoplasm, 1.5–2.0 mM in mitochondria and 0.15–0.4 mM in glyoxysomes. They also showed the NADH/NAD percentage to be less than 2% in glyoxysomes. If one calculates the concentration of NADH in the glyoxysome from these values, a glyoxysomal concentration of between 3 and 8 μM NADH would be present. These values are similar to the apparent $K_m$ values reported for NADH for HPR in numerous species. The data suggest that, within microbodies, HPR functions with a NADH concentration close to the concentration described by the apparent $K_m$. Because most enzymes operate under non-saturating substrate concentrations, Hausladen and Alscher (1994) concluded that substrate binding would be of greater importance than $V_{max}$ for the characterization of enzyme function under physiological conditions. In their evaluation of temperature responses of glutathione reductases (GRs) from spruce needles, they calculated that *in vivo* the GRs operate near the concentration providing half-maximal saturation with oxidized glutathione (GSSG). These findings suggest that HPR and GR exemplify 'good' enzymes in that their apparent $K_m$–[substrate] ratios would permit significant adjustments in activity in response to stress.

## 5.4 Modification of the thermal dependence of metabolism by changes in the concentration of existing enzymes

Changing the concentration of an enzyme is another way of achieving temperature adaptive changes in metabolic systems. Hochachka and Somero (1984) suggested that adjustments of this nature seem particularly important in seasonal adjustments of metabolic rates in a temperature-compensatory fashion. They cite the studies of Hazel and Prosser (1974) and Shaklee *et al.* (1977) as examples that, in most cases, organisms did not appear to use seasonal isozyme forms, but rather altered the concentrations of many enzymes. Davidson and Simon (1983) examined the temperature dependency of 11 ecotypic populations of *Spirodela polyrhiza* grown at either 18, 23 or 28°C. Adaptive shifts in the thermal dependence of the $K_m$ were not detected and only slight acclimation changes were observed. Specific activity of MDH was not modified in a pattern suggesting adaptive shifts, but plants that had acclimated to higher temperatures (28°C) had a tendency to produce elevated MDH levels. The $K_m$ values of MDH in all of the *S. polyrhiza* clones increased as a positive function of temperature. They suggest that positive thermal modulation is an efficient rate-compensating mechanism and is an adequate regulatory mechanism for catalytic efficiency within a narrow temperature range of the aquatic environment of *S. polyrhiza.*

## 5.5 Modification of the thermal dependence of metabolism by synthesis of isozymes

A significant body of literature exists on isozyme changes in response to temperature stresses. Some representative examples of isozyme changes and their relationship (or lack thereof) to acclimation of the apparent $K_m$ to the temperature stress are discussed below. Changes in GR activity in response to cold acclimation of spinach were reported by Guy and Carter (1984). An acclimation period of 4 weeks resulted in a 66% increase in GR activity. Similar increases in GR activities have been identified for other species (de Kok and Oosterhuis, 1983; Esterbauer and Grill, 1978; Guy *et al.*, 1984). Guy and Carter (1984) investigated whether the changes in GR activity resulted from increased enzyme content or low temperature-induced synthesis of isozymes with greater catalytic activity. They found that the kinetic characteristics, heat inactivation, freezing inactivation and electrophoretic mobilities of GR isolated from hardened spinach leaves were different from GR isolated from non-hardened spinach leaves. As the assay temperature decreased, the $K_m$ of GSSG for GR decreased in both the non-hardened and hardened spinach. At 25°C, non-hardened GR had a lower $K_m$ for GSSG than did the GR from hardened spinach. At 5°C the GR from hardened spinach had a lower $K_m$ than did the non-hardened GR. Guy and Carter suggest that growth temperature influences the affinity of GR for GSSG in such a way that the enzyme from warm-grown plants functions better at moderate temperatures and the enzyme from cold-grown plants functions better at low temperatures. They point out that these results are analogous to findings reported for Rubisco from rye and potato (Huner and Macdowall, 1979; Huner *et al.*, 1981) and PEP carboxylase (PEPC) from wheat (Graham *et al.*, 1979).

Not all investigations of the temperature characteristics of the apparent $K_m$ of isozymes have observed adaptive changes in apparent $K_m$. Simon and Vairinhos (1991) investigated the thermal stability and kinetic properties of NADP$^+$-MDH isomorphs in two populations of the $C_4$ weed species *Echinohloa crus-galli* (Barnyard grass) from sites of contrasting climate. They found that the enzyme from the Quebec population showed one isomorph, and that three isomorphs were detected in all plants from the Mississippi population. They reported that the single NADP$^+$-MDH isomorph of Quebec plants possesses good acclimatory potential for maintaining catalytic efficiency under a wide range of temperature conditions. *In vitro* thermal and kinetic data did not support the hypothesis that the multiple NADP$^+$-MDH isomorphs found in Mississippi plants may have been selected to optimize the thermal and catalytic efficiency of NADP$^+$-MDH under warm temperature conditions.

Hausladen and Alscher (1994) determined the thermal dependence in purified or partially purified preparations of a cold-hardiness-specific GR (GR-1H) and non-hardened-specific GR (GR-1NH) isolated from red spruce (*Picea rubens* Sarg.) needles to investigate a possible functional adaptation of these isozymes to environmental temperature. They found no differences in the temperature dependence of $K_m$ values for NADPH between hardiness-specific isozymes GR-1NH and GR-1H. They did find that GR-1H had higher $K_m$ (NADPH) values than GR-1NH, but discounted any physiological significance to this finding because the difference was rather small. No difference in $K_m$ (GSSG) between GR-1NH and GR-1H was observed, but a significant temperature effect on $K_m$ (GSSG) was reported. They postulate that the temperature independence of $K_m$ (NADPH) for GRs in red spruce ensures adequate enzyme function at all temperatures commonly experienced during the course of a single season (–20 to +30°C, Sheppard *et al.*, 1989). Teeri (1978) made a similar finding in a study of *Lathyrus japonicus, Arabidopsis thaliana* and *Potentilla glandulosa*, reporting that $K_m$ values of MDH and glucose-6-phosphate dehydrogenase were less dependent on temperature in a plant population native to an environment with large temperature changes than in a population growing in a mild climate with relatively constant temperatures.

## 5.6 Modification of the thermal dependence of apparent $K_m$ by modulators and effectors

In a study recently reported by Chardot and Wedding (1992), the regulation of *Crassula argentea* PEPC was evaluated in relation to changes in temperature. The effect of temperature on the $K_m$ of the CAM PEPC showed little change below 25°C. Above that temperature the $K_m$ increased until at 35°C it had doubled in value. The presence of glucose-6-phosphate substantially lowered the $K_m$ below the control values and resulted in a temperature insensitivity up to 35°C, thus indicating that glucose-6-phosphate causes a change in the enzyme that makes its kinetic behaviour more tolerant to temperature changes. They report that for the CAM enzyme the optimal night temperature is probably around 10–15°C. However, the data obtained in this study indicate that the range favourable for PEPC action can be broadened by a number of factors. They concluded that if we view malate as the overall product of PEPC, the peak in concentration of malate at low temperatures suggests that a low $K_m$ may be more effective than a high $V_{max}$ in accomplishing the accumulation of malate.

Another modulator of the temperature characteristics of the apparent $K_m$ of PEPC from a CAM plant was reported by Buchanan-Bollig *et al.* (1984). They found that the $K_m$ values for PEP were nearly temperature independent from 10 to 30°C, but increased at higher temperatures from 0.12 up to 1.0 mM PEP. With malate-inhibited enzyme, the $K_m$ values were always higher and the loss in substrate affinity could be observed above 20°C when the $K_m$ for PEP increased rapidly with increasing temperature from 0.4 up to 5 mM. Glucose-6-phosphate activated the enzyme at all temperatures tested in the usual manner by increasing the substrate affinity of the enzyme (Nott and Osmond, 1982; Pays *et al.*, 1980) and in the presence of glucose-6-phosphate (2 mol $m^{-3}$ ) the $K_m$ was less affected by temperature. They suggest that *in vivo*, where often a temperature optimum of 15–20°C for $CO_2$ dark fixation could be observed, the cytosolic pH and/or metabolite concentrations, especially the malate concentration, change with increasing temperature so that the PEPC becomes inhibited. The influence of temperature on PEPC of CAM plants is also discussed in Chapter 3.

## 5.7 Thermal kinetic windows: metabolic correlations and avenues for modification

The concept of TKWs arose from a desire to investigate temperature stresses in plants and the realization that there is a lack of knowledge about how to identify the optimal temperatures for metabolism. TKWs of optimal enzyme function were defined as the temperature range over which the value of the apparent $K_m$ was within 200% of the minimum apparent $K_m$ value observed for the enzyme (Burke *et al.*, 1988). The 200% cut-off value was used because previous studies (Somero and Low, 1976; Teeri, 1978) had reported that enzymes could function optimally with $K_m$ values below 200% of the minimum $K_m$ value. The purpose of the TKW was to provide a general indicator of the range of temperatures in which the optimal temperature for metabolism was located. The temperature ranges comprising the TKWs for wheat and cotton were 17.5–23°C and 23.5–32°C, respectively. Although the TKWs were 5–8°C in breadth, it was shown that these plants were only within the optimal temperature range of their TKWs for a fraction of the growing season (Burke *et al.*, 1988). These initial observations called for a re-evaluation of our understanding of the temperature stresses experienced by plants in the field. To date,TKWs have been reported for numerous species (Anderson *et al.*, 1992; Burke, 1990; Burke and Oliver, 1993; Burke *et al.*, 1988; Ferguson and Burke, 1991; Kidambi *et al.*, 1990; Mahan *et al.*, 1990).

In light of the reported effects of pH, activators and inhibitors of enzyme activity on the temperature dependence of the apparent $K_m$, what evidence exists that the optimal temperature ranges defined by the TKW have any relationship to metabolic responses *in vivo*? The best correlative evidence of the validity of the TKWs comes from the determination of the temperature dependence of the reappearance of photosystem II variable fluorescence following illumination (Burke, 1990; Ferguson and Burke, 1991). The chlorophyll fluorescence functions as a natural indicator of the *in vivo* temperature characteristics of the plant. A representative example of the relationship between the temperatures providing maximum reappearance of variable fluorescence and the temperatures providing minimum apparent $K_m$ values is shown in Figure 5.3. In this example, the minimum apparent $K_m$ for NADH for HPR from Norgold M potatoes occurred at 20°C, with the TKW falling between 15 and 25°C (Ferguson and Burke, 1991). The optimum temperature for variable fluorescence (*F*v) reappearance (expressed as the ratio *F*v/*F*o where *F*o is the initial fluorescence) is defined as the temperature providing the maximum *F*v/*F*o ratio and the minimum time in darkness required to reach this ratio. The optimal temperature identified from the fluorescence reappearance is also 20°C for the Norgold M

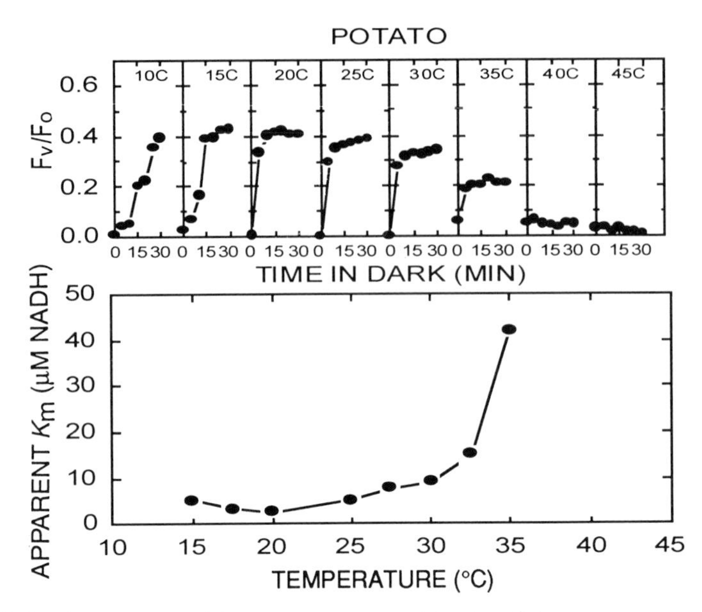


***Figure 5.3*** *Comparison of the temperature sensitivities of the reappearance of photosystem II variable fluorescence following illumination and the apparent* $K_m$ *for NADH of hydroxypyruvate reductase.*

potato. Similar correlations between the temperatures of the TKW and the temperatures providing maximum fluorescence reappearance have been reported for cucumber, wheat, cotton, soybean, tomato, petunia and bell pepper (Burke, 1990; Burke and Oliver, 1993; Ferguson and Burke, 1991).

An approach to alter the temperature characteristics of an enzyme pool based upon the concept of TKWs has been developed by Oliver, Ferguson and Burke (personal communication). A chimeric cucumber NADH-HPR gene under the control of a cauliflower mosaic virus 35S promoter was constructed and introduced into the genome of tobacco. The root system of the R1 generation of the resultant transgenic plants expressed only the cucumber enzyme (the native tobacco HPR gene is light regulated and only found in the aerial portions of the plant). Enzyme isolated from the transgenic root tissues exhibited a TKW centred at 32.5°C, characteristic of cucumber. The pool of HPR in the shoots, containing both tobacco and cucumber enzymes, exhibited a broad TKW consistent with an equal mix of the two forms. Thus, not only can sufficient enzyme be produced from the 35S chimeric gene construct to produce a phenotype, but enough is synthesised to alter the temperature characteristics of the total enzyme pool *in vitro*. This suggests that the temperature characteristics of plant biochemical pathways can be broadened by metabolic engineering to suit changing thermal environments.

## 5.8 Conclusion

This chapter has highlighted several aspects of enzyme adaptation to temperature. Special attention has been paid to the relationship between the conservation of apparent $K_m$s within the temperature range in which organisms develop. Its significance is supported by the finding that long-term exposure to changes of a few degrees Celsius in average body temperature can shift the temperature characteristics of the apparent $K_m$ presumably to optimize metabolism. Because cells have several mechanisms for *in vivo* modification of the thermal dependence of apparent $K_m$ values of existing enzymes, the potential impact of their presence must be considered in any interpretation of *in vitro* analysis of temperature responses of apparent $K_m$s. The correlation between the temperature characteristics of chlorophyll fluorescence reappearance following illumination and the TKWs determined from the temperature characteristics of the apparent $K_m$ of enzymes aids in the identification of optimal temperatures for plant metabolism. Having identified temperature limitations of organisms within an environment, molecular engineering can be employed to assist in broadening of the temperature characteristics of the limiting enzyme(s).

## References

Aducci, P., Federico, R., Carpinelli, G. and Podo, F. (1982) Temperature dependence of intracellular pH in higher plant cells. *Planta* **156**, 579–582.

Anderson, J.V., Chevone, B.I. and Hess, J.L. (1992) Seasonal variation in the antioxidant system of eastern white pine needles. Evidence for thermal dependence. *Plant Physiol.* **98**, 501–508.

Blier, P.U. and Guderley, H.E. (1993) Effects of pH and temperature on the kinetics of pyruvate oxidation by muscle mitochondria from rainbow trout (*Oncorhynchus mykiss*). *Physiol. Zool.* **66**, 474–489.

Buchanan-Bollig, I.C., Kluge, M. and Müller, D. (1984) Kinetic changes with temperature of phosphoenolpyruvate carboxylase from a CAM plant. *Plant, Cell Environ.* **7**, 63–70.

Burke, J.J. (1990) Variation among species in the temperature dependence of the reappearance of variable fluorescence following illumination. *Plant Physiol.* **93**, 652–656.

Burke, J.J. and Oliver, M.J. (1993) Optimal thermal environments for plant metabolic processes (*Cucumis sativis* L.) *Plant Physiol.* **102**, 295–302

Burke, J.J., Mahan, J.R. and Hatfield, J.L. (1988) Crop-specific thermal kinetic windows in relation to wheat and cotton biomass production. *Agron. J.* **80**, 553–556.

Burt, C.T., Cohen, S.M. and Barany, M. (1979) Analysis of intact tissue with $^{31}$P-NMR. *Annu Rev. Biophys. Bioeng.* **8**, 1–25.

Chardot, T.P. and Wedding, R.T. (1992) Regulation of *Crassula argentea* phosphoenolpyruvate carboxylase in relation to temperature. *Arch. Biochem. Biophys.* **293**, 292–297.

Coppes, Z.L. and Somero, G.N. (1990) Temperature-adaptive differences between the $M_4$-lactate dehydrogenases of stenothermal and eurythermal Sciaenid fishes. *J. Exp. Zool.* **254**, 127–131.

Dahlhoff, E. and Somero, G.N. (1993) Kinetic and structural adaptations of cytoplasmic malate dehydrogenases of eastern pacific abalone (Genus *Haliotis*) from different thermal habitats: Biochemical correlates of biogeographical patterning. *J. Exp. Biol.* **185**, 137–150.

Davidson, D. and Simon, J.P. (1983) Thermal adaptation and acclimation of Hepcotypis populations of *Spirodella polyrhiza* (L.) Schleid. (Lemnaceae): Temperature dependency of NAD malate dehydrogenase. *J. Therm. Biol.* **8**, 289–296.

de Kok, L.J. and Oosterhuis, F.S. (1983) Effects of frost-hardening and salinity on glutathione and sulfhydryl levels and on glutathione reductase activity in spinach. *Physiol. Plant.* **58**, 47–51.

Esterbauer, H. and Grill, D. (1978) Seasonal variation of glutathione and glutathione reductase in needles of *Picea abies*. *Plant Physiol.* **61**, 119–121.

Ferguson, D.L. and Burke, J.J. (1991) Influence of water and temperature stress on the temperature dependence of the reappearance of variable fluorescence following illumination. *Plant Physiol.* **97**, 188–192.

Ferguson, R.A. and Storey, K.B. (1992) Gluconeogenesis in trout (*Oncorhynchus mykiss*) white muscle: purification and characterization of fructose-1,6-bisphosphatase activity *in vitro*. *Fish Physiol. Biochem.* **10**, 201–212.

Gadian, G.G., Radda, G.K., Richards, R.E. and Seeley, P.J. (1979) $^{31}$P NMR in living tissue: the road from promising to an important tool in biology. In: *Biological Applications of Magnetic Resonance* (ed. R.G. Shulman). Academic Press, New York, pp. 463–531.

Graham, D., Hockley, D.G. and Patterson, B.D. (1979) Temperature effects on phosphoenolpyruvate carboxylase from chilling sensitive and chilling resistant plants. In: *Low Temperature Stress in Crop Plants* (eds J.M. Lyons, D. Graham and J.K. Raison). Academic Press, New York, pp. 453–461.

Graves, J.E. and Somero, G.N. (1982) Electrophoretic and functional enzymic evolution in four species of eastern Pacific Barracudas from different thermal environments. *Evolution* **36**, 97–106.

Guy, C.L. and Carter, J.V. (1984) Characterization of partially purified glutathione reductase from cold-hardened and nonhardened spinach leaf tissue. *Cryobiology* **21**, 454–464.

Guy, C.L., Carter, J.V., Yelenosky, G. and Guy, C.T. (1984) Changes in glutathione content during cold acclimation in *Cornus sericea* and *Citrus sinensis*. *Cryobiology* **21**, 443–453.

Hall, J.G. (1985) Temperature-related differentiation of glucose phosphate isomerase alleloenzymes isolated from blue mussel, *Mytilus edulis*. *Biochem. Genet.* **23**, 705–727.

Hausladen, A. and Alscher, R.G. (1994) Cold-hardiness-specific glutathione reductase isozymes in red spruce. *Plant Physiol.* **105**, 215–223.

Hazel, J.R. and Prosser, C.L. (1974) Molecular mechanisms of temperature compensation in poikilotherms. *Physiol. Rev.* **54**, 620–677.

Heisler, N. (1984) Acid–base regulation in fishes. In: *Fish Physiology,* Volume 10A (eds W.S. Hoar and D.J.Randall). Academic Press, New York, pp. 315–401.

Hochachka, P.W. and Somero, G.N. (1984) *Biochemical Adaptation.* Princeton University Press, Princeton. New Jersey.

Huey, R.B. and Kingsolver, J.G. (1993) Evolution of resistance to high temperature in ectotherms. *Am. Nat.* **142**, S21–S46.

Huner, N.P.A. and Macdowall, F.D.H. (1979) The effects of low temperature acclimation of winter rye on catalytic properties of its ribulose bisphosphate carboxylase-oxygenase. *Can. J. Biochem.* **57**, 155–164.

Huner, N.P.A., Palta, J.P., Li, P.H. and Carter J.V. (1981) Comparison of the structure and function of ribulose bisphosphate carboxylase-oxygenase from a cold-hardy and nonhardy potato species. *Can. J. Biochem.* **59**, 280–289.

Kidambi, S.P., Mahan, J.R. and Matches, A.G. (1990) Purification and thermal dependence of glutathione reductase from two forage legume species. *Plant Physiol.* **92**, 363–367.

Mahan, J.R., Burke, J.J. and Orzech, K.A. (1990) Thermal dependence of the apparent *Km* of glutathione reductases from three plant species. *Plant Physiol.* **93**, 822–824.

Malan, A., Wilson, T.L. and Reeves, R.B. (1976) Intracellular pH in cold-blooded vertebrates as a function of body temperature. *Respir. Physiol.* **28**, 29–47.

Mettler, I.J. and Beevers, H. (1980) Oxidation of NADH in glyoxysomes by a malate–aspartate shuttle. *Plant Physiol.* **66**, 555–560.

Moon, R.B. and Richards, J.H. (1973) Determination of intracellular pH by $^{31}P$ magnetic resonance. *J. Biol. Chem.* **248**, 7276–7278.

Nott, D.L. and Osmond, C.B. (1982) Purification and properties of phosphoenolpyruvate carboxylase from plants with crassulacean acid metabolism. *Aust. J. Plant Physiol.* **9**, 409–422.

Pays, A.G.G., Jones, R., Wilkins, C.A., Fewson, C.A. and Malcom, A.D.B. (1980) Kinetic analysis of effectors of phosphoenolpyruvate carboxylase from *Bryophyllum fedtschenkoi*. *Biochim. Biophys. Acta* **614**, 151–162.

Place, A.R. and Powers, D.A. (1979) Genetic variation and relative catalytic efficiencies: Lactate dehydrogenase B allozymes of *Fundulus heteroclitus*. *Proc. Natl Acad. Sci. USA* **76**, 2354–2358.

Place, A.R. and Powers, D.A. (1984) Purification and characterization of the lactate dehydrogenase ($LDH\text{-}B_4$) allozymes of *Fundulus heteroclitus*. *J. Biol. Chem.* **259**, 1299–1308.

Reeves, R.B. (1972) An imidazole alphastat hypothesis for vertebrate acid–base regulation: tissue carbon dioxide content and body temperature in bullfrogs. *Respir. Physiol.* **14**, 219–236.

Roberts, J.K.M., Ray, P.M., Wade-Jardetzky, N. and Jardetzky, O. (1980) Estimation of cytoplasmic and vacuolar pH in higher plant cells by $^{31}P$ NMR. *Nature* **283**, 870–872.

Roberts, J.K.M., Wade-Jardetzky, N. and Jardetzky, O. (1981) Extent of intracellular pH change during $H^+$ extrusion by maize root tip cells. *Planta* **152**, 74–78.

Shaklee, J.B., Christiansen, J.A., Sidell, B.D., Prosser, C.L. and Whitt, G.S. (1977) Molecular aspects of temperature acclimation in fish: Contributions of changes in enzyme activities and isozyme patterns to metabolic reorganization in the green sunfish. *J. Exp. Zool.* **201**, 1–20.

Sheppard, L.J., Smith, R.I. and Cannell, M.G.R. (1989) Frost hardiness of *Picea rubens* growing in spruce decline regions of the Appalachians. *Tree Physiol.* **5**, 25–37.

Simon, J.-P. and Vairinhos, F. (1991) Thermal stability and kinetic properties of $NADP^+$-malate dehydrogenase isomorphs in two populations of the $C_4$ weed species *Echinohloa crus-galli* (Barnyard grass) from sites of contrasting climate. *Physiol. Plant.* **83**, 216–224.

Simon, J.-P., Peloquin, M.-J. and Charest, C. (1989) Molecular forms and kinetic properties of malate dehydrogenase and glutamate oxalacetate transaminase in Glenlea and Kharkov wheat cultivars. *Environ. Exp. Bot.* **29**, 445–456.

Simon, J., Potvin, M. and Blanchard, M. (1983) Thermal adaptation and acclimation of higher plants at the enzyme level: Kinetic properties of NAD malate dehydrogenase and glutamate oxaloacetate transaminase in two genotypes of *Arabidopsis thaliana* (Brassicaceae). *Oecologia* **60**, 143–148.

Somero, G.N. (1986) Protons, osmolytes and fitness of internal milieu for protein function. *Am. J. Physiol.* **251**, R197–R213.

Somero, G.N. and Low, P.S. (1976) Temperature: A "shaping force" in protein evolution. *Biochem. Soc. Symp.* **41**, 33–42.

Somero, G.N., Siebenaler, J.F. and Hochachka, P.W. (1983) Biochemical and physiological adaptations of deep-sea animals. In: *The Sea,* Volume 8 (ed. G.T. Rowe). Wiley, New York, pp. 261–330.

Tang, Y. and Boutilier, R.G. (1991) White muscle intracellular acid–base and lactate status following exhaustive exercise: A comparison between freshwater- and seawater-adapted rainbow trout. *J. Exp. Biol.* **156**, 153–171.

Teeri, J.A. (1978) Adaptation of malate dehydrogenase to temperature variability. In: *Adaptation of Plants to Water and High Temperature Stress* (eds N.C. Turner and P.J. Kramer). Wiley, New York, pp. 251–260.

Teeri, J.A. and Peet, M.M. (1978) Adaptation of malate dehydrogenase to environmental temperature variability in two populations of *Potentilla glandulosa* Lindl. *Oecologia* **34**, 133–141.

Titus, D.E., Hondred, D. and Becker, W.M. (1983) Purification and characterization of hydroxypyruvate reductase from cucumber cotyledons. *Plant Physiol.* **72**, 402–408.

Walsh, P.J. and Somero, G.N. (1982) Interactions among pyruvate concentration, pH, and $K_m$ of pyruvate in determining *in vivo* Q10 values for lactate dehydrogenase reaction. *Can. J. Zool.* **60**, 1293–1299.

Wilson, T.L. (1977) Interrelations between pH and temperature for the catalytic rate of the M4 isozyme of lactate dehydrogenase (EC 1.1.1.27) from goldfish (*Carassius auratus* L.) *Arch. Biochem. Biophys.* **179**, 378–390.

Yacoe, M.E. (1986) Effects of temperature, pH and $CO_2$ tension on the metabolism of isolated hepatic mitochondria of iguana, *Dipsosaurus dorsalis. Physiol. Zool.* **59**, 263–272.

Yancey, P.H. and Somero, G.N. (1978) Temperature dependence of intracellular pH: its role in the conservation of pyruvate apparent $K_m$ values of vertebrate lactate dehydrogenase. *J. Comp. Physiol.* **125**, 129–134.

# 6

# Adaptations of plant respiratory metabolism to nutritional phosphate deprivation

M.E. Theodorou and W.C. Plaxton

## 6.1 Introduction

### 6.1.1 *The critical role of Pi in plant metabolism*

Phosphorus is an essential element in plant nutrition. For plants, the orthophosphate anion ($H_2PO_4^{2-}$) is the preferentially absorbed form of phosphorus. After being taken up by plant cells, it either remains as inorganic phosphate (Pi) or is esterified to a carbon backbone (e.g. sugar phosphate). Pi may also be esterified to another Pi to form the energy-rich pyrophosphate bond of compounds such as ATP, inorganic pyrophosphate (PPi) and polyphosphates. The importance of Pi in plant metabolism can be attributed to a variety of factors including: (i) the unique physical characteristics of the P–P bond and its role in energy transduction (see Sections 6.2.2 and 6.2.4), (ii) the vital role of Pi in the 'fine' control of metabolism as an allosteric effector of enzymatic activity (Iglesias *et al.*, 1993; Preiss, 1984) or by the reversible phosphorylation of key regulatory enzymes (Budde and Chollet, 1988; Nimmo, 1993; Plaxton, 1990; see Section 6.7), (iii) its function as an important constituent of macromolecules such as phospholipids and nucleic acids (Rawn, 1989), and (iv) its vital role in regulating the exchange of triose phosphate between the plastid and cytosol via the Pi translocator (Preiss, 1984; Walker and Sivak, 1986).

### 6.1.2 *Pi availability in the environment*

Despite its importance to plant growth and metabolism, Pi is the least available nutrient in many aquatic and terrestrial environments (Bieleski, 1973). Almost all of the Pi compounds occurring in nature exist in insoluble forms. Soils often contain large amounts of both organic and bound (insoluble) mineral phosphorus, however the level of Pi in the solution phase rarely exceeds 1.5 μM and is often below the level of many micronutrients (Bieleski, 1973). Vast areas of potentially fertile land are agriculturally poor because of Pi deficiency. Costly fertilizer applications needed to enrich such soils and to replenish Pi removed by growing crops account for more than 90% of total phosphorus use in the world (Bieleski and Ferguson, 1983). In lakes, Pi is the single most important factor limiting plant growth. Lake waters (pH 6) typically contain Pi concentrations of less than 40 nM, and soluble Pi levels decrease with increasing acidity (Bieleski, 1973). The disposal of Pi-rich materials such as detergents, sewage and processing wastes into inland waters removes the existing limitation and the resultant burst of plant growth is what we see as eutrophication (Bieleski and Ferguson, 1983). The insufficient supply of environmental Pi is magnified if one considers the relatively large demand for phosphorus by living organisms. The cytoplasm of a nutrient-sufficient plant typically contains millimolar concentrations of Pi (see Section 6.2.1). The apparent contradiction between supply of and demand for Pi, as well as the economic and environmental impact of nutritional Pi deficiency, has prompted a wide array of investigations into the effects of Pi status on various aspects of plant growth and function.

### 6.1.3 *Metabolic adaptations to Pi starvation*

Plants exhibit numerous morphological, physiological and metabolic adaptations to Pi deprivation. At the metabolic level, much attention has been devoted to the effect of Pi status on the secretion of acid phosphatases, Pi uptake rates, as well as other mechanisms which may serve to increase the availability of Pi to Pi-deprived cells (Duff *et al.*, 1991a,b, 1994; Goldstein *et al.*, 1989; Lefebvre *et al.*, 1990). The effects of Pi nutrition on photosynthesis and photosynthate partitioning in leaves have also been well documented (Crafts-Brander, 1992; Duchein *et al.*, 1993; Freeden *et al.*, 1990; Jacob and Lawlor, 1992; Paul and Stitt, 1993; Qui and Israel, 1992; Rao *et al.*, 1993; Sawada *et al.*, 1992; Topa and Cheeseman, 1992a,b; Usuda and Shimogawara, 1991a,b, 1992, 1993; Walker and Sivak, 1986). By contrast, relatively few studies have examined the influence of Pi nutrition on plant secondary metabolism (Fischer *et al.*, 1993; Knobloch, 1982; Knobloch *et al.*, 1981; Okazaki *et al.*, 1982; Sasse *et al.*, 1982). Likewise, investigations on the adaptations of plant respiratory metabolism to Pi deficiency have been limited. In this chapter, we summarize recent findings which indicate that plants respond adaptively to Pi deprivation through the induction of alternative pathways of glycolysis and mitochondrial electron transport. These respiratory bypasses are proposed to facilitate respiration by Pi-deficient plant cells because they negate the necessity for adenylates and Pi, both pools of which are severely depressed following nutritional Pi deprivation.

## 6.2 Effect of Pi nutrition on the Pi, nucleotide phosphate and pyrophosphate pools

### 6.2.1 *Pi pools*

Phosphate-sufficient plants and algae are capable of taking up excess Pi, even when immediate growth and metabolic demands for phosphorus have been satisfied. This process is known as luxury consumption (Cembella *et al.*, 1984). In higher plants, the vast majority of excess phosphorus is stored in the vacuole as Pi (Lee *et al.*, 1990). It has been estimated that, in nutrient-sufficient plants, 90% of leaf Pi can be sequestered in the vacuole (Bielski, 1973). Algae are capable of storing Pi as either free Pi in their vacuoles or as large insoluble Pi polyesters called polyphosphate bodies which exist primarily as cytoplasmic granules (reviewed by Bental *et al.*, 1988; Kuhl, 1974; Lundberg *et al.*, 1989; Sianoudis *et al.*, 1984). This method of packaging Pi allows algal cells to store large quantities of Pi without disrupting their osmotic balance. The disappearance of polyphosphates (Lundberg *et al.*, 1989) and vacuolar Pi stores (discussed below) under conditions of Pi starvation imply that they function as Pi reserves (Kuhl, 1974). There is also the possibility that the phosphate ester bonds of polyphosphates act as energy stores. This possibility seems likely, considering the recent recognition of PPi as an important energy donor in plant cytosolic metabolism (Davies *et al.*, 1993).

$^{31}$P-Nuclear magnetic resonance (NMR) studies have revealed that higher plant cells selectively distribute Pi between cytoplasmic ('metabolic') and vacuolar ('storage') pools, with cytoplasmic Pi being maintained essentially constant at the expense of large fluctuations in vacuolar Pi. Studies from a variety of tissues and species, including pea root tips, soybean leaves and maize roots, suggest that cytosolic Pi homeostasis, at the expense of vacuolar Pi fluctuations, may be a ubiquitous feature of plant cells (Lauer *et al.*, 1989; Lee and Ratcliffe, 1983, 1993; Lee *et al.*, 1990). During Pi starvation, vacuolar Pi is released into the cytoplasm in a controlled manner correlated to the severity of Pi limitation. It is only when vacuolar Pi stores have been completely exhausted that cytoplasmic Pi levels begin to decline. Such selective preservation of cytoplasmic Pi would ensure constant Pi concentration in the metabolically active compartments (i.e. cytosol, plastids, mitochondria) during short-lived fluctuations in environmental Pi availability. This might prevent perturbations of metabolism resulting from transient changes in Pi status and could possibly spare the cell the unnecessary (and energetically expensive) induction of Pi starvation rescue mechanisms. Regulation of Pi efflux across the tonoplast membrane as part of the mechanism to maintain the cytoplasmic Pi homeostasis is not yet fully understood (Mimura *et al.*, 1990). Similarly, $^{31}$P-NMR studies from a variety of algal species (Lundberg, 1989 and references therein) reveal that, under Pi-limited conditions, free cytoplasmic Pi levels are maintained essentially at constant levels at the expense of polyphosphate esters. In the macroalga *Ulva lacta*, the polyphosphate pool was reduced by 75% after 72 h of Pi starvation while the cytoplasmic free Pi pool remained at about the same level.

Although $^{31}$P-NMR is capable of distinguishing vacuolar and cytoplasmic Pi pools, it cannot discriminate definitively between Pi sequestered in cytoplasmic organelles versus the Pi contained in the cytosol (Roberts, 1984). Investigations of intracellular Pi compartmentation have also been performed using organelles isolated by the non-aqueous fractionation technique (Mimura *et al.*, 1990; Stitt *et al.*, 1985). However, Pi may leak from organelles, and cross-contamination of organelles (particularly by vacuolar material) may result in an overestimation of Pi when non-aqueous techniques are used (Dietz and Heber, 1984). For these reasons, there are few studies providing accurate measurements of cytosolic Pi concentrations in plant cells. Estimates of cytosolic Pi using the non-aqueous fractionation technique range from 5 to 10 mM (Stitt *et al.*, 1985). $^{31}$P-NMR estimates of cytoplasmic Pi from a variety of species and tissues range from 0.6 to 21 mM (Bligney *et al.*, 1989, 1990; Loughman *et al.*, 1989; Roby *et al.*, 1987). Cytoplasmic Pi concentrations from soybean leaves, as measured using $^{31}$P-NMR, range from 0 to 0.23 mM in Pi-starved cells, compared to 5–8 mM for Pi-sufficient leaves (Lauer *et al.*, 1989). Lee and Ratcliffe (1993) also reported a decline of cytoplasmic Pi from 8.8 to 4.8 mM upon prolonged Pi starvation of mature maize roots. Because the cytosol constitutes a large proportion of the cytoplasm, it is probable that the cytosolic concentration of Pi is reduced dramatically following Pi starvation.

Hentrich and co-workers (1993) recently have presented evidence for (i) detectability of chloroplastic Pi *in vivo* by means of $^{31}$P-NMR and (ii) for assignment of most of the intracellular Pi of the green alga, *Chlamydomonas reinhardtii*, to the chloroplast. They report stromal and cytosolic Pi concentrations of about 5 and 0.9 mM, respectively in Pi-sufficient cells (Hentrich *et al.*, 1993). However, these results are in contradiction to various reports concerning other plant species which attribute the majority of the cytoplasmic Pi signal to the cytosol.

### 6.2.2 *Nucleotide phosphate pools*

As a consequence of the significant decline in cytoplasmic Pi, large decreases in intracellular concentrations of nucleotide phosphates also occur following prolonged Pi deprivation. For example, Ashihara and co-workers (1988) reported that levels of ATP, CTP, GTP and UTP in Pi-deficient *Catharanthus roseus* suspension cells were only 20–30% of the levels found in nutrient-sufficient cells and that levels of all other nucleotide phosphates were also reduced. In Pi-starved *Brassica nigra* suspension cells, ATP and ADP levels decreased to 26 and 9% of the nutrient-sufficient values, respectively (Duff *et al.*, 1989b). Parallel findings were reported for *Glycine max* (Freeden *et al.*, 1990) and *Selenastrum minutum* (Theodorou *et al.*, 1991) but differ from those obtained with *Chenopodium rubrum* suspension cells (Dancer *et al.*, 1990), maize leaves (Jacob and Lawlor, 1992) and *Scenedesmus obtusiusculus* (Tillberg and Rowley, 1989), in which the reduction in total adenylates with Pi deprivation was attributed to a larger decrease in ATP relative to ADP. In Pi-deficient sunflower leaves, ATP and ADP levels were both reduced to approximately 45% of the nutrient-sufficient controls (Jacob and Lawlor, 1992).

### 6.2.3 *Adenylate and Pi limitation of respiratory metabolism*

The dramatic reductions in the metabolic Pi and adenylate pools that accompany long-term Pi stress have important implications with respect to respiratory metabolism. These arise from the adenylate and/or Pi substrate dependence of the phosphorylating (cytochrome) pathway of mitochondrial electron transport (Figure 6.1) and several enzymes of cytosolic glycolysis (ATP: fructose-6-phosphate 1-phosphotransferase (PFK), NAD-glyceraldehyde-3-phosphate dehydrogenase (G3PDH), 3-phosphoglycerate (3-PGA) kinase, and pyruvate kinase (PK)). Moreover, because Pi relieves the potent allosteric inhibition of plant PFKs by phosphoenolpyruvate (PEP) (Iglesias *et al.*, 1993), any significant depletion of metabolic Pi could further attenuate cytosolic PFK activity owing to a lowered cytosolic ratio of [Pi]:[PEP].

### 6.2.4 *Pyrophosphate: an autonomous energy donor of the plant cytosol*

It has been suggested (Dancer *et al.*, 1990) that three different energy donor systems may operate in the cytosol of higher plants: adenine nucleotides, uridine nucleotides and PPi. As described above, adenine and uridine nucleotide levels are extremely responsive to Pi nutrition. By contrast, the PPi level of plant cells appears to be unaffected by Pi deprivation (Dancer *et al.*, 1990; Duff *et al.*, 1989b; Stitt, 1990). To assess the relative importance of PPi as an energy donor in the plant cytosol, Davies *et al.* (1993) computed the standard and free energy changes for PPi and ATP under a variety of cytosolic conditions. The results indicate that the standard and free energies of hydrolysis favour PPi as a phosphoryl donor under conditions of decreased pH (a condition which dramatically decreases the free $HPO_4^-$ species in the cytosol), low free $Mg^{2+}$ and reduced cytosolic ATP levels (Davies *et al.*, 1993). These findings suggest increased potential of PPi to act as a phosphoryl donor, relative to ATP, under cytosolic conditions known to accompany Pi stress. These results also demonstrate the importance of PPi as an autonomous energy donor in the cytosol of Pi-starved higher plants.

PPi could be utilized as an energy donor in three different cytosolic reactions: (a) the PPi-dependent phosphorylation of fructose-6-phosphate catalysed by pyrophosphate: fructose-6-phosphate 1-phosphotransferase (PFP) (Figure 6.1), (b) the conversion of UDP-glucose to UTP and glucose 1-phosphate catalysed by UDP-glucose pyrophosphorylase, and (c) the PPi-dependent proton pump of the tonoplast. This $H^+$-PPiase is one of two primary pumps residing at the vacuolar membrane, the other being an $H^+$-ATPase (Rea and Poole, 1993). Various studies have at one time or another implicated each of the three PPi-utilizing

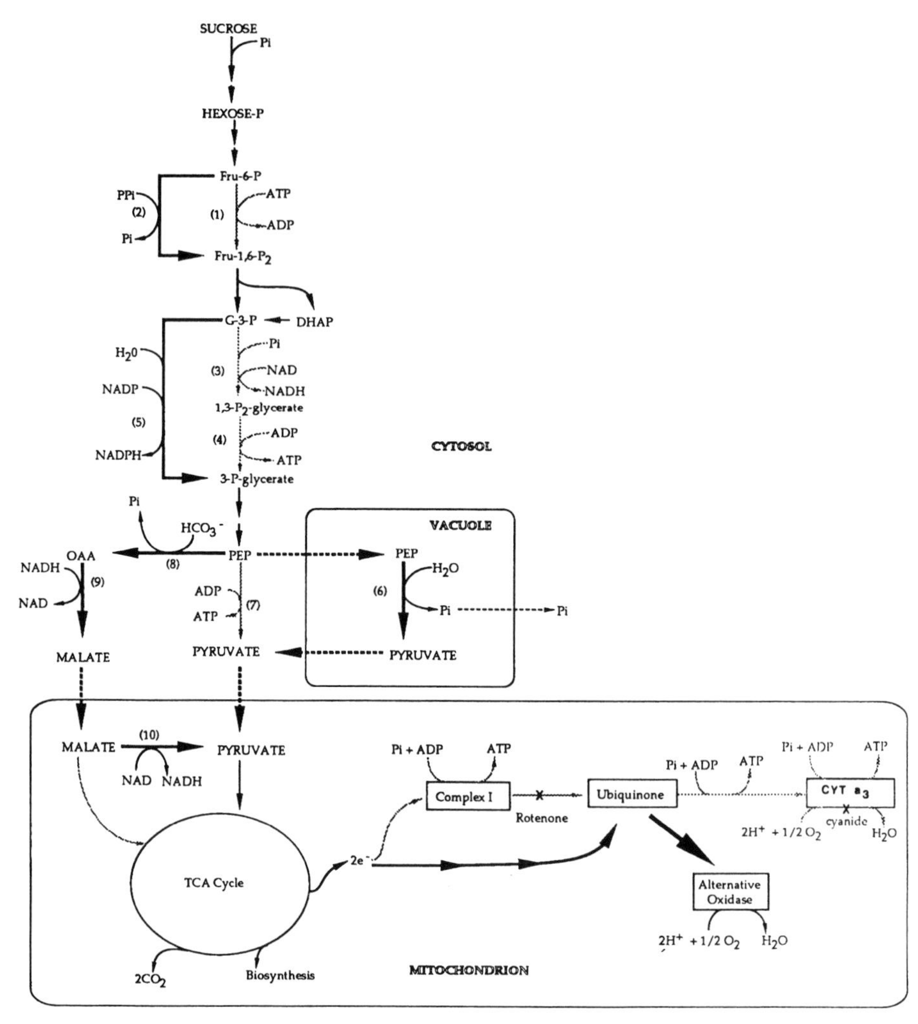


***Figure 6.1.*** *A model depicting alternative pathways of glycolytic carbon flow and mitochondrial respiration (indicated by heavy arrows) during nutritional Pi deprivation of higher plants. This metabolic flexibility permits plants to circumvent the adenylate- and Pi-dependent reactions of respiration (indicated by stippled arrows), thus facilitating respiration by Pi-deficient plants. The designations are: (1), PFK; (2) PFP; (3) phosphorylating NAD-G3PDH ; (4) 3-PGA kinase; (5) non-phosphorylating NADP-G3PDH; (6) PEP phosphatase; (7) cytosolic PK; (8) PEPC; (9) malate dehydrogenase; (10) NAD-malic enzyme. Reproduced from Theodorou and Plaxton (1993), © American Society of Plant Physiologists, with permission.*

enzymes listed above in the regulation of cytosolic PPi levels (Dancer and ap Rees, 1989; Davies *et al.,* 1993; Ekkehard-Neuhaus and Stitt, 1991; Hill and ap Rees, 1994; Rowntree and Kruger, 1992). Another possible factor regulating cytosolic PPi levels was recently reported by Lunn and Douce (1993) who have discovered evidence of a specific, high affinity PPi translocator on the spinach chloroplast envelope. At the estimated concentrations of PPi in the spinach cytosol (200–300 μM; Weiner *et al.,* 1987) this translocator would be fully active. Once inside the chloroplast, PPi would be cleaved rapidly to Pi by an alkaline PPiase which is restricted to the stroma (Weiner *et al.,* 1987). It was proposed that, in the light, cytosolic PPi is taken up into the spinach chloroplast to replenish the stromal Pi pool.

## 6.3 The remarkable flexibility of plant phosphoenolpyruvate metabolism

In 1989, Duff and co-workers (1989a) reported the purification and characterization of a PEP phosphatase from heterotrophic *B. nigra* suspension cells. Although its substrate specificity was not absolute, this acid phosphatase was designated a PEP phosphatase owing to a specificity constant ($V_{max}/K_m$ ) for PEP that was at least sixfold greater than the value obtained for any other of 14 non-synthetic substrates that were identified (Duff *et al.*, 1989a). Furthermore, the $K_m$ (PEP) of the enzyme was well within estimated physiological concentrations of this compound and is equivalent to $K_m$ (PEP) values reported for various plant cytosolic PKs. These data suggested that PEP phosphatase could compete with cytosolic PK for a common intracellular pool of PEP. Notably, *B. nigra* PEP phosphatase also demonstrated potent feedback inhibition by Pi (Duff *et al.*, 1989a), and its specific activity was increased more than 10-fold following Pi deprivation (Duff *et al.,* 1989b). Similar results have been reported for the PEP phosphatase of a Pi-limited green alga, *S. minutum* (Theodorou *et al.*, 1991). Nagano and Ashihara (1993) have also observed a significant induction of an acid phosphatase having significant hydrolytic activity against PEP following Pi starvation of heterotrophic *C. roseus* suspension cells. It was postulated (Duff *et al.*, 1989b; Theodorou *et al.*, 1991): (a) that the ADP-dependent cytosolic PK is functionally eliminated from cellular metabolism during severe Pi stress (owing to >10-fold reductions in intracellular ADP levels) and (b) that PEP phosphatase functions as a Pi starvation-inducible bypass to the cytosolic PK of *B. nigra* and *S. minutum* (Figure 6.1).

Although energetically wasteful, PEP phosphatase could allow Pi-deficient plants and algae to utilize the Pi provided by adenylate catabolism, without impairing the conversion of PEP to pyruvate. Immunoquantification studies have revealed that the significant induction of *B. nigra* PEP phosphatase activity during Pi stress arises from *de novo* synthesis of PEP phosphatase protein (Duff *et al.*, 1991b). *B. nigra* PEP phosphatase has been shown to be localized in the cell vacuole (Duff *et al.*, 1991a), a location consistent with the enzyme's acidic pH optimum (Duff *et al.*, 1989a). The 'glycolytic bypass' theory for PEP phosphatase therefore requires that Pi-depleted *B. nigra* suspension cells transport PEP into, and pyruvate out of, the cell vacuole (Figure 6.1). As no data are presently available concerning this possibility, future studies must evaluate if the tonoplast of Pi-starved plant cells is permeable to PEP and pyruvate.

Another route of primary PEP catabolism is through the enzyme PEP carboxylase (PEPC) (Figure 6.1). When compared to nutrient-sufficient controls, PEPC activity was found to be fivefold greater in extracts of Pi-deficient *B. nigra* (Duff *et al.*, 1989b), and 30% greater in extracts of Pi-limited *S. minutum* (Theodorou *et al.*, 1991). The combined activities of PEPC, malate dehydrogenase and NAD-malic enzyme may function as an alternative route of pyruvate supply to the mitochondrion during Pi limitation (Figure 6.1). In *S. minutum*, the dark $CO_2$ fixation rate (i.e. representing *in vivo* PEPC activity) was found to increase with Pi limitation, while the rate of dark respiration declined (Theodorou *et al.*, 1991). This suggested that the *in vitro* and *in vivo* elevation of PEPC activity caused by Pi limitation of this green alga was in response to increased demands for pyruvate and/or Pi recycling rather than strictly by demand for the anaplerotic replenishment of the tricarboxylic acid (TCA) cycle intermediates. Induction of PEPC activity in response to Pi deprivation was also observed in *C. roseus* suspension cells (Nagano and Ashihara, 1993), and leaves of *Brassica napus* and *Sisymbrium officinale* (Hoffland *et al.*, 1992). In the latter study, it was suggested that, under Pi deficiency, citrate produced in the leaves (as a consequence of elevated PEPC activity) is translocated to the roots, where it is excreted as malic acid for the solubilization of rock phosphates in the rhizosphere. In contrast to studies of $C_3$ plants, Usuda and Shimogawara (1992) reported a steady decline of the PEPC activity of maize leaves with Pi starvation which corresponded to a decline in the level of PEPC protein to 27% of that in control plants.

The presence of pathways that can potentially bypass the reaction catalysed by cytosolic PK affords plants remarkable flexibility at the level of PEP metabolism. Nevertheless, the physical absence of cytosolic PK could be expected to have a major impact on plant metabolism. For example, PK-deficient mutants of yeast are unable to respire glucose (as cited by McHugh Gottlob *et al.*, 1992). Thus, it was somewhat unexpected when morphologically normal transgenic tobacco plants that completely lacked cytosolic PK in their leaves were recovered (McHugh Gottlob *et al.*, 1992). The leaves had a photosynthetic rate similar to that of wild-type leaves. As well, the dark respiration rate and concentration of pyruvate were unchanged in the transgenic leaves, indicating that the supply of substrates to the mitochondrion was not significantly impeded by the physical elimination of cytosolic PK (McHugh-Gottlob *et al.*, 1992). Likewise, transgenic potato plants whose tubers expressed 17% of the wild-type PK activity were found to develop and grow normally (Dunford, 1992). These findings confirm that plants have alternative pathways to cytosolic PK. Although the activities of several potential cytosolic PK bypass enzymes were not altered in the cytosolic PK-deficient tobacco leaves, elevated levels of PEP (McHugh Gottlob *et al.*, 1992) may have initiated the flow of carbon through the alternative pathways.

## 6.4 The glycolytic network

As described above, a considerable induction of PEP phosphatase and PEPC occurred when suspension cell cultures of *B. nigra* were transferred to media lacking Pi. Subsequent analysis of the adenylate-dependent reactions of glycolysis revealed that the activities of PFK, 3-PGA kinase and PK were not altered following nutritional Pi deprivation of *B. nigra* suspension cells (Duff *et al.*, 1989b). Selective maintenance of these glycolytic kinases during Pi stress may ensure that the respiratory machinery of the plant is in place should favourable nutritional conditions be restored. By contrast, following Pi starvation, the extractable activities of two other cytosolic enzymes, PFP and non-phosphorylating (irreversible) NADP-G3PDH, were elevated at least 20-fold, whereas that of the Pi-dependent (reversible) NAD-G3PDH was reduced by about sixfold (Duff *et al.*, 1989b). PFP and non-phosphorylating NADP-G3PDH are cytosolic enzymes that catalyse parallel reactions to PFK, and 3-PGA kinase and NAD-G3PDH, respectively, but do not require adenylates or Pi as substrates (Figure 6.1). Together with PEP phosphatase and/or PEPC, these reactions may circumvent the adenylate- and Pi-requiring reactions of glycolysis, thus providing an alternate route for the conversion of hexose phosphates to pyruvate in Pi-deprived plant cells (Figure 6.1). Three of these bypass reactions (PFP, PEP phosphatase and PEPC) also constitute a Pi-recycling system that converts esterified Pi to free Pi that would be rapidly re-assimilated into the metabolism of the Pi-deficient cells.

In their reassessment of glycolysis and gluconeogenesis, Sung and co-workers (1988) concluded that these two processes are networks, not just single pathways, and that plants can direct the flow of carbon through alternative enzymes. Furthermore, they proposed that plants regulate some enzymes *adaptively* in response to developmental or environmental changes, whereas other enzymes exist primarily as *maintenance* enzymes whose activity is invariable. Therefore, according to the terminology of Sung *et al.* (1988), the induction of alternate glycolytic enzymes such as PFP and PEP phosphatase during Pi stress in *B. nigra* is an adaptive response to a nutritional deficiency, while the enzymes of 'classical' glycolysis, such as PFK and PK which displayed uniform activities regardless of cellular Pi status, are of the maintenance type.

The effect of Pi starvation on activities of several glycolytic enzymes has also been examined in maize and *C. roseus*. In maize leaves, the activities of PK, 3-PGA kinase, aldolase and phosphohexose isomerase were 30–70% lower in Pi-starved vs. Pi-fed controls (Usuda and Shimogawara, 1992). Likewise, in *C. roseus* suspension cells, the activities of several glycolytic enzymes on a fresh weight basis, including hexokinase, fructokinase, phosphoglucose isomerase, phosphorylating NAD-G3PDH, non-phosphorylating NADP-G3PDH, PFK and PFP, were unaltered or slightly decreased during the transition from Pi sufficiency to Pi deficiency (Li and Ashihara, 1990; Nagano and Ashihara, 1993). These findings contrast with the large induction of PFP activity observed in Pi-starved suspension cells and seedlings of *B. nigra* (Duff *et al.*, 1989b;

Theodorou and Plaxton, 1994; Theodorou *et al.,* 1992). This discrepancy may have been due to differences in plant material or the method of cell culture used. However, consistent with observations from *B. nigra*, the activities of PEPC as well as a PEP (acid) phosphatase were significantly elevated in the Pi-deprived *C. roseus* cells (Nagano and Ashihara, 1993).

Although there was no change in PFP activity upon Pi starvation of *C. roseus* suspension cells on a fresh weight basis, there was an increase of PFP activity when expressed in terms of protein concentration. Soluble protein concentration decreased to 68% of Pi-fed controls. Also, the activity of PFP was almost 10-fold that of PFK in the Pi-starved *C. roseus* cells (Li and Ashihara, 1990; Nagano and Ashihara, 1993). Finally, Pi and ATP levels in Pi-starved *C. roseus* suspension cells were reduced to 3 and 15% of Pi-fed levels, respectively (Li and Ashihara, 1990; Nagano and Ashihara, 1993). Owing to the potent product inhibition of PFP by Pi, the ATP substrate dependence of PFK, the decline in ATP and Pi levels, as well as the selective maintenance of PFP activity, Nagano and Ashihara (1993) have suggested that PFP may function as a bypass to the ATP-dependent PFK in Pi-deprived *C. roseus* suspension cells. An apparent increase in the $\alpha/\beta$ subunit ratio with Pi-starvation of *C. roseus* suspension cells coincident with an increased sensitivity to Fru-2,6-$P_2$ activation also support a possible function of PFP during Pi stress in this species (Figure 6.3c; Theodorou, Plaxton, and Ashihara, unpublished results).

## 6.5 PFP: an adaptive enzyme

PFP is a strictly cytosolic enzyme which occurs in plants (Carnal and Black, 1979, 1983; Kruger and Dennis, 1987) and some microorganisms (Green *et al.,* 1992, 1993; Ladror *et al.,* 1991; Peng and Mansour, 1992). PFP appears to be an adaptive enzyme whose activity is responsive to environmental stresses, such as Pi nutrition and anaerobiosis, and to developmental or tissue-specific cues (Ashihara and Sato, 1993; Blakeley *et al.,* 1992; Botha and Botha, 1991a,b, 1993a,b; Botha *et al.,* 1992; Clayton *et al.,* 1993; Duff *et al.,* 1989b; Enomoto *et al.,* 1992; Kobayashi *et al.,* 1992; Mertens, 1991; Nakamura *et al.,* 1992; Podestá and Plaxton, 1994; Sato and Ashihara, 1992; Stitt, 1990; Theodorou *et al.,* 1992). Unlike plant PFK, most plant PFPs display potent activation by nanomolar concentrations of the regulatory metabolite fructose-2,6-bisphosphate (Fru-2,6-$P_2$, Stitt, 1990). The PFP of many plants consists of two pairs of different subunits of approximately 66 ($\alpha$-subunit) and 60 ($\beta$–subunit) kDa (Kruger and Dennis, 1987; Stitt, 1990). Various evidence indicates that the $\beta$-subunit contains the catalytic site, while the $\alpha$-subunit may be involved in regulation of catalytic activity by Fru-2,6-$P_2$ (Botha and Botha, 1991a,b; Carlisle *et al.,* 1990; Cheng and Tao, 1990; Stitt, 1990; Theodorou and Plaxton, 1994; Theodorou *et al.,* 1992; Yan and Tao, 1984). Montavon and Kruger (1992, 1993) have recently identified an arginyl residue which appears to be essential for catalysis by PFP in potato tuber. This residue, which resides on the $\beta$-subunit of potato PFP, is most likely Arg115.

Chemical modification of this residue almost completely inhibits potato PFP activity in the glycolytic and gluconeogenic directions (Montavon and Kruger, 1993). This amino acid corresponds to Arg72 of bacterial PFK and is essential to the reaction mechanism of *Escherichia coli* PFK, a result which supports the proposal that the active site of PFP in plants is similar to that of the bacterial PFK (Carlisle *et al.*, 1990).

In contrast to previous reports, there is now some evidence that the α-subunit of plant PFP may also possess catalytic activity (Theodorou and Plaxton, 1994). Immunoblots of extracts from leaves of 50-day old *B. nigra* seedlings probed with: (i) antibodies raised against the potato tuber PFP holoenzyme, as well as (ii) subunit-specific antibodies raised against the α-subunit of potato tuber PFP revealed only a single immunoreactive polypeptide of 66 kDa that co-migrated with the α-subunit of homogeneous potato tuber PFP (Theodorou and Plaxton, unpublished results). By contrast, no immunoreactive polypeptides were observed on immunoblots of *B. nigra* leaf extracts probed with subunit-specific antibodies raised against the β-subunit of potato tuber PFP (Theodorou and Plaxton, unpublished results). This was the first evidence of a plant PFP comprised solely of α-subunit. The activity of this PFP displayed potent activation by Fru-2,6-$P_2$ and was elevated 2.3-fold following 50 days of Pi starvation (Theodorou and Plaxton, 1994). These results suggest that the α-subunit of PFP in leaves of *B. nigra* has both a catalytic and regulatory function.

The proposal that PFP functions as a glycolytic bypass to the ATP-dependent PFK during periods of Pi stress is supported by: (a) the potent product inhibition of PFP by Pi (Stitt, 1990), (b) the striking induction of PFP activity in Pi-deprived *B. nigra* suspension cells (Duff *et al.*, 1989b; Theodorou *et al.*, 1992) and seedlings (Theodorou and Plaxton, 1994) and (c) the selective maintenance of PPi and Fru-2,6-$P_2$ pools upon Pi starvation (Dancer *et al.*, 1990; Duff *et al.*, 1989b; Stitt, 1990). The PFP of several heterotrophic plant tissues has also been proposed to operate as a glycolytic bypass to cytosolic PFK during periods of anaerobiosis (Mertens, 1991). The use of PPi rather than ATP could confer a significant energetic advantage to plants subjected to environmental stresses, such as Pi deprivation or anoxia.

Recent work has demonstrated that synthesis of the α-subunit of PFP is tightly regulated in suspension cells of *B. nigra* and that this regulation is dependent on cellular Pi status. By contrast, the β-subunit of PFP from *B. nigra* is constitutively expressed under all nutrient regimes (Theodorou *et al.*, 1992). Quantification of immunoblots indicated that, in *B. nigra* cells experiencing a transition from Pi sufficiency to deficiency or vice versa, the relative amount of immunoreactive α-subunit correlated with the degree of activation of PFP by Fru-2,6-$P_2$ (Figure 6.2a–c; Theodorou *et al.*, 1992). Hence, the large induction of PFP activity by Pi starvation of *B. nigra* suspension cells appears to be based upon *de novo* synthesis of the enzyme's α-subunit, leading to a significant enhancement in sensitivity of the enzyme to its activator, Fru-2,6-$P_2$ (Figure 6.2a–c; Theodorou *et al.*, 1992).

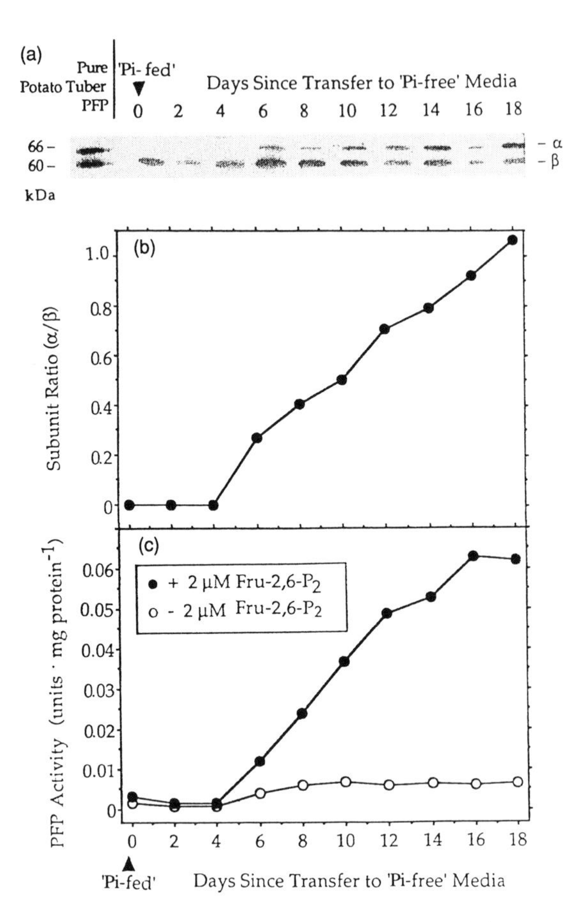


***Figure 6.2.*** *Induction of PFP in* B. nigra *suspension cells becoming Pi deficient. Cells previously cultured for 18 days in 10 mM Pi were transferred to medium containing 0 mM Pi. (a) Immunological detection of PFP. Samples were subjected to SDS–polyacrylamide gel electrophoresis and blot-transferred to a polyvinylidene difluoride membrane. Blots were probed with 1000-fold diluted rabbit anti-(potato PFP) immune serum (Moorhead and Plaxton, 1991), and immunoreactive polypeptides were detected using an alkaline phosphatase-linked secondary antibody. The first lane contains 20 ng of homogeneous potato tuber PFP (Moorhead and Plaxton, 1991); all other lanes contain 60 μg of protein from extracts of cells at days 0–18 as indicated. (b) The subunit ratios (α/β) were determined by laser densitometric immunoquantification of the immunoblot shown in a. (c) PFP activity of desalted extracts was measured in the forward direction in the presence and absence of 2 μM Fru-2,6-P$_2$. Reproduced from Theodorou* et al. *(1992) with permission from The American Society for Biochemistry and Molecular Biology.*

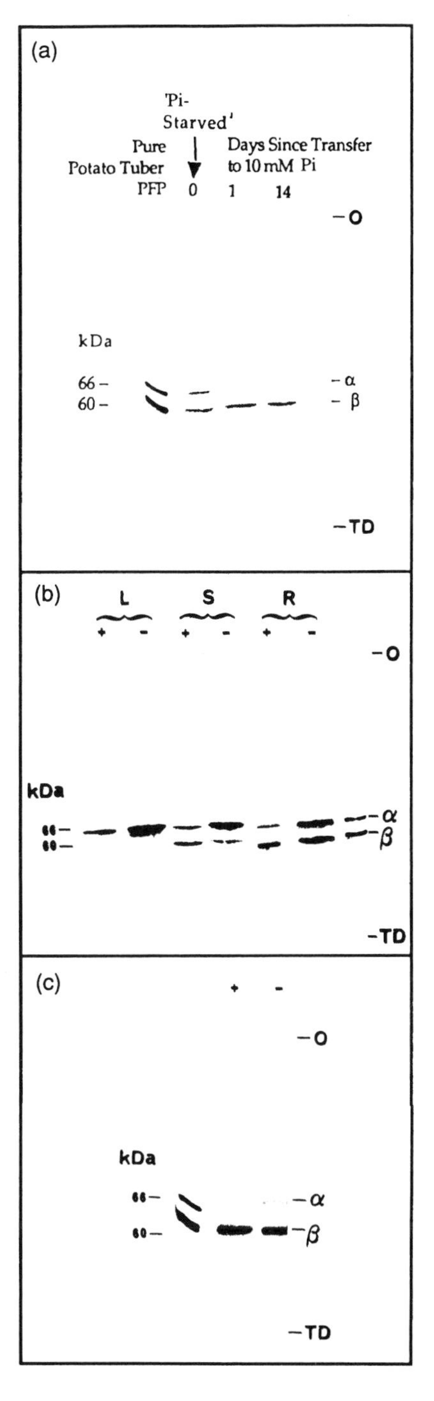


***Figure 6.3.*** *Effect of nutritional Pi status on PFP subunit composition in* B. nigra *and* C. roseus. *(a) Immunological detection of PFP from suspension cells of* B. nigra *becoming Pi sufficient. Cells were cultured for 18 days in 0 mM Pi (Pi-starved) and then transferred to medium containing 10 mM Pi. Samples were subjected to Western blotting as described in Figure 6.2a. The first lane contains 20 ng of purified potato tuber PFP (Moorhead and Plaxton, 1991); all other lanes contain 60 μg of protein from extracts prepared from cells at days 0, 1 and 14 as indicated. O, origin; TD, tracker dye front. Reproduced from Theodorou* et al., *(1992) with permission from The American Society for Biochemistry and Molecular Biology. (b) Immunological detection of PFP in leaves (L), stems (S) and roots (R) of 40 day* B. nigra *seedlings grown in medium containing 0 mM (–) or 5 mM (+) Pi. Samples were subjected to Western blotting as described in Figure 6.2a. The last lane contains 40 ng of homogeneous potato tuber PFP (Moorhead and Plaxton, 1991); all other lanes contain the equivalent of 8 mg fresh weight from extracts of leaves, stems, and roots as indicated. Reproduced from Theodorou and Plaxton (1994), with permission from Blackwell Scientific Publications Ltd. (c) Immunological detection of PFP from* C. roseus *suspension cells. Cells were cultured for 7 days in medium containing 1.25 mM Pi, and then transferred to medium containing either 1.25 mM Pi (+) or 0 mM Pi (–). Cells from both treatments were harvested after 7 days and extracts were subjected to Western blotting as described in Figure 6.2a. The first lane contains homogeneous potato tuber PFP (Moorhead and Plaxton, 1991). Lanes 2 and 3 contain crude extracts prepared from the Pi-fed (+) and Pi-starved (–) treatments, respectively.*

Similarly, the large reduction in PFP activity that occurs when Pi-deprived *B. nigra* cells become Pi sufficient appears to arise from an inhibition of synthesis and/or enhanced degradation of the α-subunit (Figure 6.3a; Theodorou *et al.*, 1992). It is evident that some form of proteolytic specificity towards the α-subunit must exist to facilitate the selective disappearance of this polypeptide upon Pi refeeding. Overall, these findings provide additional evidence that the α-subunit may function as a regulatory protein in controlling the catalytic activity of PFP and its regulation by Fru-2,6-$P_2$. The concentration of Fru-2,6-$P_2$ in the cytosol of Pi-deprived *B. nigra* was previously estimated to be about 0.5 μM (Duff *et al.*, 1989b), a level that almost fully saturates the PFP from Pi-starved cells (Theodorou *et al.*, 1992).

A subsequent study (Theodorou and Plaxton, 1994) has demonstrated that the induction of PFP by Pi starvation of *B. nigra* suspension cells is also relevant to intact seedlings of the same species. Roots, stems and leaves of 50-day old Pi-deficient seedlings displayed two- to fourfold greater PFP activity than nutrient-sufficient controls. This induction was based primarily upon an increased susceptibility of PFP from the Pi-starved tissues to activation by Fru-2,6-$P_2$. The ratio of PFP:PFK was at least twofold greater in the various Pi-deprived organs. Immunoblots probed with potato PFP antibodies revealed that the induction of PFP in Pi-starved seedlings was coincident with an elevation in the amount of PFP α-subunit in the leaves (see above) as well as an increase in the α:β subunit ratio in the stems and roots (Figure 6.3b). Time course studies revealed that the various responses to Pi stress in these seedlings were significantly delayed in the leaves as compared to the roots and stems, suggesting that Pi may be sequestered preferentially to the leaves during Pi starvation (Theodorou and Plaxton, 1994).

Although induction of PFP by Pi deprivation does not appear to be a ubiquitous feature of higher plants, several species which are closely related to *B. nigra* also appear to possess this adaptation. Roots from Pi-limited seedlings of *Sinapis arvensis, Sinapis alba, Brassica oleracea* and *Brassica rapa* were found to contain significantly greater PFP activity than did nutrient sufficient controls (Murley, Theodorou, Warwick and Plaxton, unpublished results). Like *B. nigra*, all five of these species are: (i) weeds of the Crucifer family which occur in nature on nutrient-poor soils such as old fields and roadsides (Rollins, 1993), and (ii) contain extremely low PFP activity under conditions of nutrient sufficiency. Presumably, in nutrient-sufficient plants of such species which normally contain very low PFP activity, glycolytic carbon flux is initiated by PFK. ATP limitation of PFK during Pi deprivation may necessitate increased participation, and hence increased activity of PFP. In contrast to these findings, PFP activity declined with Pi stress in suspension cells of *Brassica napus, Nicotiana silvestris* (Theodorou and Plaxton, unpublished results), and *C. roseus* (Li and Ashihara, 1990; Nagano and Ashihara, 1993). Similarly, PFP activity was reduced or unaltered in Pi-deprived tissues of *Nicotiana tabacum* (Paul and Stitt, 1993) and in bean roots (Rychter and Randall, 1994). A common feature of these five species is very high PFP activity during nutrient sufficiency. Therefore, despite the fact that PFP activity declined or was unchanged with Pi stress, activity in Pi-limited tissues of all five of these species (0.19–0.91 μmol $g^{-1}$ fresh weight $min^{-1}$) was still greater than PFP activity in the Pi-limited tissues of the species where PFP induction has been observed (0.025–0.18 μmol $g^{-1}$ fresh weight $min^{-1}$). We therefore suggest that in such species which already contain substantial PFP activity under nutrient sufficiency, there is no need for induction of PFP in order for this enzyme to support glycolytic carbon flow during Pi stress.

In several instances where PFP activity is markedly affected by Pi status (i.e. *B. nigra* suspension cells and seedlings, and *C. roseus* suspension cells), altered PFP activity appears to be correlated with modified levels of the α-subunit and a corresponding change in the enzyme's sensitivity to its allosteric activator, Fru-2,6-$P_2$ (Figures 6.2 and 6.3). Similarly, Podestá and Plaxton (1994) reported that PFP activity of germinating *Ricinus communis* cotyledons depends on differential expression of the enzyme's α-subunit during different stages of seedling development. Increased α-subunit content was correlated directly to increased PFP activity and sensitivity to Fru-2,6-$P_2$ (Podestá and Plaxton, 1994). Differential expression of the genes for the α- and β-subunits of PFP in the endosperm of germinating and developing seeds from *R. communis* has also been detected (Blakeley *et al.*, 1992). However, an increased level of mRNA for the α-subunit relative to the β-subunit during germination was not reflected directly in elevated levels of PFP α-subunit protein.

Few precedents exist in the literature demonstrating regulation of enzymatic activity by selective synthesis/degradation of one subunit while another subunit is expressed constitutively. However, Loulakakis and Roubelakis-Angelakis (1992) recently reported increased aminating activity of NAD(H)-glutamate dehydrogenase (GDH) in *Vitus vinifera* callus upon transfer to ammonium-containing medium. This ammonium-induced increase in GDH activity was attributed to *de novo* synthesis of the enzyme's α-subunit while there was no detectable alteration in β-subunit synthesis (Loulakakis and Roubelakis-Angelakis, 1992). Likewise, in *Schizosaccaromyces pombe*, mitosis is induced by activation of the enzyme M-phase-promoting factor (MPF, Glotzer *et al.*, 1991). MPF is composed of a catalytic subunit, $p34^{cdc2}$, which is constitutively expressed, and a regulatory subunit, cyclin, whose abundance fluctuates throughout the cell cycle. Rapid degradation of cyclin signals the end of mitosis, whereas accumulation of cyclin during interphase leads to the activation of MPF, and another round of mitosis ensues (Glotzer *et al.*, 1991).

## 6.6 Dark respiration and mitochondrial electron transport during Pi stress

Various investigations using different plant species and tissues have yielded conflicting results with respect to the effect of Pi deficiency on the overall rate of dark $O_2$ consumption and respiratory $CO_2$ release. Generally, in most plant and algal species examined, lower rates of respiration have been measured with suboptimal supply of Pi. In the green algae *C. reinhardtii* (Weger and Dasgupta,

1993) and *S. minutum* (Theodorou *et al.*, 1991) as well as in the higher plant *Lemna gibba* (Thorsteinsson and Tillberg, 1987), respiration rates declined with decreasing nutritional Pi in a manner closely correlated to a declining relative growth rate. Decreased rates of respiration in response to Pi deficiency have also been reported in maize leaves (Usuda and Shimogawara, 1993), sugar beet (Terry and Ulrich, 1973), pea and barley leaves (Thorsteinsson and Tillberg, 1990), barley roots (Bingham and Farrar, 1989) and heterotrophic suspension-cultured *C. roseus* (Li and Ashihara, 1990). The latter study contradicts the findings of Hoefnagel *et al.* (1993) who observed: (i) no difference in respiration rate between Pi-deprived and nutrient-sufficient *C. roseus* cells during the exponential growth phase, and (ii) a significant increase in the respiration of Pi-deprived *C. roseus* during the stationary phase of growth. Elevated respiration with Pi stress was also reported in the green alga *S. obtusiusculus* (Tillberg and Rowley, 1989), whereas in sycamore cell cultures (Rebeille *et al.*, 1984), pea root tips (Lee and Ratcliffe, 1983) and bean roots (Rychter *et al.*, 1992) the rate of respiration was not affected by Pi starvation. These opposing results suggest that the overall rate of respiration may not be as important an indicator of mitochondrial adaptations to Pi limitation as is the pathway of electron transport by which $O_2$ is consumed.

Respiratory $O_2$ consumption in plants may be mediated via the phosphorylating cytochrome pathway or the non-phosphorylating alternative pathway (Figure 6.1). The branch point between these pathways is the mobile ubiquinone pool. Since operation of the alternative pathway is coupled to ATP synthesis at only one site (complex I), effectively bypassing production of ATP at complexes II and III of the cytochrome pathway, the physiological significance of this apparently wasteful route has been the subject of debate. It has been suggested that the alternative pathway may function as an 'energy overflow' mechanism which becomes engaged only when the cytochrome pathway is working at full capacity, or when electron flux via the cytochrome pathway is restricted by low availability of adenylates and/or Pi (Bahr and Bonner, 1973; Lambers, 1982; Lambers *et al.*, 1983; Sideow and Berthold, 1986). The 'energy overflow' hypothesis has found credence in the recent finding that engagement of the alternative pathway is dependent upon the reduction state of the ubiquinone pool (Dry *et al.*, 1989). The alternative pathway is only engaged at a high ubiquinone reduction state (i.e. when the cytochrome pathway is working at its capacity or is in some way restricted). According to this view, the alternative pathway may allow oxidation of respiratory reductant during adenylate restriction of the cytochrome pathway, thereby maintaining respiratory carbon flow for the provision of biosynthetic intermediates (Møller and Palmer, 1982).

Increased alternative pathway activity has been implicated in maintenance of electron transport chain activity accompanying a wide variety of physiological stresses. These include low temperature (Collier and Cummins, 1990), high temperature (Weger and Guy, 1991; Weger and Dasgupta, 1993), osmotic stress (Jolivet *et al.*, 1990; Schmitt and Dizengremel, 1989; Weger and Dasgupta, 1993), nitrogen limitation (Hoefnagel *et al.*, 1993, 1994) and Pi limitation (discussed below). Several techniques have been employed to assess both engagement and capacity of either pathway. These include the use of: (i) specific inhibitors of the cytochrome pathway such as potassium cyanide (an inhibitor of cytochrome $a_3$) and rotenone (an inhibitor of the NADH:ubiquinone oxido-reductase of complex I); (ii) specific inhibitors of the alternative pathway such as propyl gallate or the more commonly used inhibitor salicylhydroxamic acid which inhibits electron flux between the ubiquinone pool and the alternative oxidase; or (iii) isotope discrimination using $^{18}O$ (Bryce *et al.*, 1990; Guy *et al.*, 1992; Møller and Palmer, 1982; Rychter *et al.*, 1992; Weger and Dasgupta, 1993). The use of respiratory inhibitors to measure partitioning of respiratory electron flow between the alternative and cytochrome pathways has been described in detail elsewhere (Lambers *et al.*, 1983; Møller *et al.*, 1988).

From the limited number of studies of the mechanisms of regulatory control of electron transport in plant mitochondria, it appears that the absolute concentrations of ADP and Pi are amongst the most important factors involved in the regulation of mitochondrial respiration (Bryce *et al.*, 1990). The significant reductions in cellular ADP and Pi levels that follow nutritional Pi limitation (see Section 6.2) could restrict electron flow through the cytochrome pathway at the sites of coupled ATP synthesis. Recent findings suggest that plants may respond adaptively to Pi stress and adenylate control of respiration by increased engagement and/or capacity of the non-phosphorylating (i.e. rotenone- and potassium cyanide-insensitive) alternative pathways of mitochondrial electron transport (Figure 6.1). Rychter and co-workers (Rychter and Mikulska, 1990; Rychter *et al.*, 1992) have examined the effect of Pi deficiency on mitochondrial electron transport in isolated bean root mitochondria as well as in intact roots. In both systems, Pi nutrition had no effect on the overall rate of $O_2$ consumption (Rychter and Mikulska, 1990; Rychter *et al.*, 1992). However, an increased participation of the cyanide-resistant pathway and a decreased involvement of the cytochrome pathway was observed during prolonged Pi deficiency. Cytochrome oxidase activity was 20% lower in the Pi-deficient bean roots, which may partially explain the lowered involvement of the cytochrome pathway (Rychter *et al.*, 1992). Reduced cytochrome oxidase activity was also observed in barley roots deprived of inorganic nutrients, but the alternative pathway did not become engaged (Bingham and Farrar, 1989). $O_2$ consumption rates of mitochondria isolated from Pi-deficient bean roots were also found to be less sensitive to rotenone inhibition, suggesting an increased participation of the rotenone-insensitive (bypass of complex I) alternative pathway of mitochondrial electron transport (Figure 6.1).

Similar results were reported for heterotrophic suspension-cultured *C. roseus*. During nutrient-sufficient culture of *C. roseus*, the alternative pathway activity was negligible, although there was substantial alternative respiration capacity (Hoefnagel *et al.*, 1987). With Pi starvation, in combination with excess sugar, the overall respiration rate increased and this extra respiratory activity was associated exclusively with engagement of the cyanide-resistant pathway (Hoefnagel *et al.*, 1993). Pi starvation does not always result in engagement of the alternative pathway. In the green alga *C. reinhardtii*, Pi starvation resulted in a large increase in alternative pathway capacity relative to cytochrome pathway activity, without resulting in engagement of the alternative pathway (Weger and Dasgupta, 1993). The physiological significance of increased alternative pathway capacity without increased engagement is unclear.

The studies described above provide strong evidence that increased involvement of the cyanide-resistant pathway and/or the rotenone-insensitive bypass could significantly alter the extent to which adenylates or Pi control the rate of mitochondrial electron transport. The operation of these alternative, non-phosphorylating pathways when the cytochrome pathway activity is restricted or working at capacity might allow functioning of the TCA cycle and limited ATP production, thus contributing to the survival of plants during prolonged periods of Pi deprivation (Rychter *et al.*, 1992).

## 6.7 Pi re-supply to Pi-limited cells: short-term respiratory adaptations

Pi re-supply studies have provided some insight into the nature of the restrictions on mitochondrial electron flux accompanying Pi stress. Two recent investigations examining short-term Pi re-feeding of unicellular green algae suggest that, during Pi limitation, respiratory electron flow is either limited by ADP and/or Pi availability, or restricted by substrate supply to the TCA cycle, thereby limiting supply of reductant to electron transport pathways (Gauthier and Turpin, 1994; Weger, 1993).

In *S. minutum*, an algal species which does not possess the cyanide-resistant pathway of mitochondrial electron transport, dark respiration of Pi-deprived cells appears to be primarily ADP limited (Gauthier and Turpin, 1994). Pi enrichment of Pi-limited *S. minutum* caused a 2.5-fold increase in dark $O_2$ consumption within 1 min of Pi addition. In this and other plant species, Pi uptake is postulated to involve $H^+$ co-transport which produces decreased cytosolic pH (Gauthier and Turpin, 1994; Sakano, 1990; Sakano *et al.*, 1992; Ulrich and Novacky, 1990). The resulting decrease in cytosolic pH is believed to stimulate a plasmalemma $H^+$-ATPase which pumps protons out of the cells (Gauthier and Turpin, 1994; Ulrich-Eberius *et al.*, 1984). The increased levels of ADP following activation of the $H^+$-ATPase releases the mitochondrial electron transport chain from adenylate control and the dark respiration rate increases. It should be noted that some control is probably exerted at the level of glycolytic carbon supply to the TCA cycle via restriction of key regulatory glycolytic enzymes such as PK and PFK whose activity may be limited by low ADP and Pi levels, respectively (Lin *et al.*, 1989; Turpin *et al.*, 1990). Although the mechanisms that initiate increased glycolytic flux in *S. minutum* following Pi re-feeding remain unclear, findings suggest that high triose phosphate levels and low PEP levels maintain long-term increases in PK and PFK activities, following Pi enrichment of Pi-limited cells (Gauthier and Turpin, 1994).

In *C. reinhardtii*, respiration by Pi-limited cells appears to be controlled primarily at the level of glycolysis (i.e. by control of the substrate supply to the TCA cycle; Weger, 1993). Metabolite analysis indicated a rapid activation of PFK and PK upon Pi re-supply. Respiration rate increased immediately and continued to increase following 10 min of Pi re-supply, and these events were associated with elevated levels of glycolytic intermediates (Weger, 1993).

Pi re-supply to Pi-starved *C. roseus* disengages the cyanide-resistant pathway of mitochondrial electron transport (Hoefnagel *et al.*, 1994). This process is probably only an indirect effect of Pi, because cellular Pi content, which rapidly increased following Pi addition, was relatively low again before the alternative pathway was fully disengaged. A much better correlation was observed between high ADP and adenylate content and disengagement of the cyanide-resistant pathway in this system (Hoefnagel *et al.*, 1994). It was therefore concluded that engagement of the alternative pathway following Pi starvation of *C. roseus* suspension cells was the result of a limited adenylate content (Hoefnagel *et al.*, 1994).

A rapid recovery of adenylate and Pi levels generally appears to be one of the most pronounced metabolic changes that initially accompanies Pi re-supply to Pi-deprived plants (Dancer *et al.*, 1990; Kubota *et al.*, 1989; Li and Ashihara, 1990). Although the pattern of changes in adenylate and Pi levels in the very short term following Pi re-feeding (i.e. seconds to minutes) varies depending on the type of Pi-starved respiratory control being exerted (Gauthier *et al.*, 1994); adenylate levels may almost fully recover within a few hours of Pi re-feeding (Dancer *et al.*, 1990; Kubota *et al.*, 1989; Li and Ashihara, 1990). In *C. roseus*, ATP levels doubled during the first 30 min of Pi re-feeding and attained nutrient-sufficient levels within 2 h of Pi re-feeding (Kubota *et al.*, 1989; Li and Ashihara, 1990).

## 6.8 Co-ordinated induction of Pi starvation rescue mechanisms: evidence of a plant Pi stimulon

Phosphate starvation-inducible synthesis of the α-subunit of PFP in suspension cells of *B. nigra* is coincident with *de novo* synthesis of PEPC (Cornel and Plaxton, unpublished results), vacuolar PEP phosphatase and a cell wall-localized non-specific acid phosphatase (Duff *et al.*, 1991b). Parallel induction of these enzymes with a simultaneous enhancement in cellular Pi absorptive capacity (Lefebvre *et al.*, 1990) points to the existence of a *B. nigra* 'Pi stimulon' (Goldstein *et al.*, 1989; Rao and Torriani, 1990).

The Pi stimulon is defined as the complete set of genes that are co-regulated by Pi. A subset of a stimulon composed of genes regulated by the same *trans*-acting factor is designated a regulon (Shinagawa *et al.*, 1987). In microbial systems, much of the rescue metabolism induced by Pi starvation is accomplished by a set of co-regulated genes known as the *pho* regulon (Goldstein *et al.*, 1989; Kaffman *et al.*, 1994; Rao and Torriani, 1990). The *pho* regulon of *E. coli* includes a minimum of 20 genes whose products function by enhancing Pi availability and uptake from the external medium (i.e. genes that encode Pi transporters and acid phosphatases) (Goldstein *et al.*, 1989; Rao and Torriani, 1990).

Based primarily on the observation that plants exhibit coordinated physiological and biochemical rescue responses to Pi starvation that are similar to those of microorganisms, Goldstein *et al.* (1989) have proposed the existence of a plant Pi stimulon. Studies of a variety of environmental perturbations of plants illustrate that stress-induced alterations in gene expression occur mostly at the transcriptional level, resulting in the synthesis of some proteins and repression of others (for review, see Sachs and Ho, 1986). Numerous recent studies report the *de novo* synthesis or repression of specific mRNAs and/or proteins in higher plants and algae in response to Pi starvation. For higher plants, these include *de novo* synthesis of a tomato secreted acid phosphatase (Goldstein *et al.*, 1988a,b), a number of tomato ribonucleases (Jost *et al.*, 1991; Löffler *et al.*, 1993; Nürnberger *et al.*, 1991), possible components of plasmalemma Pi transporters in cultured tomato roots (Hawkesford and Belcher, 1991), a vacuolar PEP phosphatase, a cell wall-localized acid phosphatase (Duff *et al.*, 1991b), PEPC (Cornel and Plaxton, unpublished results), the α-subunit of PFP (Theodorou *et al.*, 1992; Theodorou and Plaxton, 1994), a putative protein kinase (Budicky *et al.*, 1993) and four other unidentified polypeptides from *B. nigra* suspension cells (Malboobi, 1992; Malboobi and Lefebvre, 1993). It should be noted that, unlike heat shock, Pi starvation did not cause gross changes in the protein synthesis profiles of *B. nigra* suspension cells (Fife *et al.*, 1990) and tomato cell cultures (Goldstein *et al.*, 1989). In view of these findings, as well as the overall decline in protein concentration in these cells with Pi starvation, increased synthesis of specific proteins with low nutritional Pi punctuates their potential importance in the Pi starvation response. In *C. reinhardtii*, 26 Pi starvation-inducible genes were detected. Four sequences revealed a high degree of homology with enolase, pyruvate formate-lyase, P1 and L31 ribosomal proteins (Dumont *et al.*, 1993). Low Pi has also been reported to elicit specific degradation of proteins such as sucrose phosphate synthase (SPS) and PEPC in maize leaves (Usuda and Shimogawara, 1993).

## 6.9 Regulation of the Pi starvation response by protein phosphorylation

Enzyme modification by the reversible covalent incorporation of Pi is an important mechanism for the control of plant metabolism *in vivo* (Budde and Chollet, 1988; Iglesias *et al.*, 1993; Nimmo, 1993; Plaxton, 1990). Plant enzymes whose activities appear to be regulated by reversible phosphorylation include: SPS (Huber *et al.*, 1989), nitrate reductase (Huber *et al.*, 1992; MacKintosh, 1992), PEPC, the mitochondrial pyruvate dehydrogenase complex, glyoxosomal malate synthase, microsomal hydroxymethyl-glutaryl-CoA reductase, quinate:$NAD^+$ oxidoreductase, 6-phosphofructo-2-kinase and pyruvate:Pi dikinase (Budde and Chollet, 1988; Nimmo, 1993; Yang *et al.*, 1988). Since phosphorus plays a vital role in stimulus–response coupling in signal transduction pathways (via compounds such as phosphoinositides) as well as being involved directly in the ultimate covalent modification of target proteins, alterations in protein phosphorylation that occur in response to Pi starvation are of additional interest.

The importance of protein phosphorylation in the Pi starvation response of *E. coli* and yeast has already been well documented (Kaffman *et al.*, 1994). An excellent example is the regulation of acid phosphatase secretion in yeast in response to Pi status. In *Saccharomyces cerevisiae*, the PHO5 gene encodes a secreted acid phosphatase whose transcription is induced up to 1000-fold in response to Pi starvation. Transcription of PHO5 is controlled by the transcription factor PHO4. Under high Pi, hyperphosphorylation of PHO4 by a cyclin–CDK (cyclin-dependent protein kinase) complex (PHO80–PHO85) inhibits binding of the PHO4 transcription factor to the acid phosphatase gene. However, with Pi starvation, PHO4 is dephosphorylated and can thus bind to and mediate transcription of the acid phosphatase gene (Figure 6.4; Kaffman *et al.*, 1994).

Preliminary evidence for the involvement of protein phosphorylation in the Pi starvation response of plants has been obtained using *B. nigra* suspension cells (Budicky *et al.*, 1993; Malboobi, 1992). When cells were labelled *in vivo* with

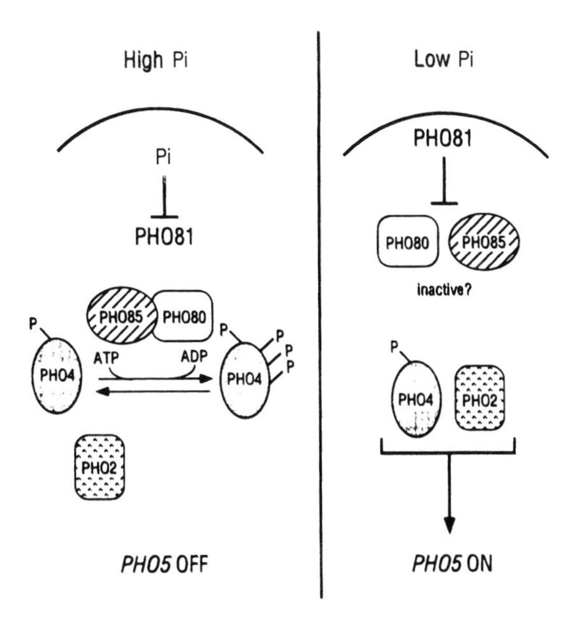


***Figure 6.4.*** *Model for regulation of Pi starvation-inducible acid phosphatase gene in* Saccharomyces cerevisiae. *When yeast are grown in high Pi (repressing conditions for the acid phosphatase gene PHO5), a cyclin–CDK complex (PHO80–PHO85) hyperphosphorylates the transcription factor PHO4, thereby inactivating it and preventing induction of the acid phosphatase gene. When yeast are Pi starved, the cyclin–CDK complex is inactivated, PHO4 is not hyperphosphorylated and can thereby activate transcription of PHO5 in conjunction with PHO2. The acid phosphatase gene is induced up to 1000-fold. Reproduced with permission from Kaffman* et al. *(1994),* Science, **263,** *1153–1154; copyright 1994 American Association for the Advancement of Science.*

$^{32}$Pi, at least 12 (de)phosphorylation conversions were detected during the transition from Pi sufficiency to Pi deficiency or vice versa. These conversions occurred within the second and third days of growing the Pi-fed cells in Pi-free medium and during the first day of transferring Pi-starved cells into nutrient-sufficient medium. The former period coincides with the inception of a variety of Pi starvation responses in these cells. Budicky *et al.* (1993) have also isolated a cDNA clone for a putative Pi starvation-inducible protein kinase from *B. nigra* suspension cells. A potential role for this kinase in modifying cellular metabolism under Pi starvation conditions has yet to be determined.

## 6.10 Concluding remarks

Figure 6.1 presents a model suggesting the major pathways of respiration which may exist in Pi-deficient plant cells. The alternative pathways of glycolytic carbon flow and mitochondrial electron transport that are depicted in Figure 6.1 may represent key facets of the survival of Pi-limited plants and exemplify the extraordinary flexibility of plant versus non-plant metabolism. Metabolic redundancy represents an essential component of the biochemical adaptations of plants that facilitates their survival in an ever-changing and frequently stressful environment.

## Acknowledgements

Research in our laboratory is supported by grants from the Natural Sciences and Engineering Research Council of Canada.

## References

Ashihara, H. and Sato, F. (1993) Pyrophosphate:fructose-6-phosphate 1-phosphotransferase and biosynthetic capacity during differentiation of hypocotyls of *Vigna* seedlings. *Biochim. Biophys. Acta* **1156**, 123–127.

Ashihara, H., Li, X.-N. and Ukaji, T. (1988) Effect of inorganic phosphate on the biosynthesis of purine and pyrimidine nucleotides in suspension-cultured cells of *Catharanthus roseus*. *Ann. Bot.* **61**, 225–232.

Bahr, J.T. and Bonner, W.D. (1973) Cyanide-insensitive respiration. I. The steady states of skunk cabbage spadix and bean hypocotyl mitochondria. *J. Biol. Chem.* **248**, 3441–3445.

Bental, M., Oren-Shamir, M., Avron, M. and Degani, H. (1988) $^{31}$P and $^{13}$C-NMR studies of the phosphorus and carbon metabolites in the halotolerant alga, *Dunaliella salina*. *Plant Physiol.* **87**, 320–324.

Bieleski, R.L. (1973) Phosphate pools, phosphate transport, and phosphate availability. *Annu. Rev. Plant Physiol.* **24**, 225–252

Bieleski, R.L. and Ferguson, I.B. (1983) Physiology and metabolism of phosphate and its compounds. In: *Encyclopedia of Plant Physiology,* Volume 15A (eds A. Lauchli and R.L. Bieleski). Springer Verlag, Berlin, pp. 422–449.

Bingham, I.J. and Farrar, J.F. (1989) Activity and capacity of respiratory pathways in barley roots deprived of inorganic nutrients. *Plant Physiol. Biochem.* **27**, 847–854.

Blakeley, S.D., Crews, L., Todd, J.F. and Dennis, D.T. (1992) Expression of the genes for the α- and β-subunits of pyrophosphate-dependent phosphofructokinase in germinating and developing seeds from *Ricinus communis. Plant Physiol.* **99**, 1245–1250.

Bligny, R., Foray, M.-F., Roby, C. and Douce, R. (1989) Transport and phosphorylation of choline in higher plant cells: phosphorus-31 nuclear magnetic resonance studies. *J. Biol. Chem.* **264**, 4888–4895.

Bligney, R., Gardestrom, P., Roby, C. and Douce, R. (1990) $^{31}$P-NMR studies of spinach leaves and their chloroplasts. *J. Biol. Chem.* **265**, 1319–1326.

Botha, A.-M. and Botha, F.C. (1991a) Effect of anoxia on the expression and molecular form of the pyrophosphate dependent phosphofructokinase. *Plant Cell Physiol.* **32**, 1299-1302.

Botha, A.-M. and Botha, F.C. (1991b) Pyrophosphate dependent phosphofructokinase of *Citrullus lanatus*: molecular forms and expression of subunits. *Plant Physiol.* **96**, 1185–1192.

Botha, A.-M. and Botha, F.C. (1993a) Induction of pyrophosphate-dependent phosphofructokinase in watermelon (*Citrullus lanatus*) cotyledons coincides with insufficient cytosolic D-fructose-1,6-bisphosphate 1-phosphohydrolase to sustain gluconeogenesis. *Plant Physiol.* **101**, 1385–1390.

Botha, A.-M. and Botha, F.C. (1993b) Effect of the radicle, and hormones on the subunit composition and molecular form of pyrophosphate-dependent phosphofructokinase in the cotyledons of *Citrullus lanatus. Aust. J. Plant Physiol.* **20**, 265–273.

Botha, F.C., O'Kennedy, M.M. and du Plessis, S. (1992) Activity of key enzymes involved in carbohydrate metabolism in *Phaseolus vulgaris* cell suspension cultures. *Plant Cell Physiol.* **33**, 477–483.

Bryce, H.J., Joaquim, A.-B., Wiskich, J.T. and Day, D.A. (1990) Adenylate control of respiration in plants: the contribution of rotenone-insensitive electron transport to ADP-limited oxygen consumption by soybean mitochondria. *Physiol. Plant.* **78**, 105–111.

Budde, J.A. and Chollet, R. (1988) Regulation of enzyme activity in plants by reversible phosphorylation. *Physiol. Plant.* **72**, 435–439.

Budicky, P.L., Malboobi, A.M. and Lefebvre, D.D. (1993) Identification of putative phosphate starvation inducible protein kinase from *Brassica nigra. Plant Physiol.* **S102**, 68.

Carlisle, S.M., Blakeley, S.D., Hemmingsen, S.M., Trevanion, S.J., Hiyoshi, T., Kruger, N.J. and Dennis, D.T. (1990) Pyrophosphate-dependent phosphofructokinase: conservation of protein sequence between the α- and β-subunits and with the ATP-dependent phosphofructokinase. *J. Biol. Chem.* **265**, 18366–18371.

Carnal, N.W. and Black, C.C. (1979) Pyrophosphate-dependent 6-phosphofructokinase, a new glycolytic enzyme in pineapple leaves. *Biochem. Biophys. Res. Commun.* **86**, 20–26.

Carnal, N.W. and Black, C.C. (1983) Phosphofructokinase activities in photosynthetic organisms. The occurrence of pyrophosphate-dependent 6-phosphofructokinase in plants. *Plant Physiol.* **71**, 150–155.

Cembella, A.D., Anita, N.J. and Harrison, P.J. (1984) The utilization of inorganic and organic phosphorus compounds as nutrients by eukaryotic microalgae: a multidisciplinary perspective, part 1. *CRC Crit. Rev. Microbiol.* **10**, 318–391.

Cheng, H.F. and Tao, M. (1990) Differential proteolysis of the subunits of pyrophosphate-dependent 6-phosphofructo-1-phosphotransferase. *J. Biol. Chem.* **265**, 2173–2177.

Clayton, H., Ranson, J. and ap Rees, T. (1993) Pyrophosphate:fructose-6-phosphate 1-

phosphotransferase and fructose-2,6-bisphosphate in the bundle sheath of maize leaves. *Arch. Biochem. Biophys.* **301**, 151–157.

Collier, D.E. and Cummins, W.R. (1990) The effects of growth and measurement of temperature on the respiratory properties of five temperate species. *Ann. Bot.* **65**, 533–538.

Crafts-Brander, S.J. (1992) Phosphorus nutrition influence on starch and sucrose accumulation, and activities of ADP-glucose pyrophosphorylase and sucrose-phosphate synthase during the grain filling period in soybean. *Plant Physiol.* **98**, 1133–1138.

Dancer, J.E. and ap Rees, T. (1989) Relationship between pyrophosphate 1-phosphotransferase, sucrose breakdown, and respiration. *Plant Physiol.* **135**, 197–206.

Dancer, J., Veith, R., Feil, R., Komor, E. and Stitt, M. (1990) Independent changes of inorganic pyrophosphate and the ATP/ADP or UTP/UDP ratios in plant cell suspension cultures. *Plant Sci.* **66**, 59–63.

Davies, J.M., Poole, R.J. and Sanders, D. (1993) The computed free energy change of hydrolysis of inorganic pyrophosphate and ATP: apparent significance of inorganic-pyrophosphate-driven reactions of intermediary metabolism. *Biochim. Biophys. Acta* **1141**, 29–36.

Dietz, K.-J. and Heber, U. (1984) Rate-limiting factors in leaf photosynthesis. 2. Carbon fluxes in the Calvin cycle. *Biochim. Biophys. Acta* **767**, 432–443.

Dry, I.B., Moore, A.L., Day, D.A. and Wiskich, J.T. (1989) Regulation of alternative pathway activity in plant mitochondria: non-linear relationship between electron flux and the redox poise of the quinone pool. *Arch. Biochem. Biophys.* **273**, 148–157.

Duchein, M.-C., Bonicel, A. and Betsche, T. (1993) Photosynthetic net $CO_2$ uptake and leaf phosphate concentrations in $CO_2$ enriched clover (*Trifolium subterraneum L.*) at three levels of phosphate nutrition. *J. Exp. Bot.* **44**, 17–22.

Duff, S.M.G., Lefebvre, D.D. and Plaxton, W.C. (1989a) Purification and characterization of a phosphoenolpyruvate phosphatase from *Brassica nigra* suspension cells. *Plant Physiol.* **90**, 734–741.

Duff, S.M.G., Moorhead, G.B.G, Lefebvre, D.D. and Plaxton, W.C. (1989b) Phosphate starvation inducible "bypasses" of adenylate and phosphate dependent glycolytic enzymes in *Brassica nigra* suspension cells. *Plant Physiol.* **90**, 1275–1278.

Duff, S.M.G, Lefebvre, D.D. and Plaxton, W.C. (1991a) Purification, characterization, and subcellular localization of an acid phosphatase from black mustard cell-suspension cultures: comparison with phosphoenolpyruvate phosphatase. *Arch. Biochem. Biophys.* **286**, 226–232.

Duff, S.M.G., Plaxton, W.C. and Lefebvre, D.D. (1991b) Phosphate-starvation response in plant cells: *de novo* synthesis and degradation of acid phosphatases. *Proc. Natl Acad. Sci. USA* **88**, 9538–9542.

Duff, S.M.G, Sarath, G. and Plaxton, W.C. (1994) The role of acid phosphatases in plant phosphorus metabolism. *Physiol. Plant.* **90**, 791–800.

Dumont, F., Joris, B., Gumusboga, A., Bruyninx, M. and Loppes, R. (1993) Isolation and characterization of cDNA sequences controlled by inorganic phosphate in *Chlamydomonas reinhardtii. Plant Sci.* **89**, 55–67.

Dunford, R. (1992) Pyruvate kinase and glycolytic control in potatoes. Ph.D. Thesis, University of Cambridge, Cambridge, UK.

Ekkehard Neuhaus, H. and Stitt, M. (1991) Inhibition of photosynthetic sucrose synthesis by imidodiphosphate, an analogue of inorganic pyrophosphate. *Plant Sci.* **76**, 49–55.

Enomoto, T., Ohyama, H. and Kodama, M. (1992) Purification and characterization of pyrophosphate: D-fructose 6-phosphate 1-phosphotransferase from rice seedlings. *Biosci. Biotech. Biochem.* **56**, 251–255.

Fife, C.A., Newcomb, W. and Lefebvre, D.D. (1990) The effect of phosphate deprivation on protein synthesis and fixed carbon storage reserves in *Brassica nigra* suspension cells. *Can. J. Bot.* **68**, 1840–1847.

Fischer, R.S., Bonner, C.A., Theodorou, M.E., Plaxton, W.C., Hrazdina, G. and Jensen, R.A. (1993) Response of aromatic pathway enzymes of plant suspension cells to phosphate limitation. *Bio. Med. Chem. Lett.* **3**, 1415–1420.

Freeden, F.L., Raab, T.K., Rao, I.M. and Terry, N. (1990) Effects of phosphorus nutrition on photosynthesis in *Glycine max* (L.) Merr. *Planta* **181**, 399–405.

Gauthier, D.A. and Turpin, D.H. (1994) Inorganic phosphate (Pi) enhancement of dark respiration in the Pi-limited green alga *Selenastrum minutum*. *Plant Physiol.* **104**, 629–637.

Glotzer, M., Murray, A.W. and Kirschner, M.W. (1991) Cyclin is degraded by the ubiquitin pathway. *Nature* **349**, 132–138.

Goldstein, A.H., Baertlein, D.A. and McDaniel, R.G. (1988a) Phosphate starvation inducible metabolism in *Lycopersicon esculentum*. I. Excretion of acid phosphatase by tomato plants and suspension-cultured cells. *Plant Physiol.* **87**, 711–715.

Goldstein, A.H., Danon, A. and Baertlein, D.A. (1988b) Phosphate starvation inducible metabolism in *Lycopersicon esculentum*. II. Characterization of the phosphate starvation inducible excreted acid phosphatase. *Plant Physiol.* **87**, 716–720.

Goldstein, A.H., Baertlein, D.A. and Danon, A. (1989) Phosphate starvation stress as an experimental system for molecular analysis. *Plant Mol. Biol. Rep.* **7**, 7–16.

Green, P.C., Latshaw, S.P., Lardor, U.S. and Kemp, R.G. (1992) Identification of critical lysyl residues in the pyrophosphate-dependent phosphofructo-1-kinase of *Propionibacterium freudenreichii*. *Biochemistry* **31**, 4815–4821.

Green, P.C., Tripathi, R.L. and Kemp, R.G. (1993) Identification of active site residues in pyrophosphate-dependent phosphofructo-1-kinase by site-directed mutagenesis. *J. Biol. Chem.* **268**, 5085–5088.

Guy, R.D., Berry, J.A., Fogel, M.L., Turpin, D.H. and Weger, H.G. (1992) Fractionation of the stable isotopes of oxygen during respiration by plants — the basis of a new technique to estimate partitioning to the alternative path. In: *Molecular, Biochemical and Physiological Aspects of Plant Respiration* (eds H. Lambers and L.H.W. van der Plas). SPB Academic Publishing, The Hague, pp. 204–216.

Hawkesford, M.J. and Belcher, A.R. (1991) Differential protein synthesis in response to sulphate and phosphate deprivation: identification of possible components of plasma-membrane transport systems in cultured tomato roots. *Planta* **185**, 323–329.

Hentrich, S., Grimme, L.H., Leibfritz, D. and Mayer, A. (1993) P-31 NMR saturation transfer experiments in *Chlamydomonas reinhardtii*: evidence for the NMR visibility of chloroplastidic Pi. *Eur. Biophys. J.* **22**, 31–39.

Hill, S.A. and ap Rees, T. (1994) Fluxes of carbohydrate metabolism in ripening bananas. *Planta* **192**, 52–60.

Hoefnagel, M.H.N, Alblas, J. and van Iren, F. (1987) Relationship between growth and alternative respiration in plant cell cultures. In: *Proceedings of the 4th European Congress in Biotechnology Vol 2* (eds O.M. Neijssel, R.R. van der Marc and K.Ch.A.M. Luyben). Elsevier Scientific Publishers, Amsterdam, pp. 421–424.

Hoefnagel, M.H.N., Van Iren, F. and Libbenga, K.R. (1993) In suspension cultures of *Catharanthus roseus* the cyanide-resistant pathway is engaged in respiration by excess sugar in combination with phosphate or nitrogen starvation. *Physiol. Plant.* **87**, 297–304.

Hoefnagel, M.H.N., Van Iren, F., Libbenga, K.R. and Van der Plas, L.H.W. (1994) Possible role of adenylates in the engagement of the cyanide resistant pathway in nutrient-starved *Catharanthus roseus* cells. *Physiol. Plant.* **90**, 269–278.

Hoffland, E., Van den Boogaard, R., Nelemans, J. and Findenegg, G. (1992) Biosynthesis and root exudation of citric and malic acids in phosphate-starved rape plants. *New Phytol.* **122**, 675–680.

Huber, J.L., Huber, S.C. and Nielsen, T.H. (1989) Protein phosphorylation as a mechanism for regulation of spinach leaf sucrose-phosphate synthase activity. *Arch. Biochem. Biophys.* **270**, 681–690.

Huber, J.L., Huber, S.C., Campbell, W.H. and Redinbaugh, M.G. (1992) Reversible light/dark modulation of spinach leaf nitrate reductase activity involves protein phosphorylation. *Arch. Biochem. Biophys.* **296**, 58–65.

Iglesisas, A.A., Plaxton, W.C. and Podestá , F.E. (1993) The role of inorganic phosphate in the regulation of $C_4$ photosynthesis. *Photosynth. Res.* **35**, 205–211.

Jacob, J. and Lawlor, D.W. (1992) Dependence of photosynthesis of sunflower and maize leaves on phosphate supply, ribulose-1,5-bisphosphate carboxylase/oxygenase activity, and ribulose-1,5-bisphosphate pool size. *Plant Physiol.* **98**, 801–807.

Jolivet, Y., Pireaux, J.-C. and Dizengremel, P. (1990) Changes in properties of barley leaf mitochondria isolated from NaCl-treated plants. *Plant Physiol.* **94**, 641–646.

Jost W., Bak, H., Glund, K., Terpstra, P. and Beintema, J. (1991) Amino acid sequence of an extracellular, phosphate-starvation-induced ribonuclease from cultured tomato (*Lycopersicon esculentum*) cells. *Eur. J. Biochem.* **198**, 1–6.

Kaffman, A., Herskowitz, I., Tjian, R. and O'Shea, E.K. (1994) Phosphorylation of the transcription factor PHO4 by a cyclin–CDK complex, PHO80–PHO85. *Science* **263**, 1153–1156.

Knobloch, K.H. (1982) Uptake of phosphate and its effect on phenylalanine ammonia-lyase activity and cinnamoyl putrescine accumulation in cell suspension cultures of *Nicotiana tabacum*. *Plant Cell Rep.* **1**, 128–130.

Knobloch, K.H., Hansen, B. and Berlin, J. (1981) Medium-induced formation of indole alkaloids and concomitant changes of interrelated enzyme activities in cell suspension cultures of *Catharanthus roseus*. *Z. Naturforsch.* **36c**, 40–43.

Kobayashi, M., Funane, K., Ohya, S., Torikoshi, H., Kurogi, M. and Ishiuchi, D. (1992) Pyrophosphate-dependent phosphofructokinase: its oligomeric forms in mature green tomato fruit. *Biosci. Biotech. Biochem.* **5**, 54–57.

Kruger, N.J. and Dennis, D.T. (1987) Molecular properties of pyrophosphate: fructose-6-phosphate phosphotransferase from potato tuber. *Arch. Biochem. Biophys.* **256**, 273–279.

Kubota, K., Li, X.-N. and Ashihara, H. (1989) The short-term effects of inorganic phosphate on the levels of metabolites in suspension-cultured *Catharanthus roseus* cells. *Z. Naturforsch.* **44c**, 802–806.

Kuhl, N.J. (1974) Phosphorus. In: *Algal Physiology and Biochemistry* (ed. W.D.P. Stewart). Blackwell Scientific, Oxford, pp. 636–654.

Lambers, H. (1982) Cyanide-resistant respiration: a non-phosphorylating electron transport pathway acting as an overflow. *Physiol. Plant.* **55**, 478–485.

Lambers, H., Day, D.A. and Azcón-Bieto, J. (1983) Cyanide-resistant respiration in roots and leaves. Measurements with intact tissues and isolated mitochondria. *Physiol. Plant.* **58**, 148–154.

Ladror, U.S., Bollapudi, L., Tripathi, R.L., Latshaw, S.P. and Kemp, R.G. (1991) Cloning, sequencing, and expression of pyrophosphate-dependent phosphofructokinase from *Propionibacterium freudenreichii*. *J. Biol. Chem.* **266**, 16550–16555.

Lauer, M.J., Blevins, D.G. and Sierzputowska-Gracz, H. (1989) $^{31}$P-Nuclear magnetic resonance determination of phosphate compartmentation in leaves of reproductive soybeans as affected by phosphate starvation. *Plant Physiol.* **93**, 504–511.

Lee, R.B. and Ratcliffe, R.G. (1983) Phosphorus nutrition and the intracellular distribution of inorganic phosphate in pea root tips: a quantitative study using $^{31}$P-NMR. *J. Exp. Bot.* **34**, 1222–1244.

Lee, R.B. and Ratcliffe, R.G. (1993) Subcellular distribution of inorganic phosphate, and levels of nucleoside triphosphate, in mature maize roots at low external phosphate concentrations: measurements with $^{31}$P-NMR. *J. Exp. Bot.* **44**, 587–598.

Lee, R.B., Ratcliffe, R.G. and Southton, T.E. (1990) $^{31}$P-NMR measurements of the cytosolic and vacuolar Pi content of mature maize roots: relationships with phosphorus status and phosphate fluxes. *J. Exp. Bot.* **41**, 1063–1078.

Lefebvre, D.D., Duff, S.M.G., Fife, C.A., Julien-Inalsingh, C. and Plaxton, W.C. (1990) Response to phosphate deprivation in *Brassica nigra* suspension cells. Enhancement of intracellular, cell surface, and secreted acid phosphatase activities compared to increases in Pi-absorption rate. *Plant Physiol.* **93**, 504–511.

Li, X.-N. and Ashihara, H. (1990) Effects of inorganic phosphate on sugar catabolism by suspension-cultured *Catharanthus roseus*. *Phytochemistry* **29**, 497–500.

Lin, M., Turpin, D.H and Plaxton, W.C. (1989) Pyruvate kinase isozymes from the green alga, *Selenastrum minutum*, II. Kinetic and regulatory properties. *Arch. Biochem. Biophys.* **269**, 228–238.

Löffler, A., Glund, K. and Irie, M. (1993) Amino acid sequence of an intracellular, phosphate-starvation-induced ribonuclease from cultured tomato (*Lycopersicon esculentum*) cells. *Eur. J. Biochem.* **214**, 627–633.

Loughman, B.C., Ratcliffe, R.G., and Southton, T.E. (1989) Observations of the cytosolic and vacuolar orthophosphate pools in leaf tissues using *in vivo* $^{31}$P-NMR spectroscopy. *FEBS Lett.* **242**, 279–284.

Loulakakis, K.A. and Roubelakis-Angelakis, K.A. (1992) Ammonium-induced increase in NADH-glutamate dehydrogenase activity is caused by *de novo* synthesis of the $\alpha$–subunit. *Planta* **187**, 322–327.

Lundberg, P., Weich, R.G., Jensén, P. and Vogel, H.J. (1989) Phosphorus-31 and nitrogen-14 NMR studies of the uptake of phosphorus and nitrogen compounds in the marine macroalga *Ulva lactuca*. *Plant Physiol.* **89**, 1380–1387.

Lunn, J.E. and Douce, R. (1993) Transport of inorganic pyrophosphate across the spinach chloroplast envelope. *Biochem. J.* **290**, 375–379.

MacKintosh, C. (1992) Regulation of spinach-leaf nitrate reductase by reversible phosphorylation. *Biochim. Biophys. Acta* **1137**, 121–126.

Malboobi, M.A. (1992) Differential gene expression as a result of phosphate starvation in *Brassica nigra* suspension cells. M.Sc. Thesis, Queen's University, Kingston, Ontario, Canada.

Malboobi, M.A. and Lefebvre, D.D. (1993) Differential gene expression at the transcriptional level in response to phosphate starvation in *Brassica nigra* suspension cells. *Plant Physiol.* **102**, S9.

McHugh-Gottlob, S.G., Sangwan, R.S., Blakeley, S.D., Vanlerberghe, G.C., Ko, K., Turpin, D.H., Plaxton, W.C., Miki, B.L. and Dennis, D.T. (1992) Normal growth of transgenic tobacco plants in the absence of cytosolic pyruvate kinase. *Plant Physiol.* **100**, 820–825.

Mertens, E. (1991) Pyrophosphate-dependent phosphofructokinase, an anaerobic glycolytic enzyme? *FEBS Lett.* **285**, 1–5.

Mimura, T., Dietz, K.J., Kaiser, W., Schramm, M.J., Kaiser, G. and Heber, U. (1990) Phosphate transport across biomembranes and cytosolic phosphate homeostasis in barley leaves. *Planta* **180**, 139–146.

Møller, I.M. and Palmer, J.M. (1982) Direct evidence for the presence of a rotenone-resistant NADH-dehydrogenase on the inner membrane of plant mitochondria. *Physiol. Plant.* **54**, 267–274.

Møller, I.M., Bérczi, A., van der Plas, L.H.W. and Lambers, H. (1988) Measurement of the activity and capacity of the alternative pathway in intact plant tissues: identification of problems and possible solutions. *Physiol. Plant.* **72**, 642–649.

Montavon, P. and Kruger, N.J. (1992) Substrate specificity of pyrophosphate:fructose 6-phosphate 1-phosphotransferase from potato tuber. *Plant Physiol.* **99**, 1487–1492.

Montavon, P. and Kruger, N.J. (1993) Essential arginyl residue at the active site of pyrophosphate:fructose 6-phosphate 1-phophotransferase from potato (*Solanum tuberosum*) tuber. *Plant Physiol.* **101**, 765–771.

Moorhead, G.B.G and Plaxton, W.C. (1991) High-yield purification of potato tuber pyrophosphate:fructose-6-phosphate 1-phosphotransferase. *Prot. Exp. Purif.* **2**, 29–33.

Nagano, M. and Ashihara, H. (1993) Long-term phosphate starvation and respiratory metabolism in suspension-cultured *Catharanthus roseus* cells. *Plant Cell Physiol.* **34**, 1219–1228.

Nakamura, N., Suzuki, Y. and Suzuki, H. (1992) Pyrophosphate-dependent phosphofructokinase from pollen: properties and possible roles in sugar metabolism. *Physiol. Plant.* **86**, 616–622.

Nimmo, H.G. (1993) The regulation of phosphoenolpyruvate carboxylase by reversible phosphorylation. *Symp. Soc. Exp. Biol.* **53**, 161–170.

Nürnberger, T., Abel, S., Jost, W. and Glund, K. (1991) Induction of an extracellular ribonuclease in cultured tomato cells upon phosphate starvation. *Plant Physiol.* **92**, 970–976.

Okazaki, M., Hino, F., Nagasawa, K. and Miura, Y. (1982) Effects of nutritional factors on formation of scopoletin and scopolin in tobacco tissue cultures. *Agric. Biol. Chem.* **46**, 601–607.

Paul, M. and Stitt, M. (1993) Effects of nitrogen and phosphorus deficiencies on levels of carbohydrates, respiratory enzymes and metabolites in seedlings of tobacco and their response to exogenous sucrose. *Plant, Cell Environ.* **16**, 1047–1057.

Paul, M., Sonnewald, U., Dennis, D. and Stitt, M. (1994) Transgenic tobacco plants with strongly decreased expression of pyrophosphate:fructose-6-phosphate 1-phosphotransferase do not differ significantly from wild type in photosynthate partitioning, plant growth or their ability to cope with limiting phosphate, limiting nitrogen and suboptimal temperatures. *Planta* in press.

Peng, Z.-Y. and Mansour, T.E. (1992) Purification properties of a pyrophosphate-dependent phosphofructokinase from *Toxoplasma gondii*. *Mol. Biochem. Parasitol.* **54**, 223–230.

Plaxton, W.C. (1990) Biochemical regulation. In: *Plant Physiology, Biochemistry and Molecular Biology* (eds D.T. Dennis and D.H. Turpin). Longman Group UK Ltd, Singapore, pp. 28–44.

Podestá, F.E. and Plaxton, W.C. (1994) Regulation of cytosolic carbon metabolism in germinating *Ricinus communis* cotyledons. I. Developmental profiles for the activity, concentration, and molecular structure of the pyrophosphate- and ATP-dependent phosphofructokinases, phosphoenolpyruvate carboxylase and pyruvate kinase. *Planta* (in press).

Preiss, J. (1984) Starch, sucrose biosynthesis and partitioning of carbon in plants are regulated by orthophosphate and triose-phosphates. *Trends Biochem. Sci.* **9**, 24–27.

Qui, J. and Israel, D.W. (1992) Diurnal starch accumulation and utilization in phosphorus-deficient soybean plants. *Plant Physiol.* **98**, 316–323.

Rao, I.M., Freeden, A.L. and Terry, N. (1993) Influence of phosphorus limitation on photosynthesis, carbon allocation and partitioning in sugar beet and soybean grown with a short photoperiod. *Plant Physiol. Biochem.* **31**, 223–231.

Rao, N.N. and Torriani, A. (1990) Molecular aspects of phosphate transport in *Escherichia coli. Mol. Microbiol.* **4**, 1083–1090.

Rawn, J.D. (1989) *Biochemistry.* Neil Patterson Publishers, Burlington, North Carolina.

Rea, P.A. and Poole, R.J. (1993) Vacuolar $H^+$-translocating pyrophosphatase. *Annu. Rev. Plant Physiol. Plant Mol. Biol.* **44**, 157–180.

Rebeille, F., Bligny, R. and Douce, R. (1984) Is the cytosolic Pi concentration a limiting factor for plant cell respiration? *Plant Physiol.* **54**, 199–206.

Roberts, J.K.M. (1984) Study of plant metabolism *in vivo* using NMR spectroscopy. *Annu. Rev. Plant Physiol.* **35**, 375–386.

Roby, C., Martin, J.-B., Bligny, R. and Douce, R. (1987) Biochemical changes during sucrose deprivation in higher plant cells: phosphorus-31 nuclear magnetic resonance studies. *J. Biol. Chem.* **262**, 5000–5007.

Rollins, R.C. (1993) *The Cruciferae of Continental North America. Systematics of the Mustard Family from the Arctic to Panama.* Stanford University Press, Stanford, CA, pp. 233–236.

Rowntree, E.G. and Kruger, N.J. (1992) Inhibition of pyrophosphate:fructose-6-phosphate 1-phosphotransferase by imidodiphosphate. *Plant Sci.* **86**, 183–189.

Rychter, A.M. and Mikulska, M. (1990) The relationship between phosphate status and cyanide-resistant respiration in bean roots. *Physiol. Plant.* **7**, 663–667.

Rychter, A.M. and Randall, D.D. (1994) The effect of phosphate deficiency on carbohydrate metabolism in bean roots. *Physiol. Plant.* **91**, 383–388.

Rychter, A.M., Chauveau, M., Bomsel, J.-L. and Lance, C. (1992) The effect of phosphate deficiency on mitochondrial activity and adenylate levels in bean roots. *Physiol. Plant.* **84**, 80–86.

Sachs, M.M. and Ho, T.-H.D. (1986) Alteration of gene expression during environmental stress in plants. *Annu. Rev. Plant Physiol.* **37**, 363–376.

Sakano, K. (1990) Proton/phosphate stoichiometry in uptake of inorganic phosphate by cultured cells of *Catharanthus roseus* (L.) G. Don. *Plant Physiol.* **93**, 479–483.

Sakano K., Yazaki, Y. and Mimura, T. (1992) Cytosolic acidification induced by inorganic phosphate uptake in suspension cultured *Catharanthus roseus* cell. *Plant Physiol.* **99**, 672–680.

Sasse, F., Heckenberg, U. and Berlin, J. (1982) Accumulation of β-carboline alkaloids and serotonin by cell cultures of *Peganum harmala. Plant Physiol.* **69**, 400–404.

Sato, F. and Ashihara, H. (1992) Pyrophosphate: fructose-6-phosphate phosphotransferase and gluconeogenic capacity in germinated peanut seeds. *Biochem. Physiol. Pflanzen* **188**, 145–151.

Sawada, S., Usuda, H. and Tsukkui, T. (1992) Participation of inorganic orthophosphate in regulation of the ribulose-1,5-bisphosphate carboxylase activity in response to changes in the photosynthetic source–sink balance. *Plant Cell Physiol.* **33**, 943–949.

Schmitt, N. and Dizengremel, P. (1989) Effect of osmotic stress on mitochondria isolated from etiolated mung bean and sorghum seedlings. *Plant Physiol. Biochem.* **27**, 17–26.

Shinagawa, H., Makino, M., Amemura, M. and Nakata, A. (1987) Structure and function

of the regulatory genes for the phosphate regulon in *E. coli*. In: *Phosphate Metabolism and Cellular Regulation in Microorganisms* (eds A. Torriani, F.G. Rothman, S. Silver, A. Wright and E. Yagil). ASM Washington, DC, pp. 20–25.
Sianoudis, J., Mayer, A. and Leibfritz, D. (1984) Investigation of intracellular phosphate pools in the green alga *Chlorella fusca* using $^{31}P$ nuclear magnetic resonance. *Org. Magn. Reson.* **22**, 364–368.
Sideow, J.N. and Berthold, D.A. (1986) The alternative oxidase: a cyanide resistant pathway in higher plants. *Physiol. Plant.* **66**, 569–573.
Stitt, M. (1990) Fructose-2,6-bisphosphate as a regulatory molecule in plants. *Annu. Rev. Plant Physiol. Mol. Biol.* **41**, 153–185.
Stitt, M., Wirtz, W., Gerhardt, R., Heldt, H.W., Spencer, C., Walker, D.A. and Foyer, C.H. (1985) A comparative study of metabolite levels in plant leaf material in the dark. *Planta* **166**, 354–364.
Sung, S.-J.S., Xu, D.-P., Galloway, C.M. and Black, C.C. Jr (1988) A reassessment of glycolysis and gluconeogenesis in higher plants. *Physiol. Plant.* **72**, 650–654.
Terry, N. and Ulrich, A. (1973) Effects of phosphorus deficiency on the photosynthesis and respiration of leaves of sugar beet. *Plant Physiol.* **51**, 43–47.
Theodorou, M.E. and Plaxton, W.C. (1993) Metabolic adaptations of plant respiration to nutritional phosphate deprivation. *Plant Physiol.* **101**, 339–344.
Theodorou, M.E. and Plaxton, W.C. (1994) Induction of pyrophosphate-dependent phosphofructokinase by phosphate starvation in seedlings of *Brassica nigra. Plant, Cell Environ.* **17**, 287–294.
Theodorou, M.E., Elrifi, I.R., Turpin, D.H. and Plaxton, W.C. (1991) Effects of phosphorus limitation on respiratory metabolism in the green alga *Selenastrum minutum. Plant Physiol.* **95**, 1089–1095.
Theodorou, M.E., Cornel, F.A., Duff, S.M.G. and Plaxton, W.C. (1992) Phosphate starvation-inducible synthesis of the $\alpha$-subunit of pyrophosphate-dependent phosphofructokinase in black mustard suspension cells. *J. Biol. Chem.* **267**, 21901–21905.
Thorsteinsson, B. and Tillberg, J.-E. (1987) Carbohydrate partitioning, photosynthesis and growth in *Lemna gibba* G3. II. Effects of phosphorus limitation. *Physiol. Plant.* **71**, 271–276.
Thorsteinsson, B. and Tillbeg, J.-E. (1990) Changes in photosynthesis/respiration ratio and levels of carbohydrates in leaves of nutrient depleted barley and pea. *J. Plant Physiol.* **136**, 532–537.
Tillberg, J.-E. and Rowley, J.R. (1989) Physiological and structural effects of phosphorus starvation on the unicellular green alga *Scenedesmus. Physiol. Plant.* **75**, 315–324.
Topa, M.A. and Cheeseman, J.M. (1992a) Effects of root hypoxia and low P supply on relative growth, carbon dioxide exchange rates and carbon partitioning in *Pinus serotina* seedlings. *Physiol. Plant.* **86**, 136–144.
Topa, M.A. and Cheeseman, J.M. (1992b) Carbon and phosphorus partitioning in *Pinus serotina* seedlings growing under hypoxic and low-phosphorus conditions. *Tree Physiol.* **10**, 195–207.
Turpin, D.H., Botha, F.C., Smith, R.G., Feil, R., Horsey, A.K. and Vanleberghe, G.C. (1990) Regulation of carbon partitioning to respiration during dark ammonium assimilation by the green alga *Selenastrum minutum. Plant Physiol.* **93**, 166–175.
Ulrich, C.I. and Novacky, A.J. (1990) Extra- and intracellular pH and membrane potential changes induced by $K^+$, $Cl^-$, $H_2PO_4^-$, and $NO_3^-$ uptake and fusicoccin in root hairs of *Limnobium stoloniferum. Plant Physiol.* **94**, 1561–1567.

Ulrich-Eberius, C.I, Novacky, C. and van Bell A.J.E. (1984) Phosphate uptake in *Lemna gibba* G1: energetics and kinetics. *Planta* **161**, 45–52.

Usuda, H. and Shimogawara, K. (1991a) Phosphate deficiency in maize. I. Leaf phosphate status, growth, photosynthesis and carbon partitioning. *Plant Cell Physiol.* **32**, 497–504.

Usuda, H. and Shimogawara, K. (1991b) Phosphate deficiency in maize. II. Enzyme activities. *Plant Cell Physiol.* **32**, 1313–1317.

Usuda, H. and Shimogawara, K. (1992) Phosphate deficiency in maize. III. Changes in enzyme activities during the course of phosphate deprivation. *Plant Physiol.* **99**, 1680–1685.

Usuda, H. and Shimogawara, K. (1993) Phosphate deficiency in maize. IV. Changes in amounts of sucrose phosphate synthase and PEP carboxylase during the course of phosphate deprivation. *Plant Cell Physiol.* **34**, 767–770.

Walker, W.A. and Sivak, M.N. (1986) Photosynthesis and phosphate: a cellular affair? *Trends Biochem. Sci.* **11**, 176–179.

Weger, H.G. (1993) Alternative pathway respiration in *Chlamydomonas reinhardtii. Plant Physiol.* **S102**, 128.

Weger, H.G. and Dasgupta, R. (1993) Regulation of alternative pathway respiration in *Chlamydomonas reinhardtii* (Chlorophyceae). *J. Phycol.* **29**, 300–308.

Weger, H.G. and Guy, R.D. (1991) Cytochrome and alternative pathway respiration in white spruce (*Picea glauca*) roots. Effects of growth and measurement of temperature. *Physiol. Plant.* **83**, 675–681.

Weiner, H., Stitt, M. and Heldt, H.W. (1987) Subcellular compartmentation of pyrophosphate and alkaline pyrophosphatase in leaves. *Biochim. Biophys. Acta* **893**, 13–21.

Yan, T.F.J. and Tao, M. (1984) Multiple forms of pyrophosphate:D-fructose-6-phosphate 1-phosphotransferase from wheat seedlings. *J. Biol. Chem.* **259**, 5087–5092.

Yang, Y.P., Randall, D.D. and Trelease, R.N. (1988) Phosphorylation of glyoxysomal malate synthase from castor oil seeds *Ricinus communis* L. *FEBS Lett.* **234**, 275–279.

# 7

# Metabolic aspects of the anoxic response in plant tissue

R.G. Ratcliffe

## 7.1 Introduction

The availability of oxygen and the likelihood of oxygen deprivation are major factors in determining the survival of plants and the resulting distribution of plant species. Tolerance to anoxia appears to depend on a combination of morphological and metabolic adaptations that are both species and tissue specific, and the success of these strategies is very variable. Thus while flooding-intolerant species such as maize (*Zea mays* L.) and pea (*Pisum sativum* L.) succumb easily to anoxia, some plants, notably rice (*Oryza sativa* L.) and several *Echinocloa* species, can germinate in the complete absence of oxygen. Ultimately even the most resistant tissues, for example the rhizomes of the wetland plant *Schoenoplectus lacustris* L., which can grow new shoots in the absence of oxygen after several weeks of anoxia, will die without oxygen and without access to the relatively abundant supplies of energy that can be derived from aerobic respiration. Thus, at best, plants are only equipped to cope with an anoxic interlude in an otherwise aerobic existence, with the result that much of the research effort in this field is directed at understanding the mechanisms that enable plants to minimize the duration and metabolic consequences of such episodes.

The complexity of the anoxic response necessitates investigations at several different levels on plants that span the whole range of anoxia tolerance. Important themes in current work include: (i) studies of the development of aerenchyma and its role in the avoidance of anoxia in root tissues and rhizomes; (ii) analyses of the changes in gene expression that occur under anoxia; and (iii) studies of the effect of anoxia on the principal pathways of metabolism. Steady, rather than spectacular, progress is being made in each of these areas and a good impression of the field can be obtained by considering some of the most recent papers. At the whole plant level, the correlation between oxygen deprivation, ethylene biosynthesis and the development of the extended intercellular air spaces that make up the aerenchyma has been studied in the primary roots of maize seedlings (Brailsford *et al.*, 1993; Jackson and Hall, 1993). This work has confirmed that ethylene production is stimulated under hypoxic conditions in maize roots and it has also shown that putrescine, an endogenous regulator of plant development, limits the formation of aerenchyma by suppressing the action of ethylene, rather than by reducing its synthesis. At the genetic level, representative papers include: (i) the detection of multiple, anaerobically induced pyruvate decarboxylase (PDC; EC 4.1.1.1) genes in maize seedlings (Peschke and Sachs, 1993); (ii) the characterization of two anaerobically induced genes in maize that, in contrast to all the other known genes of this type, do not appear to encode proteins with a glycolytic function (Peschke and Sachs, 1994); and (iii) the clear demonstration of regulation at the levels of transcription, translation and post-translation in the expression of the genes for alcohol dehydrogenase (ADH; EC 1.1.1.1) and PDC in the anoxia-tolerant plant *Acorus calamus* L. (Bucher and Kuhlemeier, 1993). Finally, at the metabolic level, there has been continuing interest in the effects of hypoxic pre-treatments on the subsequent anoxic response of plant tissues (Andrews *et al.*, 1993) and the central role of lactate in the switch from aerobic to anaerobic metabolism (Rivoal and Hanson, 1993a).

There is no shortage of reviews in this field and the availability of several comprehensive accounts makes it unnecessary to attempt anything similar here. The impact of oxygen deprivation on the ecology, physiology and biochemistry of plants was reviewed definitively relatively recently (Jackson *et al.*, 1991) and numerous other reviews are available on the anoxic response, including those of Drew (1990), Crawford (1992), Kennedy *et al.* (1992) and Perata and Alpi (1994). Accordingly, it seems more appropriate to focus on just one aspect of the anoxic response, namely the production of glycolytic end-products and to review some of the recent evidence relating to the commonly observed switch from lactate to ethanol production under anoxia.

The continued oxidation of carbohydrates during anaerobiosis requires a mechanism for recycling the NADH produced by glycolysis and, in principle, this can be achieved either by the reduction of pyruvate to lactate using lactate dehydrogenase (LDH; EC 1.1.1.27) or by the decarboxylation of pyruvate to acetaldehyde by PDC and the subsequent reduction of acetaldehyde to ethanol using ADH. While lactate is often formed at the onset of anaerobiosis, ethanol is usually the main fermentation product during prolonged anoxia and, according to the biochemical pH stat model (Davies, 1980; Davies *et al.*, 1974), it is the initial fall in the cytoplasmic pH that is responsible for the switch from lactate to ethanol. The relative activity of LDH, which has an alkaline pH optimum and a pH-dependent sensitivity to ATP, and PDC, with an acidic pH optimum, is expected to change significantly as the pH falls from a typical aerobic value of 7.5, and this suggests that a fall in pH could simultaneously inhibit the production of lactate and promote the synthesis of ethanol (Davies *et al.*, 1974). The pH-dependent properties of PDC are consistent with the model (Morrell *et al.*, 1990) and direct evidence for the key role of the cytoplasmic pH was obtained in some early *in vivo* nuclear magnetic resonance (NMR) experiments

(Roberts *et al.*, 1984a,b, 1985). In the NMR work, $^{13}C$-NMR measurements on the incorporation of [1-$^{13}C$]glucose into lactate, ethanol and alanine were correlated with $^{31}P$-NMR measurements of the cytoplasmic pH during the switch from aerobic to anaerobic conditions, and it was concluded: (i) that the fall in the cytoplasmic pH under hypoxia triggered the switch to ethanol production (Roberts *et al.*, 1984a); and (ii) that cytoplasmic acidosis, caused by a loss of pH control under anoxia and the breakdown of the pH gradient across the tonoplast, would cause cell death and could be used as an indicator of flooding intolerance (Roberts *et al.*, 1984b, 1985).

Thus a picture has emerged in which a transient burst of lactate production gives way to an ethanolic fermentation as a result of an initial fall in the cytoplasmic pH. However, it would be incorrect to assume that all the available evidence is consistent with this model, and it has been argued that the model lacks generality as a unifying theory for flooding tolerance (Kennedy *et al.*, 1992). Two aspects of the model are of particular interest in this context, namely the extent to which it is correct to consider the production of lactate as a transient phenomenon under anoxia and the origin and significance of the observed changes in the cytoplasmic pH, and it is these metabolic aspects of the anoxic response that will be reviewed here.

## 7.2 Lactate production under anoxia

The question of immediate interest is whether the accumulation of ethanol is always preceded by the transient production of lactate. The experimental evidence on this point is contradictory, with different systems showing either no lactate production, transient lactate production or long-term lactate production, and with the transient formation of lactate either preceding or running in parallel with the production of ethanol. The situation expected by the Davies model is best illustrated by the data reported by Roberts *et al.* (1984a) for excised maize root tips. In these experiments, *in vivo* $^{13}C$-NMR was used to show that the lactate level reached a maximum within 20 min of the abrupt imposition of anoxia and that the formation of lactate preceded the production of ethanol in the expected manner. Other investigations, using both *in vivo* and *in vitro* methods of analysis, have been less unambiguous. For example, working with the same tissue, Xia and Saglio (1992) observed a similar lag in the switch to ethanol but this did not coincide with the complete cessation of lactate synthesis. In contrast, in pea root tips (Smith and ap Rees, 1979), rice shoots (Fan *et al.*, 1986) and wheat roots (Menegus *et al.*, 1991) it was not possible to detect any lag between the production of lactate and ethanol.

In some systems there is no formation of lactate at all and the tissue responds to anoxia with the immediate formation of ethanol (Andreev and Vartapetian, 1992). For example, in rice shoots, the onset of anoxia caused negligible (Menegus *et al.*, 1991; Rivoal *et al.*, 1989) or at best limited (Menegus *et al.*, 1989) production of lactate. These observations have important implications for the Davies model, since it is often assumed that it is the production of lactate that is responsible for the fall in the cytoplasmic pH. In the absence of lactate, the origin of the pH change must lie elsewhere and this question is discussed further in the section on intracellular pH.

It is also possible to reduce, and in some cases eliminate, the initial production of lactate by tissues that would otherwise show the expected transient formation of the metabolite before the switch to ethanol (Roberts *et al.*, 1984a; Xia and Saglio, 1992). In one procedure, maize root tips were incubated with a permeant weak acid before the onset of anoxia with the aim of artificially reducing the pH of the cytoplasm (Roberts *et al.*, 1984a). This pre-treatment abolished both the lag in the production of ethanol and the transient accumulation of lactate, showing that both processes were regulated by the cytoplasmic pH. It follows that speculations about the reasons for the production or non-production of lactate in anoxic tissues can only be conducted usefully with a knowledge of the cytoplasmic pH of the tissue.

Another experimental procedure that can have a major effect on lactate production (Xia and Saglio, 1992), and which also has the great advantage that it mimics the gradual depletion of the oxygen supply that characterizes many naturally occurring anaerobic episodes, is the use of a hypoxic pre-treatment prior to the imposition of anoxia (Hole *et al.*, 1992; Johnson *et al.*, 1989; Saglio *et al.*, 1988). This technique has been used extensively with maize root tips and typically involves switching from a 40% v/v oxygen supply to a 4% v/v supply for a period of 16–18 h before the switch to total anoxia. The hypoxic pre-treatment leads to improved viability under anoxia (Johnson *et al.*, 1989; Saglio *et al.*, 1988) and an increased glycolytic rate (Hole *et al.*, 1992). ADH induction has also been shown to be greater and more persistent under hypoxic rather than anoxic conditions (Andrews *et al.*, 1993) and the conclusion seems to be that exposure to a reduced oxygen partial pressure can have a substantial effect on the subsequent response to anoxia.

When anoxia was imposed on maize root tips without a hypoxic pre-treatment, lactate formation preceded ethanol production and there was relatively little leakage of lactate into the suspending medium over the first hour of anoxia (Xia and Saglio, 1992). In contrast, when anoxia was imposed on hypoxically pre-treated root tips, ethanol was formed immediately, without the usual lag phase, and approximately 50% of the lactate produced in the first hour was released to the suspending medium. It is possible that the differences in the kinetics of lactate and ethanol production in the two tissues arise from differences in the response of the cytoplasmic pH to the switch to anoxia. Labelled 5,5-dimethyloxazolidine-2,4-dione (DMO) was used to estimate the cytoplasmic pH in the two tissues under anoxia and it was concluded that the pH was 0.09 units lower in the absence of exogenous glucose, and 0.14 pH units lower in the presence of 100 mM glucose, in the root tips that were not given the hypoxic pre-treatment. While these measurements are consistent with the lower production and retention of lactate by the hypoxically pre-treated tissue, they do not provide an explanation for the altered metabolic response at the onset of anoxia.

It should also be noted that, while the hypoxic pre-treatment abolished the lag in ethanol production, it did not eliminate the lactate fermentation completely and this contrasts with the earlier data on acid-pre-treated maize root tips (Roberts *et al.*, 1984a). While this discrepancy may reflect differences in the effect of the pre-treatments on the cytoplasmic pH – and this will only be resolved by using *in vivo* $^{31}$P-NMR to obtain more detailed time courses for the cytoplasmic pH than can be obtained with the DMO method – it may also reflect changes in the activities of the fermentation enzymes during the hypoxic pre-treatment. Changes in the levels of PDC (Kelley, 1989; Wignarajah and Greenway, 1976), ADH (Andrews *et al.*, 1993; Wignarajah and Greenway, 1976) and LDH (Good and Paetkau, 1992; Xia and Saglio, 1992), involving both transcriptional and translational regulation, are known to occur in maize during oxygen deprivation and there is also the likelihood of changes in the kinetic properties of PDC during a hypoxic pre-treatment (Morrell *et al.*, 1990).

Further evidence that the hypoxic pre-treatment has a profound effect on the physiology of the tissue is provided by the observation that lactate was released into the suspending medium much more rapidly by the pre-treated tissue (Xia and Saglio, 1992). This effect was abolished by the protein synthesis inhibitor cycloheximide, suggesting that the hypoxic pre-treatment induced the synthesis of a lactate carrier system. Despite this further complication, it seems clear that the use of hypoxic pre-treatments leads to a greater understanding of the interaction between the different fermentation pathways under physiologically relevant conditions and that the technique will be exploited in a wider range of species in subsequent work.

Although the Davies model predicts that lactate formation should be transient, it seems unlikely that there would be a sharp transition between the lactate and ethanol pathways and so the long-term production of lactate might be expected even after the metabolism of a tissue has switched to the predominant synthesis of ethanol and alanine. Equally, in situations where there might be a gradient in oxygen deficiency, for example in maize roots exposed to a poor oxygen supply (Thomson and Greenway, 1991), some cells will have made the transition to ethanol production while others, in a less hypoxic state, will be producing lactate. Thus the long-term production of low levels of lactate does not necessarily contradict the Davies model and is observed quite often (Good and Muench, 1993; Hoffman *et al.*, 1986). However, recent work on hypoxically pre-treated roots of various members of the halophytic genus *Limonium* has revealed tissues in which lactate is a major long-term fermentation product (Rivoal and Hanson, 1993a). Some species responded to anoxia in the usual way, with a much greater production of ethanol than lactate; while others produced substantial quantities of lactate and, in two cases, the flux into this metabolite actually exceeded the production of ethanol. No measurements of cytoplasmic pH were reported in this work and so it is possible that the prolonged synthesis of lactate merely reflects the operation of an efficient pH regulation system that prevented the usual fall in pH. If this is the case, then it would appear that lactic acid efflux makes an important contribution to the regulation of the pH since much of the lactate was transported into the suspending medium (Rivoal and Hanson, 1993a). Equally, the kinetic properties of LDH could be significantly different in the species showing a high flux to lactate, with the result that lactate synthesis is no longer self-limiting, making lactate efflux essential if the tissue is to avoid a lethal fall in the cytoplasmic pH.

These observations on the long-term accumulation of lactate may shed some light on the anaerobic induction of LDH activity. This phenomenon has been observed in a range of plant tissues, including barley roots, which showed a 20-fold increase in LDH activity within the first week of hypoxia (Hoffman *et al.*, 1986); maize root tips, which showed a twofold increase in activity over a 2-day hypoxic pre-treatment (Xia and Saglio, 1992); a rice cell culture (Mohanty *et al.*, 1993); and rice seedlings (Rivoal *et al.*, 1991). The increase in LDH activity in barley roots appears to be due to the preferential expression of one of the two *Ldh* genes (Hoffman *et al.*, 1986) and it is associated with increases in the levels of both translatable LDH mRNA and LDH protein (Hondred and Hanson, 1990). While these observations establish LDH as one of the small group of anaerobic proteins that are known to be synthesised under anoxia, they do little to establish the function of the enzyme during anaerobiosis. It is difficult to reconcile the induction of LDH with the transient production of lactate expected by the Davies model, and the view persists that LDH induction may reflect the existence of some other function for the enzyme under anoxia. Genetic manipulation, using either antisense technology to generate transgenic plants deficient in LDH or transformation techniques that lead to the over-expression of LDH, may help to identify this function, but the initial results of an over-expression experiment (Rivoal and Hanson, 1993b) were not very revealing. Transformation of tomato with a barley LDH cDNA led to the production of root clones with LDH activities that were up to 50 times higher than in the wild-type control, but this had no effect on the balance between lactate and ethanol production under anoxia either with or without a hypoxic pre-treatment (Rivoal and Hanson, 1993b). Whether an antisense approach will be any more successful remains to be seen, but it should provide a critical test of the role of lactate in triggering the switch to the production of ethanol under anoxia.

The physiological significance of the anaerobic induction of LDH is analogous to the uncertainty that surrounds the anaerobic induction of ADH (Roberts *et al.*, 1989). In the case of ADH, it has been established that the normal aerobic level of the enzyme is sufficient to catalyse the production of ethanol under anoxia without limiting the capacity for energy production via fermentation, and this makes it difficult to assign a function for the induction process in the survival of the tissue (Roberts *et al.*, 1989). However, in the case of LDH, there is the tempting hypothesis that the long-term production of lactate under anoxia serves to conserve fixed carbon (Rivoal and Hanson, 1993a). This proposal calls to mind the well-known Cori cycle in animal tissues, whereby the lactate produced by anaerobic glycolysis in muscle and other tissues is transferred to the liver for gluconeogenesis, but at present there is no evidence that something similar, perhaps involving the export of lactate to the leaves, occurs in plants. While the proposal is attractive, the reality is that the long-term production of lactate as a major fermentation product is uncommon and that lactate efflux appears to be the immediate fate of the metabolite in tissues that produce large quantities under anoxia (Hanson and Jacobsen, 1984; Rivoal and Hanson, 1993a).

## 7.3 Cytoplasmic pH changes under anoxia

Although the early NMR experiments on maize root tips (Roberts *et al.*, 1984a) provided good evidence that the switch to ethanol production under anoxia is indeed triggered by a fall in the cytoplasmic pH, subsequent work has shown that the pH response can be more complicated than originally observed and thus raises a number of questions about the origin and metabolic significance of the observed pH changes. Firstly, there is the question of the origin of the pH changes that occur under anoxia and the extent to which these changes can be attributed to the production of lactate. Secondly, there is the equally important question of the relevance of the observed pH behaviour to the *in vivo* pH dependence of the key enzymes and thus the extent to which a change in pH under anoxia can alter the metabolism of the tissue. The recent papers that relate to these questions will be discussed and it will become apparent that there are still some unsolved problems in this field, despite the general validity of the idea that underlies the Davies model. It will also become apparent that much of the progress in this area can be attributed to the existence of a technique, *in vivo* NMR spectroscopy, that can combine the measurement of cytoplasmic and vacuolar pH with non-invasive measurements of metabolism (Loughman and Ratcliffe, 1984; Ratcliffe, 1987, 1994; Roberts, 1984, 1987).

The close correlation between the accumulation of lactate and the fall in cytoplasmic pH in anoxic maize root tips (Roberts *et al.*, 1984a) reinforced the idea that lactate accumulation is responsible for the acidification of the cytoplasm under anoxia. While it is still possible to find this view expressed in recent papers (e.g. Rivoal and Hanson, 1993a), there appear to be three reasons for avoiding this rather misleading oversimplification. Firstly, it assumes that the correlation between the two events is a general feature of the anoxic response, whereas there is increasing evidence, summarized below, that it is not. Secondly, it suggests that pH regulation under anoxia can be understood without reference to the many other biochemical and biophysical processes that determine the normal aerobic pH and which can be expected to respond to the perturbation caused by the switch to anoxia. Thirdly, and at the risk of splitting hairs, it perpetuates the myth that the endogenous production of lactate by glycolysis acidifies the cytoplasm in much the same way as an acid load imposed by the uptake of a permeable weak acid. In fact, as discussed by Hochachka and Mommsen (1983), the 1:1 stoichiometry between the anaerobic production of lactate from glucose and the production of protons reflects a pH-dependent balance between the protons generated in the conversion of glucose to lactate and the protons released in the subsequent utilization of the ATP produced by glycolysis. Since these two processes are of approximately equal importance at pH values close to 7.0, it would seem best to avoid the oversimplification of assuming that it is merely the accumulation of lactate that causes acidification.

Turning to the apparent correlation between lactate production under anoxia and the decline in the cytoplasmic pH, there is increasing evidence that the situation originally reported in maize root tips (Roberts *et al.*, 1984a), and confirmed by more recent experiments on a different cultivar (Roberts *et al.*, 1992), may be unrepresentative. For example, Saint-Ges *et al.* (1991), also working with maize root tips, observed relatively little lactate during the initial acidification of the cytoplasm under anoxia and there was a substantial accumulation of lactate after the cytoplasmic pH had stabilized. Moreover, in rice shoots (Menegus *et al.*, 1991), anoxia caused an immediate decrease in the cytoplasmic pH even though the lactate production in this tissue was very low (Menegus *et al.*, 1989, 1991; Rivoal *et al.*, 1989). It would be interesting to have pH measurements for the *Limonium* spp. that produce lactate as a major fermentation product (Rivoal and Hanson, 1993a), but it seems unlikely that there would be any correlation with the cytoplasmic pH, especially given that a substantial proportion of the lactate is released into the surrounding medium. Thus, in general, it appears that there is no necessary correlation between the accumulation of lactic acid and the time dependence of the cytoplasmic pH under anoxia.

The contribution that lactate makes to the initial pH change can be calculated on the basis of the buffering capacity of the cytoplasm if it is assumed firstly that all the lactate is confined to this compartment and secondly that there is tight coupling between lactate synthesis and ATP hydrolysis (Hochachka and Mommsen, 1983). Such calculations are at best semiquantitative, because of the uncertainty in the buffering capacity of the cytoplasm, and they probably only serve to draw attention to the underlying complexity in the pH regulatory processes. Thus, working with maize root tips, and using a cytoplasmic buffering capacity of 10.2 μmol $H^+$ $g^{-1}$ fresh weight per pH unit, Reid *et al.* (1985) concluded that the predicted pH change during the first 1000 s (16 min 40 s) of anoxia was significantly greater than the observed change and that the excess $H^+$ production would have to be offset by other mechanisms. The calculated $H^+$ production included the small contributions from the net hydrolysis of ATP (unrealistically ignoring the small effect of magnesium complexation) and $H^+$ influx from the suspending medium, as well as the dominant effect of the lactate, and it was concluded that the observed pH fall must be limited by the metabolic consumption of protons. The proton-consuming synthesis of γ-aminobutyric acid (GABA) was suggested as a possible response, but a more recent study (Roberts *et al.*, 1992) indicates that this is insignificant during the first 30 min of anoxia. Roberts *et al.* (1992), using a cytoplasmic buffering capacity of approximately 7 μmol $H^+$ $g^{-1}$ fresh weight per pH unit, concluded that the pH fall observed during the first 10 min of anoxia could be explained quantitatively in terms of the accumulation of lactate and alanine, but that a proton-consuming reaction, identified as the decarboxylation of malate rather than the proposed synthesis of GABA, and $H^+$ transport to the vacuole became significant after 10 min and thus explained the partial recovery of the pH towards a less acidic value. Finally, while Saint-Ges *et al.* (1991), who observed that the bulk of the lactate accumulation in their maize tips occurred after the initial acidification, did not calculate the contribution of lactate to the fall in pH, there is sufficient information in their paper to show that the lactate produced in the first 10 min (2.35 nmol per root tip) would have caused a pH drop of less than 0.1 pH units (assuming an implied cytoplasmic buffering capacity of 15 μmol $H^+$ $g^{-1}$ fresh weight per pH unit) which was much less than the observed fall of 0.6.

Thus, even when lactate is produced as an immediate response to an oxygen deficit, different investigators, working with nominally the same tissue, have found that the production of lactate is either insufficient (Saint-Ges *et al.*, 1991), just right (Roberts *et al.*, 1992) or too great (Reid *et al.*, 1985) to explain the initial fall in the cytoplasmic pH. This indicates that attempts to understand the time dependence of the cytoplasmic pH in an anoxic tissue need to take into account the full range of biochemical and biophysical events that potentially could influence pH. These processes include proton-generating reactions, such as lactate synthesis and ATP hydolysis, proton-consuming reactions, such as malate decarboxylation and GABA synthesis, ion transport processes at the tonoplast and plasma membrane, and processes that redistribute the acidic products of metabolism, such as compartmentation in the vacuole or export to either aerobic tissues or the external medium. In this respect, tissues that produce little or no lactate during the switch to anoxia and tissues that produce large quantities of lactate over an extended period are of particular interest. The origin of the pH change in the absence of lactate production is unclear (e.g Menegus *et al.*, 1991), but an interesting suggestion, stemming from the similarity of the pH and ATP time courses observed in anoxic maize root tips (Saint-Ges *et al.*, 1991), is that it could arise from substrate limitation of the $H^+$-ATPases in the tonoplast and plasma membrane. At the other extreme, in tissues that produce lactate over an extended period, lactic acid efflux appears to be an important mechanism for avoiding cytoplasmic acidosis and this has been observed in the lactate-accumulating *Limonium* spp. (Rivoal and Hanson, 1993a), as well as in hypoxically pre-treated maize root tips (Xia and Saglio, 1992).

Irrespective of the pH-regulating mechanisms that operate under anoxia, it is the fall in pH that apparently triggers the switch to ethanol production (Davies *et al.*, 1974; Roberts *et al.*, 1984a). However, a major shortcoming in the current understanding of this effect is that it is unclear how far the cytoplasmic pH must actually fall in order to activate the ethanolic fermentation. For example, early NMR data for maize root tips (Roberts *et al.*, 1984a) and mature maize root segments (Fan *et al.*, 1988) showed falls to pH 6.8 and 7.15, respectively. This difference could have arisen because of differences in the experimental conditions or because of the difference in metabolic activity of the two tissues; but equally the observations could reflect the existence of a different threshold for the activation of PDC. Similarly, ethanol production appears to be associated with a smaller drop in pH in rice shoots than in wheat roots

(Menegus *et al.*, 1991) and again it is difficult to decide whether this reflects a difference in the ability of the tissue to resist the fall in pH or whether it indicates a difference in the kinetic properties of PDC.

A further complication becomes apparent when it is noted that the cytoplasmic pH often passes through a minimum (Menegus *et al.*, 1991; Roberts *et al.*, 1992; Saint-Ges *et al.*, 1991). In some cases, this undershoot is corrected rapidly (Roberts *et al.*, 1992; Saint-Ges *et al.*, 1991) and is reminiscent of the effects that can arise from a temporary loss of pH control (Fox and Ratcliffe, 1990), while in other cases the cytoplasmic pH can apparently drift upwards towards a more normal aerobic value over a time scale of hours (Menegus *et al.*, 1991). The question of immediate interest is whether the partial recovery of the cytoplasmic pH has any effect on the production of ethanol. For example, does ethanol production slow down as the pH increases and does this provide an explanation for the long-term production of lactate in some tissues? Alternatively, does ethanol production continue after the pH has recovered above the threshold for the activation of PDC and how can this be reconciled with the properties of the enzyme?

Significant progress towards solving this problem has been made by examining the anoxic response of maize root tips under conditions that either reduce the initial fall in the cytoplasmic pH or promote its recovery under anoxia (Fox *et al.*, 1994). *In vivo* $^{31}$P-NMR was used to measure the cytoplasmic pH following the onset of anoxia (Figure 7.1) and in the key experiment (Figure 7.2) methylamine, a permeant weak base, was used to combat the acidification of the cytoplasm. Correlating the cytoplasmic pH (Figure 7.2a) with ethanol production (Figure 7.2b) under these conditions provided unequivocal support for the critical role of pH in the switch to an ethanolic fermentation by showing that ethanol production ceased as the pH recovered to a normal aerobic value under anoxia. Thus, these experiments demonstrate the reversibility of the pH effect on ethanol production under anoxia and they indicate that external factors can have an important effect on the nature of the fermentation end-products. Moreover, with further development, the approach adopted in this work should allow the cytoplasmic pH to be held at a defined value under anaerobic conditions and thus allow the pH-dependent balance between lactate and ethanol production to be measured directly.

In principle, the fall in cytoplasmic pH under anoxia could trigger other metabolic changes, and a potential candidate is the onset of GABA synthesis. This metabolite often accumulates under anoxia (Roberts *et al.*, 1992; Tesnière *et al.*, 1994) and it has often been suggested that the accumulation of GABA could be the result of a change in the relative activities of glutamate decarboxylase (GDC; EC 4.1.1.15) and $\gamma$-aminobutyric acid: $\alpha$-oxo acid transaminase (EC 2.6.1.19) following a fall in the cytoplasmic pH (Streeter and Thomson, 1972; Wallace *et al.*, 1984). Moreover, since GDC activity increases as the pH falls and since glutamate decarboxylation is a proton-consuming reaction, GABA production could contribute to the regulation of pH under anoxia (Menegus *et al.*, 1989; Reid *et al.*, 1985; Roberts *et al.*, 1992). Proof that a fall in cytoplasmic pH can

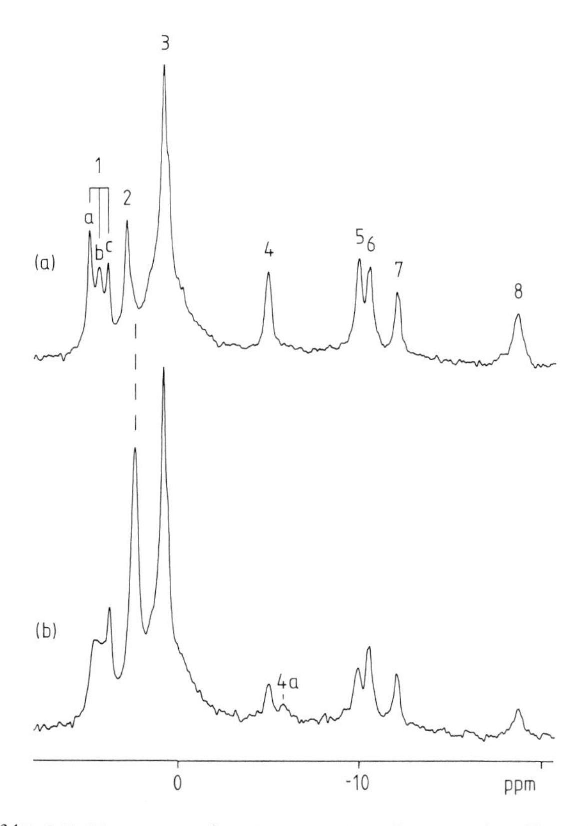


***Figure 7.1.*** *$^{31}$P-NMR spectra of maize root tips showing the effect of a switch to anoxia. Spectrum (a) was recorded with the tissue bathed in an oxygenated 2-(N-morpholino)-ethane sulphonic acid (MES) buffer at pH 6 and spectrum (b) was recorded after switching the gas supply from oxygen to nitrogen. Each spectrum was accumulated over a period of 30 min and spectrum (b) was recorded after 30-60 min under anoxia. The numbered peaks can be assigned to: 1, several phosphomonoesters, including glucose-6-phosphate (1a) and phosphocholine (1c); 2, cytoplasmic inorganic phosphate (Pi); 3, vacuolar Pi, with underlying contributions from phosphodiesters such as glycerophosphocholine; 4, 5 and 8, the γ-, α- and β-phosphates respectively of nucleoside triphosphate; 6, UDP-glucose and NAD(P)(H); and 7, UDP-glucose. Peak 4a, which is only detectable under anoxia, can be assigned to the β-phosphate of nucleoside diphosphate. The shift in the position of the cytoplasmic Pi signal, emphasized by the vertical line between the two spectra, is caused by the acidification of the cytoplasm and the pH change can be measured by calibrating the pH dependence of the Pi signal. In a typical experiment, the accumulation time might be reduced to 5 min (Roberts* et al.*, 1992) or even 2 min (Saint-Ges* et al.*, 1991) and the resulting good time resolution, coupled with the ability to monitor signals from several bioenergetically important metabolites simultaneously, explains why the technique has been used extensively in studies of anoxia. (Fox, McCallan and Ratcliffe, unpublished spectra.)*

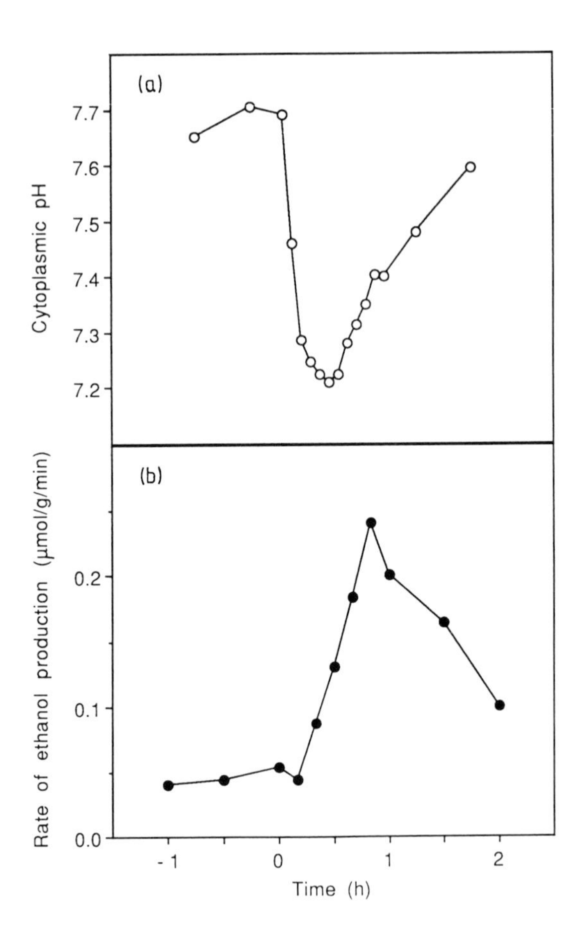


***Figure 7.2.*** *Graphs showing the effect of a switch to anoxia on (a) the cytoplasmic pH and (b) the rate of ethanol production for maize root tips in a 3-(cyclohexylamino)-1-propane sulphonic acid (CAPS) buffer at pH 10. The buffer also contained 20 mM methylamine and, under these conditions, the cytoplasmic pH recovered to a normal aerobic value following the switch from oxygen to nitrogen at time zero. The weak base also had a marked effect on the vacuolar pH (data not shown) and so the root tissue was pre-treated with manganese to eliminate the vacuolar Pi signal from the $^{31}$P-NMR spectrum, thus avoiding any overlap in the cytoplasmic and vacuolar Pi signals. (Adapted from data presented in Fox* et al.*, 1994.)*

trigger the synthesis of GABA *in vivo* has been obtained recently from a series of NMR experiments on carrot (*Daucus carota*, L.) cells (Carroll *et al.*, 1994). *In vivo* $^{15}$N-NMR was used to follow ammonium assimilation under different oxygenation conditions (Figure 7.3) and the mild hypoxia induced by the use of air, rather than pure oxygen, in the aeration system was sufficient to trigger the accumulation of GABA. *In vivo* $^{31}$P-NMR measurements showed that ammonium assimilation caused a transient fall in the cytoplasmic pH under less well oxygenated conditions (Figure 7.4) and the significance of this effect was demonstrated by varying the external pH in such a way as to control the cytoplasmic pH during ammonium assimilation. By this means, it was possible to show that the metabolic effect of a reduced oxygen supply, that is increased GABA production, could be mimicked by reducing the cytoplasmic pH of well oxygenated cells and that GABA production in mildly hypoxic cells could be abolished by preventing the fall in cytoplasmic pH (Carroll *et al.*, 1994). These results provide direct and unequivocal evidence that GABA accumulation can be

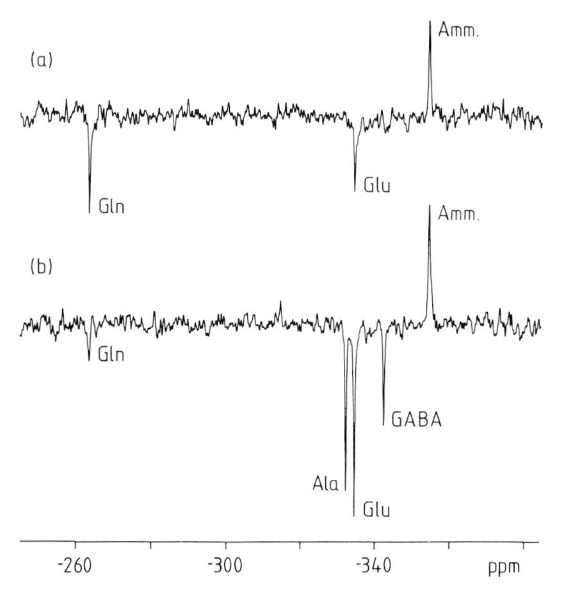


***Figure 7.3.*** *$^1$H-decoupled $^{15}$N-NMR spectra of (a) well oxygenated and (b) mildly hypoxic carrot cells following the addition of 20 mM [$^{15}$N]ammonium chloride to the suspending medium. Each spectrum was accumulated over a 2-h period 8–10 h after the addition of the ammonium. The labelled peaks can be assigned to the amide nitrogen of glutamine (Gln), the amino nitrogens of alanine (Ala), glutamate (Glu) and GABA, and ammonium (Amm). As with the $^{31}$P-NMR spectra in Figure 7.1,* in vivo *$^{15}$N-NMR spectra can be recorded sequentially from the same sample over an extended time scale. This allows time courses to be constructed more rapidly and efficiently than by extraction and it also permits the direct measurement of the metabolic fluxes into specific compounds. (Fox and Ratcliffe, unpublished spectra.)*

triggered by a fall in cytoplasmic pH and that the proton-consuming decarboxylation of glutamate can be important in the regulation of cytoplasmic pH. However, this process does not contribute to the regulation of pH in maize root tips during the first 30 min of anoxia (Roberts *et al.*, 1992), suggesting that the activation of GDC is relatively slow.

## 7.4 Conclusion

While there is considerable evidence in favour of the biochemical pH stat model for the anoxic response of plant tissues, neither the role of lactate production during anoxia nor the role of the cytoplasmic pH in mediating the transition to anaerobic metabolism is understood fully. Further progress in this area is likely to involve both the continued use of *in vivo* NMR spectroscopy and the exploitation of the potential, as yet underutilized in this field, of transgenic material. The application of these techniques to plants with widely differing suscept-

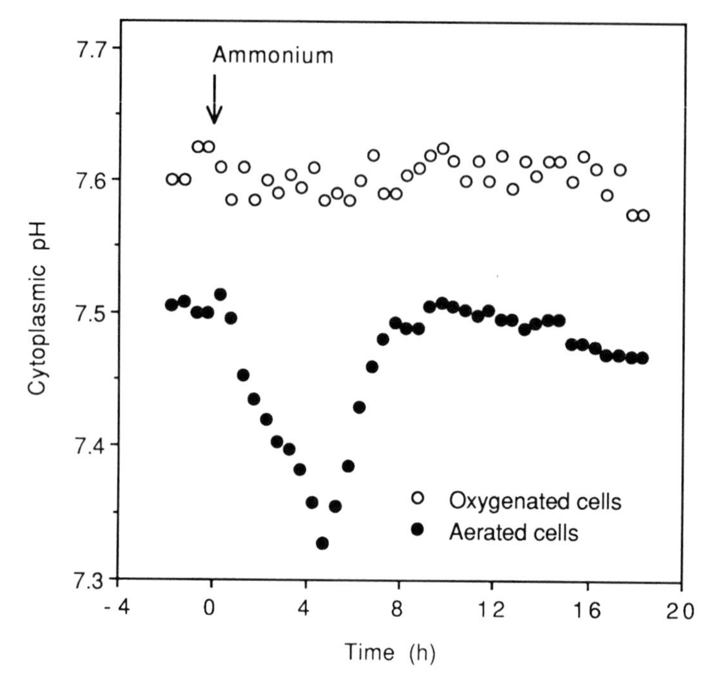


***Figure 7.4.*** *Graph showing the effect of the oxygen supply on the cytoplasmic pH of carrot cells during the assimilation of ammonium. Oxygen was supplied as either pure oxygen (○; oxygenated cells) or air (●; aerated cells) and the 20 mM [$^{15}$N]ammonium chloride was added to the suspending medium at time zero. The cytoplasmic pH was measured from the position of the cytoplasmic Pi signal in $^{31}$P-NMR spectra recorded every 30 min. (Adapted from data presented in Carroll* et al., *1994.)*

ibilities to anoxia would be particularly welcome and would assist in establishing the generality or otherwise of the current model of the anoxic response.

## Acknowledgements

The author acknowledges the financial support of the Agricultural and Food Research Council and is indebted to Mr N.R. McCallan and Dr G.G. Fox for their contribution to the experiments outlined in the figures. The author is also grateful for a critical reading of the manuscript by Dr N.J. Kruger.

## References

Andreev, V.Y. and Vartapetian, B.B. (1992) Induction of alcoholic and lactic fermentation in the early stages of anaerobic incubation of higher plants. *Phytochemistry* **31**, 1859–1861.

Andrews, D.L., Cobb, B.G., Johnson, J.R. and Drew, M.C. (1993) Hypoxic and anoxic induction of alcohol dehydrogenase in roots and shoots of seedlings of *Zea mays*. *Adh* transcripts and enzyme activity. *Plant Physiol.* **101**, 407–414.

Brailsford, R.W., Voesenek, L.A.C.J., Blom, C.W.P.M., Smith, A.R., Hall, M.A. and Jackson,M.B. (1993) Enhanced ethylene production by primary roots of *Zea mays* L. in response to sub-ambient partial pressures of oxygen. *Plant, Cell Environ.* **16**, 1071–1080.

Bucher, M. and Kuhlemeier, C. (1993) Long-term anoxia tolerance. Multi-level regulation of gene expression in the amphibious plant *Acorus calamus* L. *Plant Physiol.* **103**, 441– 448.

Carroll. A.D., Fox, G.G., Laurie, S., Phillips, R., Ratcliffe, R.G. and Stewart, G.R. (1994) Ammonium assimilation and the role of γ-aminobutyric acid in pH homeostasis in carrot cell suspensions. *Plant Physiol.* in press.

Crawford, R.M.M. (1992) Oxygen availability as an ecological limit to plant distribution. *Adv. Ecol. Res.* **23**, 93–185.

Davies, D.D. (1980) Anaerobic metabolism and the production of organic acids. In: *The Biochemistry of Plants,* Volume 2 (ed. D.D. Davies). Academic Press, New York, pp. 581–611.

Davies, D.D., Grego, S. and Kenworthy, P. (1974) The control of the production of lactate and ethanol by higher plants. *Planta* **118**, 297–310.

Drew, M.C. (1990) Sensing soil oxygen. *Plant, Cell Environ.* **13**, 681–693.

Fan, T.W.-M., Higashi, R.M. and Lane, A.N. (1986) Monitoring of hypoxic metabolism in superfused plant tissues by *in vivo* $^1$H NMR. *Arch. Biochem. Biophys.* **251**, 674–687.

Fan, T.W.-M., Higashi, R.M. and Lane, A.N. (1988) An *in vivo* $^1$H and $^{31}$P NMR investigation of the effect of nitrate on hypoxic metabolism in maize roots. *Arch. Biochem. Biophys.* **222**, 592–606.

Fox, G.G. and Ratcliffe, R.G. (1990) $^{31}$P NMR observations on the effect of the external pH on the intracellular pH values in plant cell suspension cultures. *Plant Physiol.* **93**, 512–521.

Fox, G.G., McCallan, N.R. and Ratcliffe, R.G. (1994) Manipulating cytoplasmic pH under anoxia: a critical test of the role of pH in the switch from aerobic to anaerobic metabolism. *Planta* in press.

Good, A.G. and Muench, D.G. (1993) Long term anaerobic metabolism in root tissue. Metabolic products of pyruvate metabolism. *Plant Physiol.* **101**, 1163–1168.

Good, A.G. and Paetkau, D.H. (1992) Identification and characterisation of a hypoxically induced maize lactate dehydrogenase gene. *Plant Mol. Biol.* **19**, 693–697.

Hanson, A.D. and Jacobsen, J.V. (1984) Control of lactate dehydrogenase, lactate glycolysis and α-amylase by $O_2$ deficit in barley aleurone layers. *Plant Physiol.* **75**, 566–572.

Hochachka, P.W. and Mommsen, T.P. (1983) Protons and anaerobiosis. *Science* **219**, 1391–1397.

Hoffman, N.E., Bent, A.F. and Hanson, A.D. (1986) Induction of lactate dehydrogenase isozymes by oxygen deficit in barley root tissue. *Plant Physiol.* **82**, 658–663.

Hole, D.J., Cobb, B.G., Hole, P.S. and Drew, M.C. (1992) Enhancement of anaerobic respiration in root tips of *Zea mays* following low oxygen (hypoxic) acclimation. *Plant Physiol.* **99**, 213–218.

Hondred, D. and Hanson, A.D. (1990) Hypoxically inducible barley lactate dehydrogenase: cDNA cloning and molecular analysis. *Proc. Natl Acad. Sci USA* **87**, 7300–7304.

Jackson, M.B. and Hall, K.C. (1993) Polyamine content and action in roots of *Zea mays* L. in relation to aerenchyma development. *Ann. Bot.* **72**, 569–575.

Jackson, M.B., Davies, D.D. and Lambers, H. (eds) (1991) *Plant Life Under Oxygen Deprivation: Ecology, Physiology and Biochemistry.* SPB Academic Publishing, The Hague, The Netherlands.

Johnson, J., Cobb, B.G. and Drew, M.C. (1989) Hypoxic induction of anoxia tolerance in root tips of *Zea mays*. *Plant Physiol.* **91**, 837–841.

Kelley, P.M. (1989) Maize pyruvate decarboxylase mRNA is induced anaerobically. *Plant Mol. Biol.* **13**, 213–222.

Kennedy, R.A., Rumpho, M.E. and Fox, T.C. (1992) Anaerobic metabolism in plants. *Plant Physiol.* **100**, 1–6.

Loughman, B.C. and Ratcliffe, R.G. (1984) Nuclear magnetic resonance and the study of plants. In: *Advances in Plant Nutrition,* Volume 1 (eds P.B. Tinker and A. Läuchli). Praeger, New York, pp. 241–283.

Menegus, F., Cattaruzza, L., Chersi, A. and Fronza, G. (1989) Differences in the anaerobic lactate–succinate production and in the changes of cell sap pH for plants with high and low resistance to anoxia. *Plant Physiol.* **90**, 29–32.

Menegus, F., Cattaruzza, L., Mattana, M., Beffagna, N. and Ragg, E. (1991) Response to anoxia in rice and wheat seedlings. Changes in the pH of intracellular compartments, glucose-6-phosphate level and metabolic rate. *Plant Physiol.* **95**, 760–767.

Mohanty, B., Wilson, P.M. and ap Rees, T. (1993) Effects of anoxia on growth and carbohydrate metabolism in suspension cultures of soybean and rice. *Phytochemistry* **34**, 75–82.

Morrell, S., Greenway, H. and Davies, D.D. (1990) Regulation of pyruvate decarboxylase *in vitro* and *in vivo*. *J. Exp. Bot.* **41**, 131–139.

Perata, P. and Alpi, A. (1994) Plant responses to anaerobiosis. *Plant Sci.* **93**, 1–17.

Peschke,V.M. and Sachs, M.M. (1993) Multiple pyruvate decarboxylase genes in maize are induced by hypoxia. *Mol. Gen. Genet.* **240**, 206–212.

Peschke, V.M. and Sachs, M.M. (1994) Characterisation and expression of transcripts induced by oxygen deprivation in maize (*Zea mays* L.). *Plant Physiol.* **104**, 387–394.

Ratcliffe, R.G. (1987) The application of NMR methods to plant tissues. In: *Methods in Enzymology,* Volume 148 (eds L. Packer and R. Douce). Academic Press, New York, pp. 683–700.

Ratcliffe, R.G. (1994) *In vivo* nuclear magnetic resonance studies of higher plants and algae. *Adv. Bot. Res.* **20**, 43–123.

Reid, R.J., Loughman B.C. and Ratcliffe, R.G. (1985) $^{31}$P NMR measurements of cytoplasmic pH changes in maize root tips. *J. Exp. Bot.* **36**, 889–897.

Rivoal, J. and Hanson, A.D. (1993a) Evidence for a large and sustained glycolytic flux to lactate in anoxic roots of some members of the halophytic genus *Limonium*. *Plant Physiol.* **101**, 553–560.

Rivoal, J. and Hanson, A.D. (1993b) The overexpression of lactate dehydrogenase in transgenic tomato roots supports the Davies hypothesis. *Plant Physiol.* **102**, S-38.

Rivoal, J., Ricard, B. and Pradet, A. (1989) Glycolytic and fermentative enzyme induction during anaerobiosis in rice seedlings. *Plant Physiol. Biochem.* **27**, 43–52.

Rivoal, J., Ricard, B. and Pradet, A. (1991) Lactate dehydrogenase in *Oryza sativa* seedlings and roots. Identification and partial characterisation. *Plant Physiol.* **95**, 682–686.

Roberts, J.K.M. (1984) Study of plant metabolism *in vivo* using NMR spectroscopy. *Annu. Rev. Plant Physiol.* **35**, 375–386.

Roberts, J.K.M. (1987) NMR in plant biochemistry. In: *The Biochemistry of Plants,* Volume 13 (ed. D.D. Davies). Academic Press, New York, pp. 181–227.

Roberts, J.K.M., Callis, J., Wemmer, D., Walbot, V. and Jardetzky, O. (1984a) Mechanism of cytoplasmic pH regulation in hypoxic maize root tips and its role in survival under hypoxia. *Proc. Natl Acad. Sci. USA* **81**, 3379–3383.

Roberts, J.K.M., Callis, J., Jardetzky, O., Walbot, V. and Freeling, M. (1984b) Cytoplasmic acidosis as a determinant of flooding intolerance in plants. *Proc. Natl Acad. Sci. USA* **81**, 6029–6033.

Roberts, J.K.M., Andrade, F.H. and Anderson, I.C. (1985) Further evidence that cytoplasmic acidosis is a determinant of flooding intolerance in plants. *Plant Physiol.* **77**, 492–494.

Roberts, J.K.M., Chang, K., Webster, C., Callis, J. and Walbot, V. (1989) Dependence of ethanolic fermentation, cytoplasmic pH regulation, and viability on the activity of alcohol dehydrogenase in hypoxic maize root tips. *Plant Physiol.* **89**, 1275–1278.

Roberts, J.K.M., Hooks, M.A., Miaullis, A.P., Edwards, S. and Webster, C. (1992) Contribution of malate and amino acid metabolism to cytoplasmic pH regulation in hypoxic maize root tips studied using nuclear magnetic resonance spectroscopy. *Plant Physiol.* **98**, 480–487.

Saglio, P.H., Drew, M.C. and Pradet, A. (1988) Metabolic acclimation to anoxia induced by low (2–4 kPa partial pressure) oxygen pre-treatment (hypoxia) in root tips of *Zea mays. Plant Physiol.* **86**, 61–66.

Saint-Ges, V., Roby, C., Bligny, R., Pradet, A. and Douce, R. (1991) Kinetic studies of the variations of cytoplasmic pH, nucleotide triphosphates ($^{31}P$ NMR) and lactate during normoxic and anoxic transitions in maize root tips. *Eur. J. Biochem.* **200**, 477–482.

Smith, A.M. and ap Rees, T. (1979) Effects of anaerobiosis on carbohydrate oxidation by roots of *Pisum sativum. Phytochemistry* **18**, 1453–1458.

Streeter, J.G. and Thompson J.F. (1972) *In vivo* and *in vitro* studies of γ-aminobutyric acid metabolism with the radish plant (*Raphanus sativus* L.). *Plant Physiol.* **49**, 579–584.

Tesnière, C., Romieu, C., Dugelay, I., Nicol, M.Z., Flanzy, C. and Robin, J.P. (1994) Partial recovery of grape energy metabolism upon aeration following anaerobic stress. *J. Exp. Bot.* **45**, 145–151.

Thomson, C.J. and Greenway, H. (1991) Metabolic evidence for stelar anoxia in maize roots exposed to low $O_2$ concentrations. *Plant Physiol.* **96**, 1294–1301.

Wallace, W., Secor, J. and Schrader, L.E. (1984) Rapid accumulation of γ-aminobutyric acid and alanine in soybean leaves in response to an abrupt transfer to lower temperature, darkness, or mechanical manipulation. *Plant Physiol.* **75**, 170–175.

Wignarajah, K. and Greenway, H. (1976) Effects of anaerobiosis on activities of alcohol dehydrogenase and pyruvate decarboxylase in roots of *Zea mays. New Phytol.* **77**, 575–584.

Xia, J.-H. and Saglio, P.H. (1992) Lactic acid efflux as a mechanism of hypoxic acclimation of maize root tips to anoxia. *Plant Physiol.* **100**, 40–46.

# 8

# The effects of water deficit on photosynthesis

D.W. Lawlor

## 8.1 Introduction

Photosynthesis is the driving force of plant productivity and the ability to maintain the rate of photosynthetic carbon dioxide and nitrate assimilation under environmental stresses is fundamental to the maintenance of plant growth and production. This is particularly so for water deficits which severely limit productivity world-wide. (Alscher and Cummings, 1990; Lawlor, 1994; Lawlor and Uprety, 1993). However, improving the stability of photosynthesis under water stress, to enable the existing photosynthetic mechanisms to attain their full capacity under limiting resources, has proved elusive (Richards, 1993). Selection of crops for increased yield might have been expected to have improved the capacity for, and stress tolerance of, photosynthesis. However, the traditional plant breeding approach has not improved tolerance of the process appreciably nor has increased photosynthetic rate *per se* yet been achieved (Richards, 1993). Currently there is much hope, indeed expectation, that the techniques of genetic engineering will allow the biochemistry of plants to be altered to better adapt them to stressful environments (Close and Bray, 1993; Kononowicz *et al.*, 1993). Photosynthetic metabolism is a prime candidate. However, a limitation to successfully selecting and, particularly, engineering plants for water stress tolerance is the current inadequate understanding of the mechanisms by which photosynthesis does respond and adapt to environmental conditions (Boyer *et al.*, 1987). Despite great progress in understanding photosynthetic metabolism and physiology, demonstrated by the extensive literature on the subject (Lawlor, 1993), and increasing understanding of the effects of water stress on fundamental processes of photosynthesis and its interaction with other environmental factors, e.g. light (see Baker and Bowyer, 1994) and oxygen (Pell and Steffen, 1991; Smirnoff, 1993), no unified concept or 'model' of the events reducing photosynthetic efficiency under stress conditions has found favour. Indeed, conflict over the importance of limitation to $CO_2$ diffusion by stomata versus metabolic regulation or damage is stronger than ever (Cornic, 1994; Lawlor and Uprety, 1993).

Lack of an acceptable model of the effects of water stress on physiology and metabolism is a consequence of the very complex systems being analysed, the range of species (possibly with differing mechanisms) and absence of standardization in plant material and environmental conditions. Despite many attempts to find 'the cause' of sensitivity to water stress, no single factor has been accepted as the limiting factor over the whole range of stress (Boyer *et al.*, 1987; Sharkey, 1990). To attempt to provide a model of photosynthesis under water stress, based on the existing information, is probably premature. Yet a model, however imperfect, would act as an hypothesis to be tested. Clearly, although analysis of parts of the complete photosynthetic system provides the basic information to understand the whole system, it is vital to know how sub-systems influence each other. In the case of water stress, for example, the relationship between capture and use of energy and carbon dioxide assimilation is complementary and these factors must be considered together. The result of modification of one part of the photosynthetic mechanism cannot easily be predicted in terms of photosynthetic rates and plant production. An aim of research is to identify how water deficits developing at different rates and of different intensity and duration affect metabolism. These variables, plus the interaction with environment, complicate the analysis: plants adjust ('harden') by increasing the mechanisms to protect the more vulnerable parts of metabolism. Such processes must be considered when the cause of the responses to water deficits are sought: differences, for example, between rapidly and slowly stressed plants (Boyer *et al.*, 1987; Ort *et al.*, 1994; Wise *et al.*, 1990) may be caused by such adjustments.

This chapter examines how photosynthetic metabolism and its regulation are affected by sub-optimal cellular water supply, concentrating on mesophytic $C_3$ plants. The aim is to give an abstracted and robust view of the way that stress affects the supply of $CO_2$ to the chloroplast, $CO_2$ assimilation there and the interactions with capture of light energy and its transduction in the thylakoids.

## 8.2 Water status of cells and its relevance to metabolism

Despite detailed understanding of cellular water relations, there is lack of clarity about the effects of the components of water balance on specific metabolic processes (Ludlow, 1987; Chapter 1) and therefore of which is the best indicator to use. Water potential (the chemical potential of water) is the most widely used measure of water status. It provides a thermodynamically based description of water in the plant and is the driving force for water movement. The water potential of cells is determined by two major components: osmotic potential (determined by the intracellular concentration of osmotically active solutes) and turgor. Turgor provides the driving force for cell expansion and may act as a signal for water status. Osmotic potential is not sensed directly by the cell contents but the solutes contributing to it have direct effects on metabolism. The other commonly used measure of water status is water content. This is usually expressed as relative water content (RWC) which is the water content as a proportion of that at full saturation. RWC is therefore a measure of relative cell volume and shows the changes in cell volume which could affect interactions between macromolecules and organelles. As a rule of thumb, a RWC of 100–90% is related to stomatal closure and decreased cell expansion and growth of organs, and 90–80% with changes in composition of tissues and some alterations in the relative rates of photosynthesis and respiration. Below 80% RWC (around water potentials of −1.5 MPa or less), changes in metabolism become marked, with cessation of photosynthesis, much increased respiration and accumulation of proline and abscisic acid (ABA). Water potential is a poor basis of comparison because the relationship between water potential, osmotic potential, turgor and RWC can vary between species and different cell types. These relationships depend on properties such as cell wall elasticity, cell size and the extent of osmotic adjustment. All these are known to be influenced by growth conditions and stage of development. Further discussion of plant water status and its measurement can be found in Ludlow (1987) and Smith and Griffiths (1993).

To improve understanding of the response of metabolism to stress, the way in which stress is applied (e.g. plants in osmotica or in drying soil), the rate of stress development and the factors which generally may affect metabolism (e.g. the temperature) or have a major interaction with water deficit (e.g. radiant energy flux), must be defined. Plant metabolism is very dynamic and the mechanism of adjustment (e.g. decreased osmotic potential) as a consequence of response to environmental stimuli are being addressed very actively at present. The contrast between plants grown in the field and in controlled environments should not be underestimated (Wise *et al.*, 1990), particularly the radiant energy which affects development of photoprotective systems (e.g. greater compliment of protective carotenoids) (Björkman and Demmig-Adams, 1994; Chapter 13). This chapter considers information from mature leaves of mainly $C_3$ plants grown in soil or solution culture and stressed over periods of hours to days by withholding water or adding osmotica, under physiologically realistic (as far as they can be defined) environmental conditions of light and temperature adequate for active photosynthesis and growth.

## 8.3 Inhibition of photosynthesis by water deficit

### 8.3.1 *Background*

Water stress decreases the rate of $CO_2$ assimilation per unit leaf area ($A$) with constant incident radiation, temperature and vapour pressure deficit, independent of the $CO_2$ supply outside the leaf ($C_a$). Figures 8.1 and 8.2 show the effect clearly. Possible reasons for the decrease in $A$ are:

(i) limited $CO_2$ diffusion to the intercellular spaces of the leaf as a consequence of reduced stomatal conductance ($g_s$);

(ii) impaired metabolism by direct inhibition of biochemical processes caused by ionic, osmotic or other conditions induced by loss of cellular water.

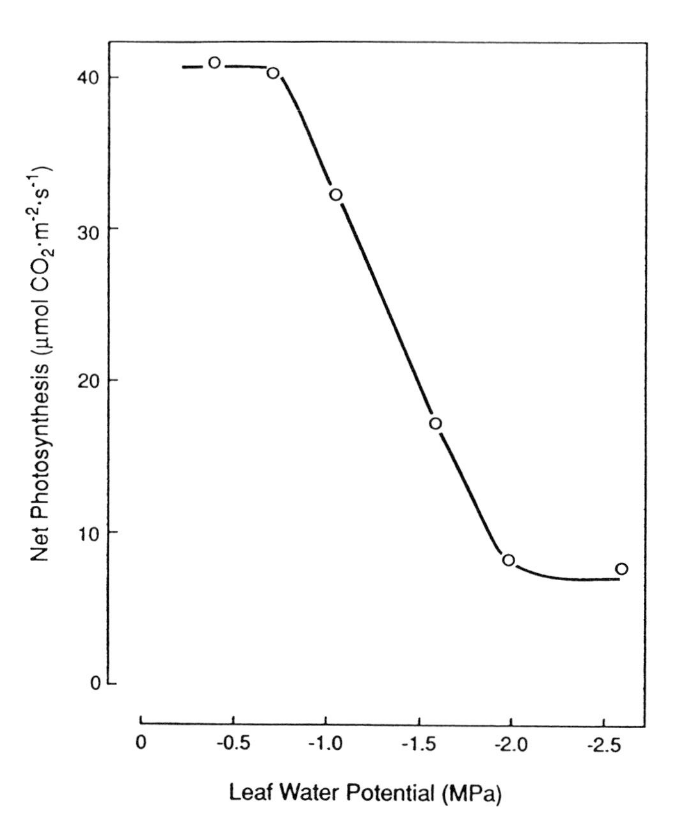


***Figure 8.1.*** *The relationship between leaf water potential and net photosynthesis in attached sunflower (*Helianthus annuus *L.) leaves exposed to increasing water deficit over 6 days. Measurements were made at 25°C with saturating irradiance (1600 µmol m$^{-2}$ s$^{-1}$) in air containing 3 mmol $CO_2$ mol$^{-1}$ air. Reproduced from Ortiz-Lopez* et al. *(1991) with permission from the American Society of Plant Physiologists.*

The decrease in *A* correlates with the decrease in stomatal conductance (Lawlor, 1976b; Lawlor and Khanna-Chopra, 1984; Matthews and Boyer, 1984; Ort *et al.*, 1994). However, this does not exclude the possibility of metabolic effects, as indicated by assimilation decreasing in parallel with reduction in RWC in spinach leaves exposed to hyperosmotic solutions in which stomatal limitations are negligible (Kaiser *et al.*, 1986) and measurements of *A* at given $C_a$, independent of $g_s$ (Gimenez *et al.*, 1992; Matthew and Boyer, 1984). There is a continued polarization of attitudes to regulation of *A*.

## 8.3.2 *Sub-stomatal carbon dioxide concentration*

An approach to finding out if decreased diffusion of $CO_2$ is, or is not, the factor limiting photosynthesis is to measure *A* as a function of the calculated $CO_2$ concentration inside the leaf (the $A/C_i$ function). In this method the intercellular carbon dioxide concentration ($C_i$) is calculated from the measured values of $C_a$, $g_s$ and *A* (Farquhar and Sharkey, 1982). The $A/C_i$ function assesses the rate of photosynthesis as a direct function of the intercellular $CO_2$ available to the mesophyll cells independently of $g_s$. This allows the biochemical effects (or, more strictly, mesophyll effects as physical as well as biochemical processes may be involved) of water deficit to be assessed. In leaves photosynthesising at steady-state in the ambient atmosphere ($C_a$, currently 0.35 mmol mol$^{-1}$), $C_i$ is smaller than $C_a$ due to the finite $g_s$. The $C_i/C_a$ ratio is around 0.8. The photosynthetic rate of $C_3$ leaves in air ($C_i$ = 0.28 mmol mol$^{-1}$) is not generally saturated. Increasing $C_a$ to 0.5–1 mmol mol$^{-1}$ increases $C_i$ and eventually photosynthesis reaches a plateau. Thus, stomata exert control over photosynthetic rate. If $C_a$ is reduced, then $C_i$ decreases and eventually reaches the compensation point where respiratory $CO_2$ production balances photosynthesis: for $C_3$ plants this is about 50 μmol mol$^{-1}$ at 20°C.

Many studies have shown that $A$ decreases progressively at both saturating and sub-saturating $C_i$ as water deficit develops (Bunce, 1988; Gimenez *et al.*, 1992; Gunasekera and Berkowitz, 1993; Johnson *et al.*, 1987; Lawlor, 1976b; Lawlor and Khanna-Chopra, 1984; Matthews and Boyer, 1984; Sharkey and Seemann, 1989). The interpretation is that the $CO_2$ assimilation mechanism is inhibited by processes not dependent upon the maintenance of $C_i$, i.e. biochemical processes are made less effective or damaged in some way. This interpretation has been contested: in particular uncertainty in calculation of $C_i$ is a major stumbling block to interpretation of stress effects. Measurement of small $g_s$ is particularly subject to errors in measuring water vapour fluxes and 'true' leaf temperatures. However, the problem of non-uniform distribution of stomatal closure across leaf surfaces ('patchy' stomatal apertures) during water stress (Farquhar *et al.*, 1987) is of much greater potential importance. The effect of patchiness is to reduce $A$ but not to the same proportion as the calculated $C_i$. This results in an underestimation of $A$ at a given calculated $C_i$. van Kraalingen (1990) provides a review of the literature on the occurrence and effects of patchiness. Non-uniformity of $A$ may be assessed by short-term $^{14}CO_2$ feeding (Gimenez *et al.*, 1992; Sharkey and Seemann, 1989) or measurement of chlorophyll *a* fluorescence and calculation of the non-photochemical quenching. Non-uniform stomatal conductance over the leaf surface has the greatest effect where transfer of $CO_2$ between the sub-stomatal spaces is restricted by veins. The distribution of conductance is dynamic, changing over periods of minutes to hours, and thus difficult to assess. Calculations suggest that the difference between 'closed' and 'open' stomata must be large and boundary layer conductance small to severely distort the calculated $C_i$. Stomatal patchiness is not a universal effect of stress, and it depends on the type of water deficit, rate of application and species. Leaves fed with ABA via the petiole suffer considerable non-uniformity of photosynthesis (judged by short-term measurements with radioactively labelled $CO_2$), greater than is observed for leaves droughted relatively slowly (hours to days) on the plant. Whole leaves show non-uniform $CO_2$ assimilation more clearly than do smaller areas of leaves (Gimenez *et al.*, 1992; Gunasekera and Berkowitz, 1993; Scheuermann *et al.*, 1991; Sharkey and Seemann, 1989).

By increasing $C_a$, it should be possible to overcome stomatal limitations and return $A$ to the unstressed value if metabolic processes are not affected. Very large concentrations of $CO_2$ have been used to overcome the stomatal limitation (Graan and Boyer, 1990) but, in general, with moderate to severe stress, increasing

$C_a$ does not increase $A$ to the unstressed rate. However, oxygen electrode measurements show that photosynthetic $O_2$ evolution increases with very large concentrations of $CO_2$ (Chaves, 1991; Quick *et al.*, 1992). In these studies, $CO_2$ concentrations of up to 0.2 mmol mol$^{-1}$ (20%) were employed to obtain rates of $O_2$ evolution twice those likely for $CO_2$ assimilation. At low (but still substantial) $CO_2$ concentration, $O_2$ evolution was depressed in stressed leaves. It is unlikely that the stimulation of oxygen evolution by large concentrations of $CO_2$ is simply due to overcoming stomatal diffusion limitation. In well-watered leaves ($g_s$ = 0.2 mol m$^{-2}$ s$^{-1}$), a $C_i$ of 0.5 mmol mol$^{-1}$ is needed to saturate photosynthesis. This would require a $C_a$ of 0.6 mmol mol$^{-1}$. In a water-stressed leaf ($g_s$ = 0.02 mol m$^{-2}$ s$^{-1}$), the $C_a$ to maintain $C_i$ at 0.5 mmol mol$^{-1}$ would be 1.5 mmol mol$^{-1}$. In an even more severely stressed leaf ($g_s$ = 0.002 mol m$^{-2}$ s$^{-1}$), $C_a$ would need to be 10.5 mmol mol$^{-1}$ to saturate photosynthesis. This is a very small $g_s$ yet the $C_i$ is only one-twentieth of the $CO_2$ required in the oxygen electrode to achieve saturation. Even quite modest $CO_2$ concentrations of the order of 10 mmol mol$^{-1}$ should be sufficient to overcome the limitation of even closed stomata and the small boundary layer conductance in unstirred oxygen electrode systems and provide the $CO_2$ required for saturation. This analysis suggests that very large $CO_2$ concentrations may have non-physiological effects. Also as $O_2$ evolution is not absolutely coupled to $CO_2$ assimilation, alternative sinks for electrons may, in the short-term, allow continued $O_2$ evolution. Such sinks could be reduction of metabolites as well as the de-epoxidation of violaxanthin in the xanthophyll cycle carotenoid energy dissipating mechanism (Figure 8.4; Section 8.5.5). However, these processes are probably not sufficient as the rates of $O_2$ evolution are double those expected for $CO_2$ assimilation. Despite such considerations, decreased $A$ is regarded as a consequence of small $g_s$. However, Tang *et al.* (1994), using a similar oxygen electrode system, required much smaller $CO_2$ concentrations to saturate $O_2$ evolution and found removal of the epidermis of sunflower leaves did not alter the effects of stress on $A$: they conclude that decreased $A$ results from damage to metabolism and not only decreased $g_s$.

Other approaches, for example analysis of the stable carbon and oxygen isotope exchanges of stressed plants, support the view that $A$ is regulated by $g_s$ because at mild stress $C_i$ does decrease (Thomas and André, 1982). Vassey and Sharkey (1989) concluded that decreased photosynthesis in *Phaseolus vulgaris* was due to decreased $g_s$, despite evidence from the $A/C_i$ curves. In *Lupinus albus*, photosynthesis is not affected by decreasing RWC from 100 to 80%, but below this $A$ decreases almost linearly with decreasing RWC (Chaves, 1991). In other

***Figure 8.2.*** *The effect of exposure to increasing water deficit (leaf water potential, MPa) on water relations, gas exchange and RuBP content of sunflower (*Helianthus annuus L.*) leaves with 0.35 mmol $CO_2$ mol$^{-1}$ air and 1200 mmol m$^{-2}$ s$^{-1}$ photosynthetically active radiation. (a) Osmotic pressure and turgor. (b) Stomatal conductance (mol m$^{-2}$ s$^{-1}$) and intercellular $CO_2$ concentration (µmol mol$^{-1}$). (c) Photosynthetic rate (µmol m$^{-2}$ s$^{-1}$) and ribulose bisphosphate (RuBP, µmol m$^{-2}$) (after Gimenez* et al., *1992; Gimenez and Lawlor, unpublished).*

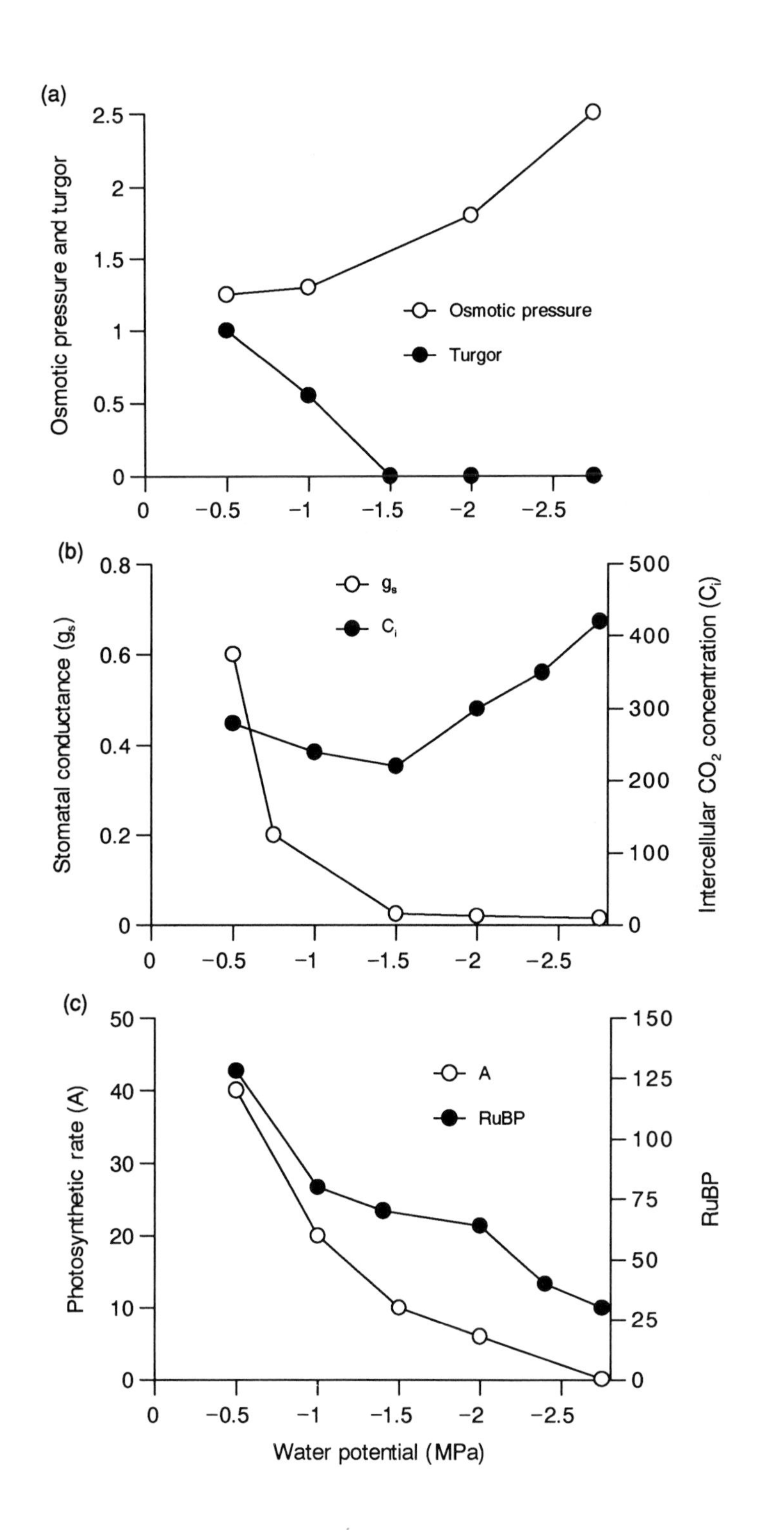
(a)
Osmotic pressure and turgor
2.5
2
1.5
1
0.5
0
0
−0.5
−1
−1.5
−2
−2.5
Osmotic pressure
Turgor
(b)
Stomatal conductance ($g_s$)
0.8
0.6
0.4
0.2
0
$g_s$
$C_i$
Intercellular $CO_2$ concentration ($C_i$)
500
400
300
200
100
0
(c)
Photosynthetic rate (A)
50
40
30
20
10
0
A
RuBP
150
125
100
75
50
25
0
Water potential (MPa)

plants, greater sensitivity of metabolism to water stress is apparent (Cornic *et al.*, 1989). Cornic and Briantais (1991) determined that allocation of electrons to $O_2$ increased as $A$ decreased with mild water stress (35% leaf water deficit), indicating that the $C_i$ decreased, despite a constant calculated $C_i$: they attributed the discrepancy to heterogeneity in the stomatal conductance distribution, supporting earlier, similar conclusions (Cornic *et al.*, 1989). Thus metabolism seems unaffected by mild water deficit (in agreement with Sharkey, 1990) until fairly small RWC and low water potentials are reached.

In contrast to the view that $g_s$ regulates $A$, a number of analyses show that metabolism is affected by water deficit. Increasing $C_a$ does not increase $A$, despite the fact that doubling $C_a$ at constant, if small, $g_s$ should increase $C_i$ and thus stimulate $A$ (Matthews and Boyer, 1984), even when patchy stomatal distribution was not considered a factor (Gimenez *et al.*, 1992; Gunesekera and Berkowitz, 1993). Thus, non-uniform stomatal closure may not affect the $A/C_i$ relationship significantly to negate the evidence that stress inhibits photosynthetic metabolism. Doubts about the validity of the oxygen electrode studies have been raised above. Such variation in observations and interpretations of the effects of water deficit on photosynthesis requires more critical analysis.

### 8.3.3 *Carbon dioxide compensation concentration*

Analysis of $CO_2$ compensation concentration indicates the relative changes in photosynthesis and respiration in stressed leaves. Generally, respiration increases with decreasing water potential; for example, measured compensation concentration of wheat leaves enclosed in an illuminated chamber were as expected for unstressed $C_3$ plants (50 μmol $mol^{-1}$), but water deficit (caused by application of polyethylene glycol) increased the compensation concentration gradually until a water potential of −1.5 MPa was reached. Below this the concentration increased to greater than 300 μmol $mol^{-1}$ (Lawlor, 1976b). This $CO_2$ is derived from increased respiration relative to photosynthesis. In unstressed or mildly stressed leaves, $CO_2$ is mainly derived from photorespiration as the compensation concentration is reduced to a few μmol $CO_2$ $mol^{-1}$ by decreasing the atmospheric oxygen from 0.21 to 0.02 mol $mol^{-1}$. However, the marked increase in compensation concentration at low water potential is not prevented by low oxygen, so 'dark' respiration (tricarboxylic acid (TCA) cycle or oxidative pentose phosphate pathway) occurring in the light is the probable cause (Lawlor, 1976b; Section 8.5.11). However, there is evidence of both decreased and increased $C_i$ during the transition from unstressed to severely stressed state. As turgor falls and $g_s$ is decreased so $C_i$ decreases. Below about −1.5 MPa water potential, the turgor loss point, $C_i$ rises to approach or exceed $C_a$. Figure 8.2 shows the changes in $A$, $C_i$ and $g_s$ of sunflower leaves as a consequence of decreasing water potential. Luo (1991) and Johnson *et al.* (1987) observe similar effects of stress on $C_i$ in *Abutilon* and wheat, respectively. Sharkey and Seemann (1989) showed a similar trend: $C_i$ drops with mild stress and rises to exceed control $C_i$ in severely stressed leaves.

### 8.3.4 *Relative control of photosynthesis by stomata and mesophyll*

Accepting that the $A/C_i$ relationship is valid, the importance of the two components may be assessed by taking the saturated rates of $A$ to be the metabolic potential and relating $A$ measured at a given $g_s$ to it (Matthews and Boyer, 1984; see Assmann, 1988 for detailed analysis). As an example, in unstressed sunflower leaves the effect of $g_s$ in decreasing photosynthesis was about 15% with no effect of stress on metabolism, so by definition all the reduction was stomatal. In moderately stressed leaves (–1.2 MPa), $g_s$ reduced photosynthesis under stress by 30% from the $CO_2$-saturated value and contributed about 40% to the total reduction in photosynthesis from the unstressed $CO_2$-saturated value. However, in severely stressed leaves (–2.6 MPa), $g_s$ reduced photosynthesis by 20% but was responsible for less than 10% of the total decrease in photosynthesis compared to the unstressed controls (Gimenez *et al.*, 1992). Therefore, the regulation of photosynthesis by $g_s$ under stress conditions is variable, decreasing with increasing water stress as metabolism is progressively inhibited. This assessment agrees with much of the evidence: it is also to be expected that, as conditions in the cell depart further from normal, then the adverse effects on photosynthetic metabolism will become greater. Thus control of $A$ by $g_s$ and mesophyll processes is more complex than a 'yes' or 'no' answer allows.

## 8.4 Conditions in chloroplasts under water deficit

Chloroplasts from unstressed leaves contain a range of ions, often at a considerable concentration (e.g. $Mg^{2+}$ and $PO_4^{3-}$), and undergo large changes in ionic composition during the light to dark transitions. Stromal $Mg^{2+}$ concentration increases from around 6 mM in the dark to 13 mM in the light and is important in the activation of ribulose-1,5-bisphosphate carboxylase-oxygenase (Rubisco) and coupling factor (CF). Phosphate is important because it is required for ATP synthesis and is imported into the chloroplasts in exchange for triose phosphate (TP) export (Lawlor, 1993). Dehydration of chloroplasts in water-stressed leaves may therefore alter the concentration of regulatory ions, for example $Mg^{2+}$, which has been implicated in the inhibition of CF (Berkowitz and Wahlen, 1985; Boyer *et al.*, 1987; Pier and Berkowitz, 1989; Section 8.5.6). Differences in the effects of ions and neutral solutes on metabolism are well established. Starch degradation and synthesis of 3-phosphoglycerate (PGA) from dihydroxyacetone phosphate were inhibited by an osmotic potential of –4 to –5 MPa, produced with neutral solutes, but by –1.5 to –2 MPa with ionic solutes (Kaiser and Heber, 1981). Also sulphate and phosphate ions have been implicated by *in vitro* studies of the inhibition of enzyme activities. Kaiser *et al.* (1986) demonstrated, in media simulating the composition of the stroma, that enzyme activities were inhibited by the concerted increase in solute concentration (and correlated with the *in vivo* inhibition of photosynthesis). Divalent inorganic ions such as sulphate and phosphate were most inhibitory. Rubisco was inhibited by sulphate and phosphate which competed with ribulose bisphosphate (RuBP) but not $CO_2$. The effects of reduced chloroplast volume on processes may, thus, be relatively unspecific although more specific sites of action of given ions are possible, for example the role of $Mg^{2+}$ in regulating photophosphorylation.

## 8.5 The effects of water deficit on the partial processes of photosynthesis

### 8.5.1 *Introduction*

The consensus scheme of photosynthetic metabolism shown in Figure 8.3 provides a framework for assessing the effects of water deficit on the partial processes:

(i) Light capture by chlorophyll–protein antenna complexes in thylakoids, energy transfer to core antennae of photosystems I and II (PSI and PSII), charge separation at reaction centres and reduction of primary acceptors of electrons ($Q_A$ and $Q_B$).
(ii) Electron transport from PSII to primary acceptor $Q_A$ plastoquinone and on to PSI and reduction of $NADP^+$ and ferredoxin.
(iii) Water splitting and release of $O_2$ in thylakoids related to (i) and (ii).
(iv) Generation of the proton gradient ($\Delta pH$) in thylakoids and photophosphorylation of ADP giving ATP.
(v) Utilization of ATP and NADPH in $CO_2$ assimilation by the Calvin cycle.
(vi) Carbohydrate (sucrose and starch) synthesis and metabolism.

### 8.5.2 *Light capture, energy transduction, photosystem function and electron transport chain*

Water stress frequently decreases the number of photons captured by leaves simply because wilted leaves are at a steeper angle to the sun's rays. Changes in absorption characteristics of leaves occur due to cell shrinkage. Initial stress may increase chlorophyll concentration but prolonged stress causes pigment loss and senescence. However, the changes at the chloroplast and thylakoid level (light capture and energy transfer in the antenna or to the reaction centres) are relatively small. Genty *et al.* (1987) did not observe changes in photon capture or energy distribution between the two photosytems in cotton. The fluorescence component *F*o remains relatively constant, showing that stress does not alter the pigment bed substantially. Neither PSII photochemistry nor photoinhibition of PSII or electron flow mediated by PSII are greatly affected by mild stress (Genty *et al.*, 1987). There are indications that more severe stress damages both photosystems (Genty *et al.*, 1987; Havaux *et al.*, 1987; Meyer and de Kouchovsky, 1993). Meyer and de Kouchovsky (1993) observed that severe stress decreased maximum rates of PSII, possibly caused by loss of active PSII centres, although the antenna and proteins of the complexes are not damaged by such severe stress. However, there is a suggestion that the oxidizing side of PSII is damaged before other processes are affected. This may be due to photoinhibition (see Baker and

Bowyer, 1994) rather than a direct water stress-related damage. This would be expected as stress increased the proportion of reduced $Q_A$ (but did not lead to complete reduction) even at low light, and slowed the rate of re-oxidation of plastoquinone. Havaux *et al.* (1987) analysed $O_2$ and heat production of water-stressed tobacco by the photoacoustic method. With slowly developing, mild to moderate stress the efficiency of photon capture by PSII was reduced by stress (in contrast to Genty *et al.*, 1987). Further dehydration damaged both photosystems and more excitation energy went to PSII.

Electron transport to the normal physiological acceptors has not been well examined in water-stressed tissues. There is evidence that the pools of reduced pyridine nucleotides are little affected (NADPH) or increased (NADH) in stressed leaves and that $NADP^+$ increased and then decreased with progressive stress and $NAD^+$ decreased, suggesting that both the supply of electrons for reduction of acceptors and the supply of oxidized nucleotides is adequate and that the process is efficient even under severe stress (Lawlor and Khanna-Chopra, 1984; Stuhlfauth *et al.*, 1991; see Section 8.5.8). Other redox components between $Q_A$, the primary electron acceptor of PSII, and $NADP^+$ are presumably also maintained in a reduced state, together with ferredoxin. There is good evidence that a larger proportion of the electron flux is diverted from $CO_2$ assimilation to $O_2$ reduction under stressed conditions with low $C_i$. From measurements of fluorescence parameters and calculation of the quantum yield of electron flow, $(Fm-Fs)/Fm$), Cornic and Briantais (1991) concluded that the allocation of electrons to $O_2$ increased, particularly at more severe stress (Cornic, 1994), as a consequence of the decrease in $C_i$ and $g_s$ (in agreement with $^{18}O_2$ uptake relative to $CO_2$ uptake measured by Thomas and André (1982)).

Electron transport therefore is considered to continue at a considerable rate even in severely stressed tissues and the electrons are passed to the normal physiological acceptors $NADP^+$ and ferredoxin (which may however be kept in an abnormally high reduction state) and to alternative acceptors such as oxygen (Mehler reaction) and eventually used in photorespiration.

### 8.5.3 *Generation of active oxygen*

The reduction of oxygen in the Mehler reaction mentioned above leads to the formation of superoxide and hydrogen peroxide (Figure 8.3). These reactive forms of oxygen, collectively known as active oxygen, are potentially dangerous. Additionally, absorption of excess light by chlorophyll results in formation of triplet chlorophyll which can then pass excitation energy to oxygen resulting in formation of singlet oxygen. This is more reactive than ground-state oxygen and can lead to peroxidation and breakdown of thylakoid lipids. Limitation of carbon dioxide fixation by water deficit appears to increase formation of active oxygen and is a potential means of damage (Björkman and Demmig-Adams, 1994; Pell and Steffen, 1991; Smirnoff, 1993). The role of active oxygen in plants exposed to water deficit and high light intensity and the nature of the antioxidant mechanisms involved in protection are discussed in Chapter 12.

### 8.5.4 *Water splitting*

There is little evidence that water splitting is inhibited by water stress in the physiological range. Cornic *et al.* (1989) observed no inhibition of $O_2$ evolution and PSII activity for mildly stressed tissue. Rates of $O_2$ evolution by stressed leaves have been measured in oxygen electrodes under elevated $CO_2$, and in some cases a decrease has been found (Havaux *et al.*, 1987; Lawlor and Khanna-Chopra, 1984). However, generally inhibition was much less than experienced by $CO_2$ assimilation. Water stress did not inhibit $O_2$ evolution at all in some studies (Chaves, 1991; Quick *et al.*, 1992; Section 8.3.2), leading to the effects of stress on $CO_2$ assimilation being ascribed to stomatal diffusion limitation.

### 8.5.5 *Energy quenching in chloroplasts*

If the energy of photons captured is not used rapidly enough to prevent accumulation of high energy states of chlorophyll then several routes of dissipation are possible. One is chlorophyll *a* fluorescence, energy re-emitted as photons from the chlorophyll matrix associated with PSII. The energy emitted as fluorescence is small but easily measured and fluorescence is an important indicator of thylakoid energetics in light–dark transitions and with changing actinic light (Björkman and Demmig-Adams, 1994). An explanation of the measurement and interpretation of chlorophyll fluorescence is provided by Björkman and Demmig-Adams (1994) and the references therein. Briefly, fluorescence is stimulated by applying a beam of weak modulated light to the leaf and measuring the fluorescence wavelengths emitted with a suitable detector. This enables the effects of normal, actinic light on photosynthesis to be analysed. The main fluorescence parameters measured are the minimum fluorescence *Fo*, and the maximum, *Fm*, of dark-adapted leaves and *Fo*' and *Fm*', the superscript ' denoting measurements made with actinic light. *Fo* is obtained when all the reaction

---

***Figure 8.3.*** *A diagrammatic scheme of photosynthetic processes in the thylakoid and its environment of chloroplast stroma and lumen illustrating the interrelationships between partial processes of photosynthesis (see Section 8.5.1). This scheme is used as a framework for discussion of the effects of water stress on photosynthesis. Under water-stressed conditions, the inhibition of $CO_2$ assimilation, the main energy-demanding process under unstressed conditions, causes accumulation of energy in the light-harvesting and energy transfer systems with increased reduction state of intermediates and products. Inhibition of ATP synthesis may also exacerbate the effects of decreased $CO_2$ supply. The excess energy and reductant are dissipated by the interconversion of the xanthophyll cycle pigments zeaxanthin, antheraxanthin and violaxanthin in an epoxidation–de-epoxidation cycle. This is shown in more detail in Figure 8.4. Abbreviations: CF0 and CF1, ATP synthase subunits; $Fd_{red/ox}$, reduced/oxidized ferredoxin; RuBP, ribulose bisphosphate; MDHA, monodehydroascorbate, SOD, superoxide dismutase; AP, ascorbate peroxidase. Further details of the role and regeneration of ascorbate are given in Chapter 12.*

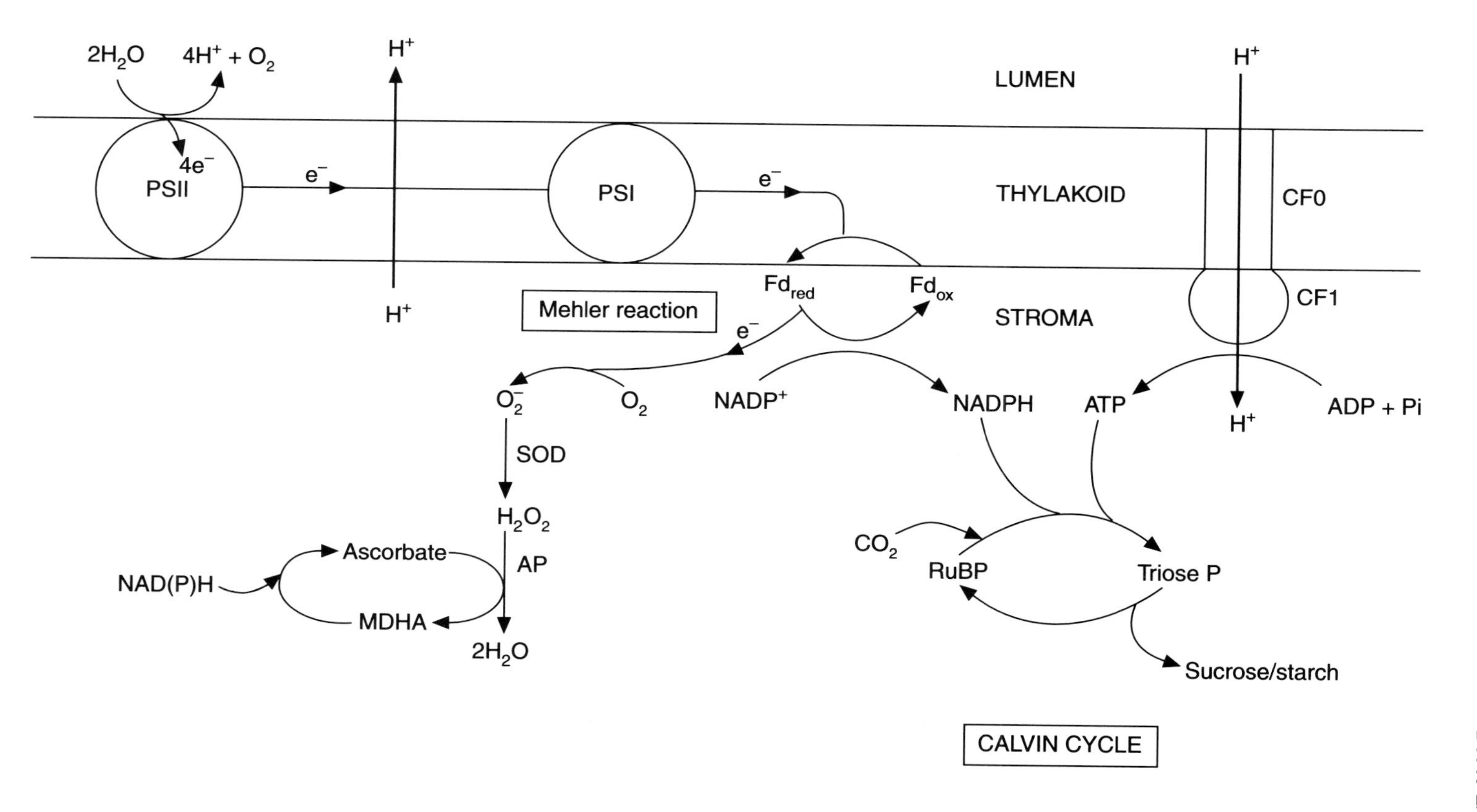

$2H_2O$
$4H^+ + O_2$
$H^+$
$H^+$
LUMEN
$4e^-$
PSII
$e^-$
PSI
$e^-$
THYLAKOID
CF0
$H^+$
Mehler reaction
$Fd_{red}$
$Fd_{ox}$
STROMA
CF1
$e^-$
$O_2^-$
$O_2$
$NADP^+$
NADPH
ATP
$H^+$
ADP + Pi
SOD
$H_2O_2$
AP
Ascorbate
NAD(P)H
MDHA
$2H_2O$
$CO_2$
RuBP
Triose P
Sucrose/starch
CALVIN CYCLE

centres of PSII are in the oxidized, open state (when excitation of PSII is minimal and the electron transport chain is drained by exciting PSI with far-red light). *Fm* and *Fm*' are measured when all the reaction centres are reduced or closed by a saturating pulse of intense light and fluorescence is maximum, the condition when quenching by non-radiative dissipation (NRD) is minimal. Variable fluorescence, *Fv* = *Fm* –*Fo*, or *Fv*' = *Fm*' –*Fo*', and the ratio *Fv*/*Fm* indicate the reduction state of the acceptors of electrons from PSII. A small *Fv*/*Fm* indicates a state in which electrons are not efficiently consumed and the primary quinone acceptor of electrons from PSII, Q, is reduced. The efficiency with which PSII converts energy is given by $\phi_{PSII}'$ = (*Fm*' –*F*)/*Fm*' under normal light conditions: when all reaction centres are open $\phi_{PSII}'$ = (*Fm*' –*Fo*')/*Fm*' = *Fv*'/*Fm*'. The ratio *Fv*/*Fm* = $\phi_{PSII}$ is the maximum efficiency of PSII when all centres are open and NRD is zero. It is also possible to estimate the fraction of the reaction centres of PSII in the oxidized state, $R_{ox}/R_{total}$, from (*Fm*' –*F*)/(*Fm*' –*Fo*') and in the reduced condition from 1 –$R_{ox}/R_{total}$.

When darkened leaves are illuminated, *Fm* decreases to *Fm*' due to NRD: this may be calculated from (*Fm* –*Fm*')/*Fm*' = *Fm*/*Fm*' –1. This quenching of fluorescence due to dissipation as heat, that is by non-photochemical quenching ($q_{NP}$), probably involves carotenoids in the pigment bed of the antennae and reaction centres. The quenching due to photochemistry is $q_P$ = (*Fm* –*Fs*)/(*Fm*' –*Fo*') = (*Fm*' –*Fv*)/*Fv*' and is related to the assimilation of $CO_2$ and nitrate ions and hence to the oxidation and reduction state of Q. Under normal conditions of illumination and with environmental conditions which allow full expression of photosynthesis, the energy captured in the light-harvesting antennae is used predominantly in photochemistry ($q_P$), with $CO_2$ assimilation (particularly under photorespiratory conditions) the largest sink in $C_3$ plants. However, as these processes decrease so $q_{NP}$ increases.

The efficiency of energy conversion is not affected by mild stress but decreases with severe stress, particularly if the thylakoids have been subjected to bright light which causes photoinhibition of the photosystems. Decreased *Fv*/*Fm* and *Fv*'/*Fm*' reflect this. Closure of the reaction centres may increase with increasing stress, particularly in bright light. There are probably large differences between species and within species resulting from differences in growth conditions and exposure to bright light and water deficit. Generally there is little change in $q_P$ with increasing stress, as energy is transferred to $O_2$ in the absence of $CO_2$, but $q_{NP}$ increases greatly in bright light and with stress. Water stress decreased $q_P$ in bean but in sunflower it increased in 0.2 mol $mol^{-1}$ $O_2$ (photorespiratory conditions) but decreased with stress in 0.02 mol $mol^{-1}$ $O_2$ (low photorespiration), indicating differences in response to photorespiration and in the state of $Q_A$ (Scheuerman *et al.*, 1991; Stuhlfauth *et al.*, 1988). Also, $q_P$ and $q_{NP}$ were unaffected by stress over the water potential range –0.7 to –2.5 MPa during steady-state photosynthesis in *Digitalis lanata*, although in the dark to light transition $q_{NP}$ increased with stress, indicating increased thylakoid membrane energization possibly as a consequence of inhibition of ATP generation or of ATP consumption (Stuhlfauth *et al.*, 1988). Other studies show that $q_{NP}$ increases substantially with severe stress at the point of turgor loss and below (Lawlor and Herrera, unpublished).

Energy dissipation by $q_{NP}$ has several components (Gilmore and Yamamoto, 1991). Energy-unrelated quenching is related to photoinhibition (it is not further considered here although of importance under stress conditions). Energy-related quenching ($q_E$), determined by the transthylakoid proton gradient (ΔpH), is rapidly reversible and correlated with the balance between ATP generation and consumption. Reduced ATP consumption increases the gradient and causes electrons to 'back up' in the electron transport system and energy to accumulate in the antennae. To avoid this, electrons are transferred to $O_2$ in the Mehler–peroxidase reaction, which uses NADPH but no ATP (Section 8.5.4 and Chapter 12). The increased ΔpH probably induces changes in the thylakoid membrane, perhaps by altering the spatial arrangement of pigments such as chlorophyll and carotenoids, allowing increased loss of energy as heat. This is a primary route of energy quenching by $q_E$. The carotenoids function both in energy dissipation and to detoxify chlorophyll triplets and singlet oxygen (Goodwin, 1980; Owens, 1994).

The xanthophylls, oxygen-containing carotenoids, are considered to play a major role in energy regulation in the xanthophyll cycle within, or associated with, the light-harvesting complexes of the thylakoids (Figure 8.4; Adams and Demmig-Adams, 1993; Björkman and Demmig-Adams, 1994). The xanthophyll cycle involves the conversion of violaxanthin, via antheraxanthin to zeaxanthin, by the activity of violaxanthin de-epoxidase which requires acid conditions (pH 5.3) in the thylakoid lumen and so is active in bright light. This removal of its oxygen increases the number of double bonds from 9 to 11. Owens (1994) suggests that protonation of the thylakoids and light-harvesting complexes may alter the environment and contact (e.g. spectral overlap and dipole character of the molecular transitions or aggregation state of light-harvesting complexes) between chlorophyll and zeaxanthin so increasing the rate of energy transfer from chlorophyll to zeaxanthin. Specifically, the ΔpH lowers the energy of the first singlet electronic state of zeaxanthin to close to that of chlorophyll singlet state, thus allowing new routes for energy dissipation. The rate of thermal dissipation from zeaxanthin is two orders of magnitude faster than chlorophyll so it can quench the excited chlorophyll.

Epoxidation occurs in the thylakoid stroma (pH 7.2) and is stimulated in dim light; the epoxidation and de-epoxidation reactions proceed simultaneously but at very different rates. Together they permit a cyclic consumption of electrons and $H^+$ thus providing a pH-dependent fluorescence quenching mechanism. When NADPH synthesis exceeds demand and ATP is limiting, then the xanthophyll cycle may regulate the balance, triggered by ΔpH. There is a strong correlation between the increased synthesis of zeaxanthin and $q_{NP}$ under conditions such as increased temperature and photon flux and the maximum capacity for $q_{NP}$ is related to the total size of the zeaxanthin, antheraxanthin and violaxanthin pool. Zeaxanthin increases whilst violaxanthin decreases during illumination, the process usually reversing in the dark. In droughted plants the xanthophylls increase and stress causes the zeaxanthin to increase relative to violaxanthin (Young and Britton, 1990). Possibly, under such conditions, the capacity of the system to remove energy would become limited by the rate of transfer and the number of zeaxanthin molecules available.

In general, $q_{NP}$, and particularly the $q_E$ component, increases substantially with increasing stress (water potential –1.2 to –1.5 MPa, RWC below 85%) and correlates with the loss of $CO_2$ assimilation. The increased $q$NP and $q_E$ would protect the photosynthetic system against photoinduced damage but only for as long as energy load is below the capacity of the xanthophyll cycle. If breakdown of the xanthophylls as a result of energy overload occurs, then re-synthesis is required. There seems to be no information about this in stressed leaves. Clearly, if photophosphorylation is inhibited (see Section 8.5.6), or indeed any other process which prevents the consumption of thylakoid energy, the xanthophyll cycle may provide only partial protection against damage. In isolated chloroplasts, Laasch (1987) observed a greater proportion of $q_{NP}$ than $q_E$ with increasing photon flux, a linear dependence of $q_E$ on $\Delta$pH with increasing photon flux and, under osmotic stress in low light, $q_E$ increased with increasing $\Delta$pH, whereas in bright light $q_E$ was decreased. He concluded that $q_E$ is regulated both by $\Delta$pH and external factors. Probably xanthophyll synthesis contributes to the regulation of energy in thylakoids. There are differences in the capacity of the systems to quench fluorescence in different species and within the same species grown in different conditions (Björkman and Demmig-Adams, 1994; Scheuermann *et al.*, 1991; Stuhlfauth *et al.*, 1988, 1991).

Increased $q_{NP}$ correlates with conditions where many of the stress-induced changes in photosynthesising cells occur and the normal mechanisms for energy consumption cannot deal with the energy loads, so that increased dissipation by alternative systems results. As the synthesis of ABA is now regarded as a consequence of the breakdown of the xanthophyll neoxanthin (Figure 8.4; Parry,

---

***Figure 8.4.*** *The xanthophyll cycle in thylakoids and its relation to the reductant (NADPH) generated in the chloroplast. Epoxidation of the xanthophyll cycle pigments (violaxanthin, antheroxanthin and zeaxanthin) uses stromal NADPH. The de-epoxidation reactions use ascorbate as reductant but it is not known if ascorbate is regenerated from the oxidized forms of ascorbate (MDHA and DHA) in the lumen or if DHA is transported to the stroma for reduction by GSH. The latter option is illustrated by dotted lines. The combined system serves to regulate the reductant and energy status of the chloroplast, but under extreme energy loads with much decreased* $CO_2$ *assimilation the system is unable to dissipate the energy. Under these conditions it is possible that the xanthophylls in the thylakoid xanthophyll cycle are partly converted to neoxanthin and broken down to form xanthoxin and ultimately abscisic acid (ABA). See Chapter 12 for further discussion of this system. Abbreviations: DHA, dehydroascorbate; DHAR, dehydroascorbate reductase; GR, glutathione reductase; GSH, glutathione; GSSG, oxidized glutathione; MDHAR, monodehydroascorbate; VE, violaxanthin epoxidase; VDE, violaxanthin de-epoxidase;* $\Delta pH$*, transthylakoid pH gradient.*

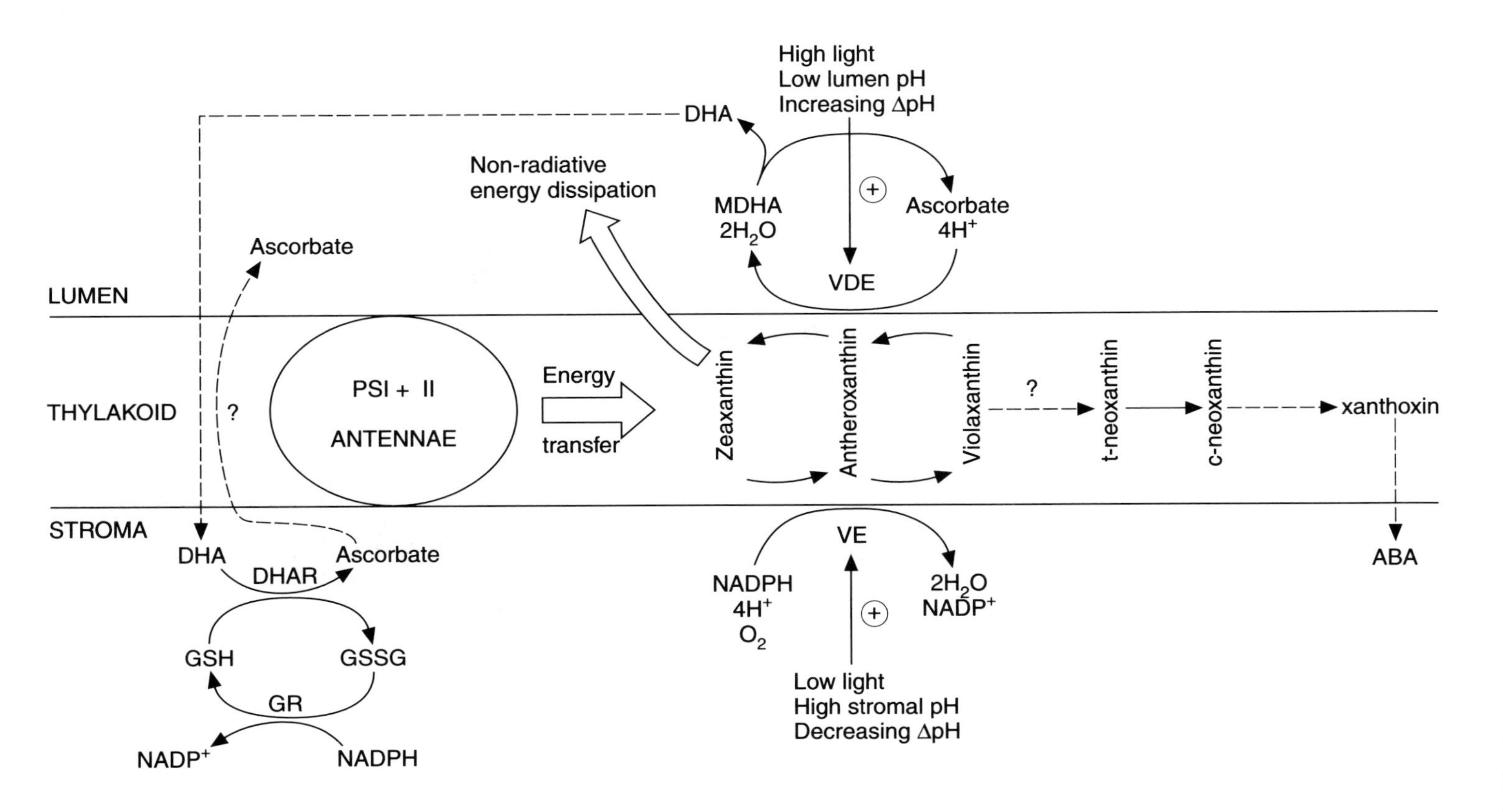
High light
Low lumen pH
Increasing ΔpH
DHA
Non-radiative
energy dissipation
MDHA
2H2O
Ascorbate
4H+
VDE
Ascorbate
LUMEN
THYLAKOID
?
PSI + II
ANTENNAE
Energy
transfer
Zeaxanthin
Antheroxanthin
Violaxanthin
?
t-neoxanthin
c-neoxanthin
xanthoxin
STROMA
DHA
DHAR
Ascorbate
VE
NADPH
4H+
O2
2H2O
NADP+
ABA
GSH
GSSG
GR
NADP+
NADPH
Low light
High stromal pH
Decreasing ΔpH

1993; Zeevart *et al.*, 1991) and ABA accumulates when turgor is lost (Pierce and Raschke, 1980), it suggests that there is a link between excess energy in thylakoids, overload of the carotenoid dissipation mechanism and their breakdown to form xanthoxin, the precursor for ABA synthesis. As the alternative systems are not complementary, energy 'backs up' in the electron transport chain, photosystems and antennae, and is seen as increased fluorescence. Large $q_{NP}$ and $q_E$ reflect an increase in the ΔpH of thylakoids as a consequence of the continuation of electron transport but inhibition of $CO_2$ assimilation.

### 8.5.6 *Photophosphorylation*

In photosynthesising tissues, photophosphorylation, the light-driven synthesis of ATP from ADP and inorganic phosphate (Pi), is crucial to the rate of $CO_2$ assimilation. Ultimately it defines the energy economy of the plant. Photophosphorylation under water-stressed conditions is therefore of great importance but poorly understood. A simple view is that water deficit does not cause decreased rates of photophosphorylation or loss of ATP (Ort *et al.*, 1994; Sharkey and Seemann, 1989; Wise *et al.*, 1990, 1992) and that it is therefore not a primary lesion. Alternatively it has been suggested that water deficit decreases the rate of photophosphorylation and the content of ATP (Boyer *et al.*, 1987). Lawlor and Khanna-Chopra (1984) showed substantial loss of ATP and decreased energy charge during water stress. Stuhlfauth *et al.* (1988, 1991) concluded, from fluorescence measurements, that changes in membrane energetics reflected damage to ATP-generating or -consuming reactions; ATP content decreased but energy charge slightly increased with stress. However, Sharkey and Badger (1982) and Sharkey and Seemann (1989) observed no change in ATP with stress.

Until recently, photophosphorylation was considered to decrease in stressed leaves as a result of stress-induced changes in the structure and function of thylakoid membranes which are not caused by structural degradation (Boyer *et al.*, 1987). The ability of chloroplast coupling factor (CF0–CF1, the ATP synthase attached to the thylakoid membrane) to bind ADP thus allowing photophosphorylation to proceed is impaired in dehydrated chloroplasts (Younis *et al.*, 1979). The cause of the damage to CF1 is increased ionic concentration, principally $Mg^{2+}$ ions, in the dehydrated chloroplast stroma. The effects of water stress on CF protein conformation and ATPase activity could be mimicked by $Mg^{2+}$. Also the recovery from stress was slow, as it is in stressed leaves. However, this conclusion has been questioned more recently by measurements of electrochromic changes *in vivo* which suggest that the non-stomatal limitation of photosynthesis in stressed leaves cannot be ascribed to the decreased efficiency of CF1 (Ortiz-Lopez *et al.*, 1991; Wise *et al.*, 1990).

However, other assessments of the role of CF0–CF1 do suggest that the CF0–CF1 activity is decreased by water deficits. Havaux *et al.* (1987) observed that photochemical energy storage (which reflects the state of a reaction not directly linked to whole-chain electron transport, for example cyclic photophosphorylation) is decreased by more than half indicating loss of photophosphorylation capacity under drought. Meyer *et al.* (1992), de Kouchkowski and Meyer (1992) and Meyer and de Kouchkowsky (1992) measured ΔpH and photophosphorylation in isolated chloroplasts; they observed that thylakoids from dehydrated tissue had lower rates of ATP synthesis for a given ΔpH than those from well watered tissues and the same was observed for ATP hydrolysis by CF1. However, artefacts due to chloroplast isolation must be considered. There was no effect of stress on the amount of CF1–CF0 or on the proton conductance of CF0. ΔpH did decrease with stress, although this was not correlated with damage to the thylakoid membranes or changes in permeability. However, it is possible that some of the CF units on the membranes are inactivated and others function normally. The cause of the decreased ATP synthesis is unexplained.

To conclude, there is good evidence of decreased ATP content and energy charge in stressed tissues and of inhibition of photophosphorylation by damage to CF in relatively rapidly and severely stressed plants, but recent *in vivo* studies on field-grown plants have contradicted this. The difference may reflect the operation of defence mechanisms protecting ATP synthesis during slow stress development: this would be expected if ATP synthesis is indeed very susceptible to stress development. The importance of the process makes it imperative that the effects of stress on photophosphorylation and on the ATP content and energy charge should be clarified as soon as possible.

### 8.5.7 *Calvin cycle activity*

As the central process in photosynthetic carbon metabolism, and the major process using the ATP and NADPH synthesised as a consequence of energy captured in the photosynthetic light reactions, the behaviour of this autocatalytic cycle is important for understanding the decrease in photosynthetic rate under water stress. The cycle is highly regulated, with many feedforward and feedback controls on the constituent enzymes, so identifying the sites of action is difficult. In common with other areas of water stress physiology and biochemistry, no clear understanding of the direct or indirect effects of stress on the response of the enzyme reactions responsible for $CO_2$ assimilation exists. Here only a few aspects of the problem are considered, concentrating mainly on the roles of Rubisco activity and the supply of its substrate RuBP.

Decreased $g_s$ and small $A$ will decrease the flux of $CO_2$ to the chloroplast and the flux of carbon through the Calvin cycle. Thus, even without direct stress effects on metabolism, carbon fluxes will be altered. Although $C_i$ may decrease, the $O_2$ concentration in the intercellular spaces (and we may assume in the chloroplast stroma) is not altered, so the $O_2/CO_2$ ratio increases and oxygenation of RuBP by Rubisco will increase relative to RuBP carboxylation. This increases the flux of carbon through the photorespiratory glycolate pathway relative to that via the phosphate translocator to sucrose. Glycine and serine synthesis associated with photorespiratory glycolate metabolism will increase (Lawlor and Fock, 1977a,b) as will the flux of $NH_4^+$ in the photorespiratory nitrogen cycle. Rubisco kinetics show that glycolate pathway carbon flux rises by

60% (relative to total flux) for a $C_i$ of 180 compared to a $C_i$ of 270 µmol mol$^{-1}$, although the absolute flux decreases (Lawlor and Keys, 1993). The energy-consuming functions of photorespiration versus $CO_2$ assimilation depend on the absolute fluxes but photorespiratory $CO_2$ recycling in stressed leaves is likely to be of major importance (André and Du Cloux, 1993; Becker *et al.*, 1986).

Water stress decreases the pools of metabolites in the Calvin cycle (PGA and RuBP; Figure 8.2) and products (sucrose and starch) (Krampitz *et al.*, 1984; Lawlor and Fock, 1977b). In general, mild water deficits have little effect on metabolism (Quick *et al.*, 1989) but greater deficits have more substantial effects (Sharkey and Seemann, 1989). Of major importance is the decrease in RuBP content, even when $C_i$ is maintained (Gimenez *et al.*, 1992; Gunesakera and Berkowitz, 1993). Gimenez *et al.* (1992) suggest that inhibition of $CO_2$ fixation is related to the decreased RuBP pool (Figure 8.5), not to the supply of $CO_2$, and so factors regulating the RuBP pool are prime candidates for the limitation of photosynthesis under stress. A marked sigmoidal relationship between photosynthesis and the RuBP content of the tissue at constant $C_i$ was observed (Figure 8.5; Gimenez *et al.*, 1992). The $K_m$ of the enzyme for RuBP is 25–40 mmol m$^{-3}$ and the estimated concentration of RuBP is 8 mol m$^{-3}$ in unstressed leaves decreasing to 1 mol m$^{-3}$ in stressed leaves. The study by Gimenez *et al.* (1992)

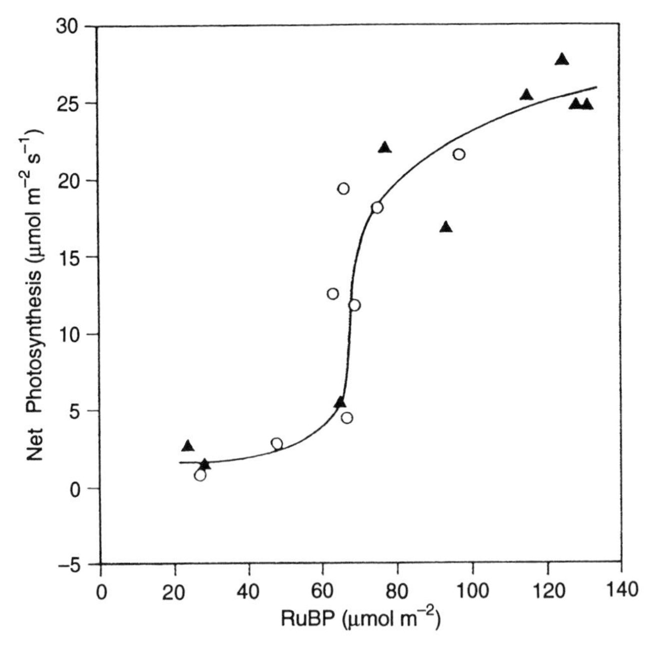


***Figure 8.5.*** *The relationship between photosynthesis and ribulose bisphosphate (RuBP) content of water-stressed sunflower (*Helianthus annuus L.*) leaves. The measurements were made during a period of increasing water deficit on attached leaves under constant internal $CO_2$ concentration. Reproduced from Gimenez* et al. *(1992) with permission from the American Society of Plant Physiologists.*

shows that the number of RuBP molecules per active site of Rubisco was about 2 at the highest rates of photosynthesis, dropping to about 1 at the lowest rate. As Rubisco is inefficient at obtaining RuBP in the conditions prevailing in the chloroplast, a large concentration of RuBP is necessary for rapid $CO_2$ assimilation, probably to maintain flux to the active sites in the protein gel of the stroma, so any inhibition of RuBP synthesis would inhibit assimilation. Mild water stress alters the fluxes of carbon through the Calvin cycle in a manner analogous to that of decreased $CO_2$ supply, but Lawlor and Pearlman (1981) were unable to model the effect of severe stress on photosynthesis and metabolites from Rubisco kinetics. If the limitation were caused by damage or loss of Rubisco, then it might be expected that RuBP would accumulate, which is not the case.

Rubisco is one of the best studied Calvin cycle enzymes in water-stressed leaves and there is a degree of agreement on its response. Stress does not substantially affect the $k_{cat}$ of the enzyme extracted rapidly and measured *in vitro*, its carbamylation state, the amount of Rubisco per unit area or its proportion to total soluble protein (Gimenez *et al.*, 1992; Sharkey and Seemann, 1989). In tobacco with different amounts of Rubisco achieved by transformation with antisense DNA techniques, the amount of the enzyme had no effect on the photosynthetic rate or the sensitivity of photosynthesis to stress nor was the activation state of the enzyme altered by stress (Gunasekera and Berkowitz, 1993). However, evidence for decreased Rubisco should not be ignored (Castrillo and Calcagno, 1989; Chaves, 1991). The type of regulation which might be expected with Rubisco, e.g. the loss of activity associated with tight binding inhibitors such as carboxyarabinitol 1-phosphate or the failure of the Rubisco activase to operate in stressed leaves, would seem not to be important. The analysis suggests that the effects of stress on $CO_2$ assimilation are operating through the RuBP pool and RuBP synthesis not via Rubisco activity.

Other enzymes of the Calvin cycle have not been analysed as extensively as Rubisco. Fructose-1,6-bisphosphatase (FBPase) has been suggested as a likely site of inhibition. The increased acidification of the chloroplast stroma which occurs with drought could inhibit the enzyme (Berkowitz and Gibbs, 1983; Berkowitz and Wahlen, 1985). This would decrease RuBP synthesis, but the evidence suggests that mild stress does not affect the activity of FBPase (Vassey and Sharkey, 1989). Activation of FBPase, phosphoribulokinase, sedoheptulose 1,7-bisphosphatase and NADP-glyceraldehyde phosphate dehydrogenase by the thioredoxin system would be expected, as the chloroplast is generally highly reduced under stress conditions (see above). The content of fructose 6-phosphate did decrease with mild stress in the study by Sharkey and Seemann (1989), suggesting that the FBPase activity might be decreased. However, as the PGA content was also much reduced, this is not surprising. In the autocatalytic cycle, in steady-state, decreased RuBP is expected to be accompanied by decreased PGA, especially as Rubisco is not affected. However, if inhibition of a particular enzymatic process resulted from stress, then the ratios of cycle metabolites should reflect it. An increased TP/RuBP ratio in mildly stressed spinach suggested that RuBP regeneration was impaired more than the synthesis of PGA (Quick *et al.*, 1989). The concentration of most metabolites increased, including a large increase in fructose-2,6-bisphosphate probably due to increased fructose 6-phosphate and Pi as the cell volume decreased (Quick *et al.*, 1989). Studies of enzymes under water stress are bedevilled by the need to assay under conditions reflecting the *in vivo* state as actual enzyme activity rather than total activity or amount of protein is important in determining the rate of photosynthesis. Care in interpretation of changes in enzyme amounts is also required because senescence involves the remobilization/breakdown of proteins and it is difficult to decide where primary events trigger secondary process such as senescence. A case in point is Rubisco, which constitutes 40–60% of the total soluble protein of leaves. Conditions such as water stress, warm temperature and nitrogen deficiency induce senescence and proteolysis and so exacerbate the loss. Therefore it is important to distinguish, where possible, the primary effects of stress on enzyme activity from those which affect the amount, and studies should be done before long-term developmental changes occur. Of course, it is possible that a primary effect of stress is on the enzyme synthesis. Appropriate studies of enzyme activities and amounts should be done to resolve these questions.

### 8.5.8 *Supply of ATP and NADPH for photosynthesis*

From the preceding analysis, a possible, indeed most likely, control in the rate of RuBP synthesis is the supply of ATP for ribulose-5-phosphate kinase (PRK). Inadequate ATP concentration and phosphorylation potential, with adequate Pi in the system, would decrease the conversion of ribulose-5-phosphate to RuBP, slowing the cycle irrespective of the effects of $CO_2$ supply or enzyme activation. Recently, comparisons of the effects of water stress on tobacco plants with normal PRK activity and plants transformed with an antisense PRK construct resulting in 90% reduction of activity have been made to assess the role of RuBP availability. The depletion of the enzyme decreased RuBP synthesis in unstressed leaves: under stress the wild-type contained less RuBP but the transgenic plant RuBP content was unaffected, i.e. the stress decreases RuBP synthesis except where the enzyme capacity is limiting (Paul, Tezara and Lawlor, unpublished). A corollary of decreased ATP content is that increased ADP would decrease the activities of PGA and PRK. ATP limitation would explain many of the effects of stress on photosynthetic metabolism. However, evidence for decreased ATP synthesis (photophosphorylation) in stressed chloroplasts is not conclusive, as discussed earlier (Section 8.5.6). In contrast, NADPH is available and indeed may be superabundant during water stress. The effects of water stress on adenylates and pyridine nucleotides are shown in Table 8.1. ATP pools decrease and ADP pools increase leading to a lower energy charge. Stuhlfauth *et al.* (1991) observed similar changes in the adenylates and the free phosphate pool dropped markedly, but the energy charge increased. Pyridine nucleotides behave differently. In general NADPH content decreased little or only slightly in the two studies, so the ratio ATP/NADPH dropped substantially and the reduced to oxidized nucleotides and reductant to ATP ratios increased greatly. The supply of ATP

***Table 8.1.*** *The effect of water stress ($\Psi$ in MPa) over 12 h on the adenylates (ATP and ADP), energy charge (EC), pyridine nucleotides (NAD(P)(H)), the ratio of reduced to oxidized pyridine nucleotides (R/O) and the ratio of ATP to total reduced pyridine nucleotides in wheat (*Triticum aestivum*) leaves. (Reproduced from Lawlor and Khanna-Chopra (1984) with permission from Kluwer Academic Publishers.)*

| $\Psi$ | ATP | ADP | EC | NADPH | $NADP^+$ | NADH | $NAD^+$ | R/O | ATP/NADPH + NADH |
|---|---|---|---|---|---|---|---|---|---|
| | (μmol $m^{-2}$) | | | | (μmol $m^{-2}$) | | | | |
| –0.4 | 170 | 60 | 0.80 | 3 | 9 | 2 | 14 | 0.22 | 34 |
| –0.7 | 90 | 100 | 0.67 | 5 | 14 | 3 | 16 | 0.27 | 11 |
| –1.3 | 60 | 100 | 0.61 | 3 | 15 | 7 | 7 | 0.45 | 6 |
| –1.8 | 50 | 160 | 0.57 | 4 | 4 | 8 | 6 | 1.20 | 4 |

and NADPH is clearly a very important aspect of metabolism and resolution of the changes in these energy metabolites under water stress is essential.

### 8.5.9 *Sucrose and starch synthesis*

Information on effects of stress on enzymes of carbon metabolism beyond the Calvin cycle is meagre, although the consumption of metabolites by growth and respiration is potentially important in feedback regulation of photosynthesis and carbon fluxes in and around the chloroplast. In mildly stressed plants, carbohydrates may accumulate: possibly this reflects slowed growth (expansion of organs is much more inhibited by slight drought than is photosynthesis). With more severe stress and stomatal closure the total carbon flux is decreased but a larger proportion goes to sucrose and a smaller proportion to starch. This is expected as starch synthesis in chloroplasts is characteristic of over-production of assimilates. ATP is needed for synthesis of ADP-glucose by ADP-glucose pyrophosphorylase, which regulates carbon flow to starch. This enzyme is controlled allosterically by Calvin cycle intermediates, e.g. PGA and fructose 6-phosphate stimulate activity while high Pi and Pi/PGA ratio inhibit. Stress decreases PGA so probably slowing the enzyme activity. Transport of TPs out of the chloroplast to the cytosol, via the phosphate translocator, in counter exchange with Pi is required for sucrose synthesis. There is no evidence of inhibition, although at severe stress loss of metabolites from chloroplasts may be unspecific and therefore damaging (Kaiser, 1987). Synthesis of sucrose is probably inhibited in moderately stressed plants, due to inadequate TP and UDP-glucose supply. UDP-glucose reacts with fructose 6-phosphate, catalysed by sucrose phosphate synthase (SPS), the product, sucrose 6-phosphate, being dephosphorylated by sucrose 6-phosphate phosphatase. Evidence about the effects of stress on SPS is confusing. Vassey and Sharkey (1989) found that enzyme activity decreased by 60% at mild water stress (–0.9 MPa) in *Phaseolus vulgaris* leaves, whereas Quick *et al.* (1989) concluded that particular kinetic forms of the enzyme from spinach increased in activity and they considered it as a primary site of response. The discrepancy is explained by the very large $CO_2$ concentrations used by Quick *et al.*, thus enabling photosynthesis to continue at substantial rates under relatively mild water stress. The pool of sucrose decreased substantially and rapidly in severely stressed leaves of sunflower (Lawlor and Fock, 1977a). Sucrose is consumed by respiration and serves as a substrate for organic acid synthesis in the TCA cycle and eventually for the synthesis of such stress metabolites as proline (Larher *et al.*, 1993; Venekamp *et al.*, 1989).

### 8.5.10 *Photorespiration*

Photorespiration consumes energy as NADPH and ATP in the synthesis of RuBP and as NADH in the glycolate pathway. (The phosphoglycolate produced in the Rubisco reaction of RuBP with oxygen is metabolized to glycerate via glycine and serine in the mitochondria and peroxisomes). Photorespiration increases relative to photosynthetic $CO_2$ fixation with increasing water stress (Lawlor and Fock, 1975, 1977a,b; Lawlor, 1976a,b) so expending relatively more reductant than ATP, particularly as the NAD(P)H produced (or NADH available outside the chloroplast as a consequence of shuttles of metabolites across the chloroplast envelope) may be oxidized by the mitochondrial respiratory chain and result in ATP synthesis. This would be an important function if photophosphorylation were partially inhibited under stress. Other mechanisms for dealing with NADH, e.g. reduction of oxaloacetic acid to malate may be responsible for the accumulation of malate in stressed leaves (Lawlor and Fock, 1977b). Reduced photorespiration slows consumption of electrons via NADPH and results in an increased reduction state of $Q_A$; this may lead to greater photoinhibition. As the consumption of NADPH increases relative to ATP then photorespiration will remove reductant and recycle $NADP^+$, thus lowering the reductant and energy status of the stroma and thylakoids. At the compensation concentration, without stress, the reductive and oxidative carbon cycles are balanced and consume energy and reductant; under these conditions the re-fixation of $CO_2$ within the mesophyll constitutes a large part of the energy demand. This is possibly the case under mild stress when photorespiration plus re-fixation of $CO_2$ play an important role in protection against over-reduction of the chloroplast (Gerbaud and André, 1987; Krampitz *et al.*, 1984; Lawlor, 1976a,b; Lawlor and Fock, 1975, 1977a,b). Photorespiration in unstressed plants in warm conditions may be 30–40% of net photosynthesis and increase to 50–70% with severe stress, but as the absolute rates of both decrease so their contribution to total energy dissipation falls. This is a more serious situation for metabolism even than reduced $CO_2$ supply, *vide* the ability of leaves at the compensation point to maintain photosynthetic competence compared to the rapid damage to leaves once 15–20% RWC is lost. At very small rates of net $CO_2$ assimilation, even with internal recycling, the protection offered is probably very much less than required to use the energy and reductant. Hence, the thylakoids are 'over-energized' and the stroma 'over-reduced'. This would be the situation if photosynthesis and therefore photorespiration are inhibited under severe stress by ATP-limited RuBP synthesis: then photorespiration can provide no protection against energy overload, so although $q_{NP}$ rises dramatically, reduced pyridine nucleotides increase, indicating malfunction of the cell. The capacity to capture light energy and use it to drive electron transport and water splitting is maintained at relatively high rates so the protective system is probably massively overloaded. The exact balance of these processes is, as yet, unclear. However, the evidence points to progressive, energy-related damage resulting from impaired RuBP synthesis. As these conditions become worse, the alternative energy and reductant dissipating processes (e.g. xanthophyll epoxidation cycles) assume greater importance and the chances of generating energized states of pigment molecules (e.g. triplet chlorophyll) and reactive oxygen species rises, placing an extra burden on protective components of the chloroplast (see Sections 8.5.2 and 8.5.3).

### 8.5.11 *Respiration*

The relative rate of photorespiration increases during water deficit (Section 8.5.10). With more severe stress, the TCA cycle ('dark respiration') contribution increases both relatively and absolutely, as indicated by the insensitivity of the $CO_2$ compensation concentration to $O_2$, and eventually becomes the dominant source of $CO_2$ production with severe stress. Another indication of the source of this $CO_2$ is seen in studies of the release of $^{14}CO_2$ from stressed leaves following photosynthesis in $^{14}CO_2$ of known specific radioactivity. Over a relatively short time the specific radioactivity of the respired $CO_2$ from unstressed leaves rises to approach the feeding gas specific activity. However, very severely stressed leaves evolve very little $^{14}CO_2$ which is of much lower specific activity than the photosynthetically fixed $CO_2$. This may be explained by consumption of unlabelled substrates in TCA cycle respiration which are not of photorespiratory origin; sucrose from a large pool would be only slowly labelled in such experiments (Lawlor and Fock, 1975). This transition from photosynthesis to respiration is very significant for metabolism, occurring at about −1.5 MPa, RWC 85% in plants grown in solution, and correlating with the marked increases in ABA (Chapter 1) and proline content (Chapter 10) of leaves and with increased damage to the photosynthetic system.

The interaction of photosynthesis, photorespiration and dark (mitochondrial) respiration has not been examined satisfactorily in water-stressed tissues. The complexity is considerable (see Amthor, 1994 for review of the interactions under non-stressed conditions). Salient points are that the fluxes of energy and metabolites between cell compartments are large and regulated, so that a very dynamic energy balance is achieved linked to fluxes of assimilates. Chloroplasts and mitochondria serve opposite, but complementary, roles in supplying both energy and substrates for synthetic mechanisms in the cell, but chloroplasts use external energy to drive the process, whereas mitochondria use photosynthates. If stress decreases the production of assimilates then inevitably the demands for energy and substrates from dark respiration must increase. Mitochondrial and photosynthetic processes are intimately linked in the energy balance of stressed cells. With the reductant status of the stressed chloroplasts increasing as described, the transfer of reducing equivalents to the cytosol via shuttle mechanisms (TP shuttle of dihydroxyacetone phosphate and PGA or oxaloacetate and malate shuttles) provides ways of regulating NADPH and NADH and also the ATP in different compartments. There is substantial evidence that plant mitochondria use NADPH directly, as well as NADH, coupling their oxidation to electron transport and ATP synthesis (Rasmusson and Moller, 1990). If ATP production in stressed chloroplasts is decreased, it is logical that excess reducing equivalents should be used in mitochondria to overcome the limitation and also 'detoxify' the system. Hence, the regulatory mechanisms normally operating between chloroplasts and mitochondria will be used, but also over-stretched, under extreme stress conditions. As with other regulatory processes, this mechanism may operate under stress to prevent or reduce metabolic damage. The role of mitochondrial and respiratory processes cannot be separated from the behaviour of photosynthesis and may be related to the metabolism of proline which accumulates during water deficit (Chapter 9).

## 8.6 Scheme of photosynthesis and respiration during water deficit

The evidence reviewed in Section 8.5 can be used to suggest the sequence of events as $C_3$ leaves are exposed to increasing water deficit.

(i) Partial loss (25%) of turgor and a small decrease in RWC (100 down to 90%), i.e. 'mild stress,' has little effect on photosynthetic metabolism, the main limitation in photosynthesis is the reduction in diffusive conductance caused by stomatal closure and decreased $C_i$. Accumulation of sucrose and starch may occur, reflecting the maintenance of a positive balance between synthesis and consumption despite reduced photosynthesis.

(ii) Further loss of turgor and reduction of RWC below 85% further decreases stomatal conductance and potential $CO_2$ assimilation rate (judged from $A/C_i$ curves after allowing for heterogeneity in stomatal conductance across the leaf surface) due to metabolic alterations, but $C_i$ may continue to decrease. The low $C_i$ leads to an increase of the *in vivo* oxygenase/carboxylase ratio of Rubisco, causing a larger relative flux of carbon through the photorespiratory glycolate pathway. Increased photorespiration, relative to photosynthesis, recycles $CO_2$, consuming relatively more NADPH than ATP.

(iii) Light-harvesting, electron transport and reduction of pyridine nucleotides and other electron acceptors are little affected in the physiological range of stress. The reduced to oxidized pyridine nucleotide ratio and the size of the reduced pyridine nucleotide pool increase with moderate to severe stress.

(iv) ATP content decreases with moderate to severe stress due to impaired synthesis of ATP by coupling factor, which is inhibited by the ionic conditions in the chloroplast. Consequently the ATP/reduced pyridine nucleotide ratio falls.

(v) Synthesis of RuBP is inhibited as a result of limited ATP supply so the potential for $CO_2$ assimilation and photorespiration decreases progressively with developing stress and control shifts from $CO_2$ availability to RuBP synthesis.

(vi) Abnormal regulation of, or damage to, enzymes is not the major lesion in photosynthetic metabolism.

(vii) $CO_2$ assimilation and synthesis of TP, sucrose and starch decreases substantially whilst consumption of sucrose by respiration continues so that total carbohydrate content of leaves falls, starch more than sucrose.

(viii) The proportion of electron flow to $O_2$ (Mehler reaction) increases, generating superoxide and hydrogen peroxide which damage membranes and enzymes. A greater proportion of the energy is dissipated by $q_{NP}$ than by

$q_P$. Dissipation mechanisms (carotenoids, including the xanthophyll cycle, and antioxidant systems) become increasingly important, removing energy and reductant and destroying toxic compounds generated as energy is transferred to unphysiological acceptors (see Chapter 12).

(ix) Severe water deficit causes substantial inhibition of photophosphorylation and a further decrease in ATP content. $CO_2$ assimilation and photorespiration almost stop, respiratory processes dominate and $C_i$ rises greatly.

(x) Excessive energy loads on the thylakoids and the detoxification systems eventually lead to membrane damage and irreversible loss of photosynthesis. These processes may be linked to accumulation of metabolites such as proline.

## References

Adams, W.W. and Demmig-Adams, B. (1993) Energy dissipation and photoprotection in leaves of higher plants. In: *Photosynthetic Responses to the Environment* (eds H.Y. Yamamoto and C.M. Smith). American Society of Plant Physiologists, Rockville, MD, pp. 27–36.

Alscher, R.G. and Cummings, J.R. (eds) (1990) *Stress Responses in Plants: Adaptation and Acclimation Mechanisms.* Wiley-Liss, New York.

Amthor, J.S. (1994) Higher plant respiration and its relationships to photosynthesis. In: *Ecophysiology of Photosynthesis* (eds E.-D. Schulze and M.M. Caldwell). Springer Verlag, Berlin, pp. 71–101.

André M. and Du Cloux, H. (1993) Interaction of $CO_2$ enrichment and water limitations on photosynthesis and water efficiency in wheat. *Plant Physiol. Biochem.* **31**, 103–112.

Assmann, S.M. (1988) Stomatal and non-stomatal limitations to carbon assimilation: an evaluation of the path-dependent method. *Plant, Cell Environ.* **11**, 577–582.

Baker, N.R. and Bowyer, J.R. (eds) (1994) *Photoinhibition of Photosynthesis. From Molecular Mechanisms to the Field.* BIOS Scientific Publishers, Oxford.

Becker, T.W., Hoppe, M. and Fock, H.P. (1986) Evidence for the participation of dissimilatory processes in maintaining high carbon fluxes through the photosynthetic carbon reduction and oxidation cycles in water stressed *Phaseolus* leaves. *Photosynthetica* **20**, 153–157.

Berkowitz, G.A. and Gibbs, M. (1983) Reduced osmotic potential inhibition of photosynthesis. Site-specific effects of osmotically induced stromal acidification. *Plant Physiol.* **72**, 1100–1109.

Berkowitz, G.A. and Wahlen, C. (1985) Leaf $K^+$ interaction with water stress inhibition of non-stomatal controlled photosynthesis. *Plant Physiol.* **79**, 189–193.

Björkman, O. and Demmig-Adams, B. (1994) Regulation of photosynthetic light energy capture, conversion and dissipation in leaves of higher plants. In: *Ecophysiology of Photosynthesis* (eds E.-D. Schulze and M.M. Caldwell). Springer Verlag, Berlin, pp. 17–47.

Boyer J.S., Ort, D.R. and Ortiz-Lopez A. (1987) Photophosphorylation at low water potentials. *Curr. Top. Plant Biochem. Physiol.* **6**, 69–73.

Bunce, J.A. (1988) Non-stomatal inhibition of photosynthesis by water stress. *Photosynth. Res.* **18**, 357–362.

Castrillo, M. and Calcagno, A.M. (1989) Effects of water stress and rewatering on ribulose

1,5-bisphosphate carboxylase activity, chlorophyll and protein contents in two cultivars of tomato. *J. Hortic. Sci. Am.* **64**, 717–724.
Chaves, M.M. (1991) Effects of water deficits on carbon assimilation. *J. Exp. Bot.* **42**, 1–16.
Close, T.J. and Bray, E.A. (eds) (1993) *Plant Response to Cellular Dehydration During Environmental Stress.* American Society of Plant Physiologists, Rockville, MD.
Cornic, G. (1994) Drought stress and high light effects on leaf photosynthesis. In: *Photoinhibition of Photosynthesis. From Molecular Mechanisms to the Field* (eds N.R. Baker and J.R. Bowyer). BIOS Scientific Publishers, Oxford, pp. 297–313.
Cornic, G. and Briantais J.-M. (1991) Partitioning of photosynthetic electron flow between $CO_2$ and $O_2$ reduction in a C3 leaf (*Phaseolus vulgaris* L.) at different $CO_2$ concentrations and during drought stress. *Planta* **183**, 178–184.
Cornic, G., Le Gouallec, J.-L., Briantais, J.M. and Hodges, M. (1989) Effect of dehydration and high light on photosynthesis of two C3 plants (*Phaseolus vulgaris* L. and *Elatostema repens* (Lour.) Hall f.) *Planta* **177**, 84–90.
de Kouchkowsky, Y. and Meyer, S. (1992) Inactivation of chloroplast ATPase by *in vivo* decrease of water potential. In: *Research in Photosynthesis,* Volume 2 (ed. N. Murata). Kluwer Academic Publishers, Dordrecht, pp. 709–712.
Farquhar, G.D. and Sharkey, T.D. (1982) Stomatal conductance and photosynthesis. *Annu. Rev. Plant Physiol.* **33**, 317–345.
Farquhar, G.D., Hubic, K.T., Terashima, I., Condon, A.G. and Richards, R.A. (1987) Genetic variation in the relationship between photosynthetic $CO_2$ assimilation rate and stomatal conductance to water loss. In: *Progress in Photosynthesis Research,* Volume 4 (ed. J. Biggins). Martinus Nijhoff Publishers, The Hague, pp. 209–212.
Genty, B., Briantais, J.-M. and Vieira Da Silva, J.B. (1987) Effects of drought on primary photosynthetic processes of cotton leaves. *Plant Physiol.* **83**, 360–364.
Gerbaud, A. and André, M. (1987) An evaluation of the recycling in measurements of photorespiration. *Plant Physiol.* **83**, 933–937.
Gilmore, A.M. and Yamamoto, H.Y. (1991) Zeaxanthin formation and energy-dependent fluorescence quenching in pea chloroplasts under artificially mediated linear and cyclic electron transport. *Plant Physiol.* **96**, 635–643.
Gimenez, C., Mitchell, V.J. and Lawlor, D.W. (1992) Regulation of photosynthetic rate of two sunflower hybrids under water stress. *Plant Physiol.* **98**, 516–524.
Goodwin, T.W. (1980) *The Biochemistry of Carotenoids. 2nd edn, Volume 1 Plastids.* Chapman and Hall, London.
Graan, T. and Boyer, J.S. (1990) Very high $CO_2$ partially restores photosynthesis in sunflower at low water potentials. *Planta* **181**, 373–384.
Gunasekera, D. and Berkowitz, G.A. (1993) Use of transgenic plants with ribulose 1,5-bisphosphate carboxylase/oxygenase antisense DNA to evaluate the rate limitation of photosynthesis under water stress. *Plant Physiol.* **103**, 629–635.
Havaux, M., Canaani, O. and Malkin, S. (1987) Inhibition of photosynthetic activities under slow water stress measured *in vivo* by the photoacoustic method. *Physiol. Plant.* **70**, 503–510.
Johnson, R.C., Mornhinweg, D.W., Ferris, D.M. and Heitholt, J.J. (1987) Leaf photosynthesis and conductance of selected *Triticum* species at different water potentials. *Plant Physiol.* **83**, 1014–1017.
Kaiser, W.M. (1987) Effects of water deficit on photosynthetic capacity. *Physiol. Plant.* **71**, 142–149.
Kaiser, W.M. and Heber, U. (1981) Photosynthesis under osmotic stress. *Planta* **153**, 423–429

Kaiser, W.M., Schröppel-Meier, G. and Wirth, E. (1986) Enzyme activities in an artificial stroma medium. *Planta* **167**, 292–299.

Kononowicz, A.K., Raghothama, K.G., Casa, A.M., Reuveni, M., Watad, A.-E., Lui, D., Bressan, R.A. and Hasegawa, A. (1993) Osmotin: regulation of gene expression and function. In: *Plant Response to Cellular Dehydration During Environmental Stress* (eds T.J. Close and E.A. Bray). American Society of Plant Physiologists, Rockville, MD, pp. 144–158.

Krampitz, M.J., Klüge, K. and Fock, H.P. (1984) Rates of photosynthetic $CO_2$ uptake, photorespiratory $CO_2$ evolution and dark respiration in water stressed sunflower and bean leaves. *Photosynthetica* **18**, 322–328.

Laasch, H. (1987) Non-photochemical quenching of chlorophyll *a* fluorescence in isolated chloroplasts under conditions of stressed photosynthesis. *Planta* **171**, 220–226.

Larher, F., Leport, L., Petrivalsky, M. and Chappart, M. (1993) Effectors for the osmoinduced proline response in higher plants. *Plant Physiol. Biochem.* **31**, 911–922.

Lawlor, D.W. (1976a) Assimilation of carbon into photosynthetic intermediates of water stressed wheat. *Photosynthetica* **10**, 431–439.

Lawlor, D.W. (1976b) Water stress induced changes in photosynthesis, photorespiration, respiration and $CO_2$ compensation concentration of wheat. *Photosynthetica* **10**, 378–387.

Lawlor, D.W. (1993) *Photosynthesis: Molecular, Physiological and Environmental Processes.* 2nd edn. Longman Scientific and Technical, Harlow, UK.

Lawlor, D.W. (1994) Physiological and biochemical criteria for evaluating genotype responses to heat and related stresses. In: *Wheat in Heat-stressed Environments: Irrigated, Dry Areas and Rice–Wheat Farming Systems* (eds D.A. Saunders and G.P. Hettel). CIMMYT, Mexico, pp. 127–142.

Lawlor, D.W. and Fock, H. (1975) Photosynthesis and photorespiratory $CO_2$ evolution of water-stressed sunflower leaves. *Planta* **126**, 247–258.

Lawlor, D.W. and Fock, H. (1977a) Photosynthetic assimilation of $^{14}CO_2$ by water-stressed sunflower leaves in two oxygen concentrations and the specific activity of products. *J. Exp. Bot.* **28**, 320–328.

Lawlor, D.W. and Fock, H. (1977b) Water stress induced changes in the amounts of some photosynthetic assimilation products and respiratory metabolites of sunflower leaves. *J. Exp. Bot.* **28**, 329–337.

Lawlor, D.W. and Keys, A.J. (1993) Understanding photosynthetic adaptation to changing climate. In: *Plant Adaptation to Environmental Stress* (eds L. Fowden, J. Stoddart and T.A. Mansfield). Chapman and Hall, London, pp. 85–106.

Lawlor, D.W. and Khanna-Chopra, R. (1984) Inhibition of photosynthesis during water stress. In: *Advances in Photosynthesis Research,* Volume 4 (ed. C. Sybesma). Martinus Nijhoff/Dr W. Junk Publishers, The Hague, pp. 379–382.

Lawlor, D.W. and Pearlman, J.G. (1981) Compartmental modelling of photorespiration and carbon metabolism of water-stressed leaves. *Plant, Cell Environ.* **4**, 37–52.

Lawlor, D.W. and Uprety, D.C. (1993) Effects of water stress on photosynthesis of crops and the biochemical mechanism. In: *Photosynthesis: Photoreactions to Plant Productivity* (eds Y.P. Abrol, P. Mohanty and Govindjee). Oxford and IBH Publ. Co. PVT. Ltd, New Delhi, pp. 419–449.

Luo, Y. (1991) Changes of $C_i/C_a$ in association with stomatal and non-stomatal limitation to photosynthesis in water stressed *Abutilon theophrasti. Photosynthetica* **25**, 273–297.

Ludlow, M.M. (1987) Defining shoot water status in the most meaningful way to relate to physiological processes. In: *Proceedings, International Conference on Measurement of Soil and Plant Water Status*, Volume 2. Utah State University, Logan, UT, pp. 47–53.

Matthews, M.A. and Boyer, J.S. (1984) Acclimation of photosynthesis to low leaf water potentials. *Plant Physiol.* **74**, 161–166.

Meyer, S. and de Kouchkowsky, Y. (1992) ATPase state and activity in thylakoids from normal and water-stressed lupin. *FEBS Lett.* **303**, 233–236.

Meyer, S. and de Kouchkowsky, Y. (1993) Electron transport, photosystem-II reaction centres and chlorophyll–protein complexes of thylakoids of drought resistant and sensitive lupin plants. *Photosynth. Res.* **36**, 49–60.

Meyer, S., Phung Nhu Hung, S., Trémolières, A. and de Kouchkowsky, Y. (1992) Energy coupling, membrane lipids and structure of thylakoids of lupin plants submitted to water stress. *Photosynth. Res.* **32**, 95–107.

Ort, D.R., Oxborough, K. and Wise, R.R. (1994) Depressions of photosynthesis in crops with water deficits. In: *Photoinhibition of Photosynthesis. From Molecular Mechanisms to the Field* (eds N.R. Baker and J.R. Bowyer). BIOS Scientific Publishers, Oxford, pp. 315–329.

Ortiz-Lopez, A., Ort, D.R. and Boyer, J.S. (1991) Photophosphorylation in attached leaves of *Helianthus annuus* at low water potentials. *Plant Physiol.* **96**, 1018–1025.

Owens, T.G. (1994) Excitation energy transfered between chlorophylls and carotenoids. A proposed molecular mechanism for non-photochemical quenching. In: *Photoinhibition of Photosynthesis. From Molecular Mechanisms to the Field* (eds N.R. Baker and J.R. Bowyer). BIOS Scientific Publishers, Oxford, pp. 95–109.

Parry, A.D. (1993) Abscisic acid metabolism. In: *Methods in Plant Biochemistry, Volume 9, Enzymes of Secondary Metabolism* (ed. P.J. Lea). Academic Press, London, pp. 381–402.

Pell, E.J. and Steffen, K.L. (eds) (1991) *Active Oxygen/Oxidative Stress and Plant Metabolism.* American Society of Plant Physiologists, Rockville, MD.

Pier, P. and Berkowitz, G.A. (1989) The effects of chloroplast envelope-$Mg^{2+}$, cation movement, and osmotic stress on photosynthesis. *Plant Sci.* **64**, 45–53.

Pierce, M. and Raschke, K. (1980) Correlation between loss of turgor and accumulation of abscisic acid in detached leaves. *Planta* **148**, 174–182.

Quick, P., Siegl, G., Neuhaus, E., Feil, R. and Stitt, M. (1989) Short-term water stress leads to a stimulation of sucrose synthesis by activating sucrose-phosphate synthase. *Planta* **177**, 535–546.

Quick, W.P., Chaves, M.M, Wendler, R., *et al.* (1992) The effect of water stress on photosynthetic carbon metabolism in four species grown under field conditions. *Plant, Cell Environ.* **15**, 25–35.

Rasmusson, A.G. and Moller, I.M. (1990) NADP utilizing enzymes in the matrix of plant mitochondria. *Plant Physiol.* **94**, 1012–1018.

Richards, R.A. (1993) Breeding crops with improved stress resistance. In: *Plant Response to Cellular Dehydration During Environmental Stress* (eds T.J. Close and E.A. Bray). American Society of Plant Physiologists, Rockville, MD, pp. 211–223 .

Scheuermann, R., Biehler, K., Stuhlfauth, T. and Fock, H.P. (1991) Simultaneous gas exchange and fluorescence measurements indicate differences in the responses of sunflower, bean and maize to water stress. *Photosynth. Res.* **27**, 189–197.

Sharkey, T.D. (1990) Water stress effects on photosynthesis. *Photosynthetica* **24**, 651.

Sharkey, T.D. and Badger, M.R. (1982) Effects of water stress on photosynthetic electron transport, photophosphorylation and metabolite levels of *Xanthium strumarium* cells. *Planta* **156**, 199–206.

Sharkey, T.D. and Seemann, J.R. (1989) Mild water stress effects on carbon-reduction-cycle intermediates, ribulose bisphosphate carboxylase activity, and spatial homogeneity of photosynthesis in intact leaves. *Plant Physiol.* **89**, 1060–1065.

Smirnoff, N. (1993) The role of active oxygen in the response of plants to water deficit and desiccation. Tansley Review No. 52. *New Phytol.* **125**, 27–58.

Smith, J.A.C. and Griffiths, H. (eds) (1993). *Water Deficits: Plant Responses from Cell to Community.* BIOS Scientific Publishers, Oxford.

Stuhlfauth, T., Sültemeyer, D.F., Weinz, S. and Fock, H.P. (1988) Fluorescence quenching and gas exchange in a water stressed C3 plant, *Digitalis lanata. Plant Physiol.* **86**, 246–250.

Stuhlfauth, T. Beckedahl, J. and Fock, H.P. (1991) The response of energy charge, NADPH and free phosphate pools to water stress. *Photosynthetica* **25**, 11–15.

Tang, A.C., Kawamitsu, Y., Kanechi, M. and Boyer, J.S. (1994) Non-stomatal limitation in sunflower at low leaf water potential. *Plant Physiol. (Suppl.)* **105**, 14.

Thomas D.A. and André, M. (1982) The response of oxygen and carbon dioxide exchanges and root activity to short term water stress in soybean. *J. Exp. Bot.* **134**, 309–405.

van Kraalingen, X. (1990) Implications of non-uniform stomatal closure on gas exchange calculations. *Plant, Cell Environ.* **13**, 1001–1004.

Vassey, T.L. and Sharkey, T.D. (1989) Mild water stress of *Phaseolus vulgaris* plants leads to reduced starch synthesis and extractable sucrose phosphate synthase activity. *Plant Physiol.* **89**, 1066–1070.

Venekamp, J.H., Lampe, J.E.M. and Koot, J.T.M. (1989) Organic acids as sources for drought induced protein synthesis in field bean plants, *Vicia faba* L. *Plant Physiol.* **133**, 654–659.

Wise, R.R, Frederick, J.R., Alm, D.M., Kramer, D.M., Hesketh, J.D, Crofts, A.R. and Ort, D.R. (1990) Investigation of the limitations to photosynthesis induced by leaf water deficit in field-grown sunflower (*Helianthus annuus* L.). *Plant, Cell Environ.* **13**, 923–931.

Wise, R.R., Frederick, J.R., Alm, D.M., Kramer, D.M., Hesketh, J.D, Crofts, A.R. and Ort, D.R. (1992) Corrigendum. Investigation of the limitations to photosynthesis induced by leaf water deficit in field-grown sunflower (*Helianthus annuus* L.). *Plant, Cell Environ.* **15**, 755–756.

Young, A. and Britton, G. (1990) Carotenoids and stress. In: *Stress Responses in Plants: Adaptation and Acclimation Mechanisms* (eds R.G. Alscher and J.R. Cummings). Wiley-Liss, New York, pp. 87–112.

Younis, H.M., Boyer, J.S. and Govindjee (1979) Conformation and activity of chloroplast coupling factor exposed to low chemical potential of water in cells. *Biochim. Biophys. Acta* **548**, 328–340.

Zeevaart, J.A.D., Rock, C.D., Fantauzzo, F., Heath, T.G. and Gage, D.A. (1991) Metabolism of ABA and its physiological implications. In: *Abscisic Acid: Physiology and Biochemistry* (eds W.J. Davies and H.G. Jones). BIOS Scientific Publishers, Oxford, pp. 39–52.

9

# Proline accumulation during drought and salinity

Y. Samaras, R.A. Bressan, L.N. Csonka, M.G. García-Ríos, M. Paino D'Urzo and D. Rhodes

## 9.1 Introduction

Proline is one of a relatively small number of naturally occurring compounds which function as 'compatible' or 'counteracting' solutes (Yancey, 1994). This chapter will consider recent progress in our understanding of the role of proline in plant adaptation to water deficits and salinity stress, and the regulation of proline biosynthesis in higher plants. The other major compatible solutes in plants, quaternary ammonium, tertiary sulphonium compounds and polyols are discussed in Chapters 10 and 11 Much of this recent progress has been made possible by advances in our knowledge of the enzymes and genes involved in proline biosynthesis in bacteria. The availability of well characterized proline-requiring mutants of *Escherichia coli* has facilitated the isolation (by complementation) of several cDNAs encoding enzymes of proline synthesis from higher plants (Delauney and Verma, 1990, 1993; Delauney *et al.*, 1993; García-Ríos *et al.*, 1994; Hu *et al.*, 1992; Verma *et al.*, 1992).

## 9.2 Proline as a compatible solute

The class of compounds known as 'compatible' or 'counteracting solutes' includes several amino acids (e.g. proline, alanine, β-alanine and taurine), quaternary ammonium compounds (e.g. glycinebetaine, prolinebetaine, β-alaninebetaine, glycerophosphorylcholine and choline-O-sulphate, Chapter 11), the tertiary sulphonium compound β-dimethylsulphoniopropionate, and certain carbohydrates (e.g. trehalose, glycerol, mannitol, sorbitol and pinitol, Chapter 11) (Yancey, 1994). These compounds tend to be uncharged at neutral pH, and are of high solubility in water (Ballantyne and Chamberlin, 1994). Osmolyte compatibility is thought to result primarily from the absence of perturbing effects of the solutes on macromolecule–solvent interactions (Low, 1985; Yancey, 1994; Yancey *et al.*, 1982). Perturbing solutes (such as inorganic ions) readily enter the hydration sphere of proteins, favouring unfolding. In contrast, compatible solutes tend to be excluded from the hydration sphere of proteins and stabilize folded protein structures (Low, 1985). Like glycinebetaine (and several other quaternary ammonium compounds), exogenously supplied proline is osmoprotective for bacteria, facilitating growth in highly saline environments (Csonka, 1989; Csonka and Hanson, 1991; Strøm *et al.*, 1983; Yancey, 1994). Accumulation of proline or glycinebetaine in the cytoplasm is accompanied by a reduction in the concentrations of less compatible solutes (e.g. $K^+$ and glutamate or trehalose) and an increase in cytosolic water volume (Cayley *et al.*, 1991, 1992). A number of studies suggest that exogenously supplied proline can also be osmoprotective for higher plant cells (Handa *et al.*, 1986; Lone *et al.*, 1987; Tal and Katz, 1980; Wyn Jones and Gorham, 1983).

*In vitro* studies indicate that proline is much less inhibitory than equivalent concentrations of NaCl to enzymes and the protein synthesis machinery (Brady *et al.*, 1984; Gibson *et al.*, 1984; Pollard and Wyn Jones, 1979) and may protect proteins against heat denaturation (Paleg *et al.*, 1981; Santoro *et al.*, 1992; Smirnoff and Stewart, 1985) and, in some cases, against the inhibitory effects of NaCl (Manetas *et al.*, 1986). Proline influences protein solvation and protects against the 'biologically unfavourable consequences of dehydration-induced thermodynamic perturbation' (Paleg *et al.*, 1984). Proline may also function as a hydroxyl radical scavenger (Smirnoff and Cumbes, 1989), and may stabilize membranes by interacting with phospholipids (Rudolph *et al.*, 1986). Proline has been shown to be cryoprotective to plant cells (Duncan and Widholm, 1987; Santarius, 1992; Songstad *et al.*, 1990; van Swaaji *et al.*, 1985; Withers and King, 1979).

As has been reviewed extensively elsewhere, the accumulation of proline is a common metabolic response of higher plants to water deficits, salinity stress and cold stress (Csonka and Baich, 1983; Delauney and Verma, 1993; Dörffling *et al.*, 1993; Hanson and Hitz, 1982; Stewart, 1981; Stewart and Larher, 1980; Thompson, 1980). In some plant tissues (e.g. apical meristems) and cell cultures adapted to growth at low water potentials, proline accumulation may play a major role in osmotic adjustment (Binzel *et al.*, 1987; Ketchum *et al.*, 1991; Voetberg and Sharp, 1991). For example, proline is accumulated to particularly high levels in leaves of certain halophytic higher plant species adapted to growth in highly saline environments (Briens and Larher, 1982; Stewart and Lee, 1974), in leaf tissues of plants experiencing water stress (Barnett and Naylor, 1966; Boggess *et al.*, 1976; Jones *et al.*, 1980), in specialized plant cells which must withstand severe desiccation (e.g. pollen) (Hong-qi *et al.*, 1982), in root apical regions growing at low water potentials (Voetberg and Sharp, 1991) and in plant cell cultures adapted to water stress (Handa *et al.*, 1986; Rhodes *et al.*, 1986; Tal and Katz, 1980), or NaCl stress (Binzel *et al.*, 1987; Katz and Tal, 1980; Rhodes and Handa, 1989; Tal and Katz, 1980; Treichel, 1975, 1986).

In the apical millimetre of maize roots, proline represents a major solute, reaching concentrations of 120 mM in roots growing at a water potential of −1.6 MPa (Voetberg and Sharp, 1991). The accumulated proline accounts for almost half the osmotic adjustment in this region (Voetberg and Sharp, 1991). Similarly, in cell cultures of tobacco adapted to 25 g $l^{-1}$ (428 mM) NaCl, proline represents over

80% of the free amino acid pool and, even assuming uniform distribution of the proline in total cell water, this amino acid is present at levels in excess of 129 mM (Binzel *et al.*, 1987). Results from several studies suggest that the proline accumulated in response to water stress or salinity stress in plants is localized primarily in the cytosol (Ketchum *et al.*, 1991; Leigh *et al.*, 1981; Pahlich *et al.*, 1983). The total proline concentration of salt-stressed *Distichlis spicata* cells (treated with 200 mM NaCl) is 28 mM, but the proline concentration of the cytosol is estimated to be greater than 230 mM in these cells (Ketchum *et al.*, 1991). This is in accordance with the hypothesis on cytoplasmic osmoregulation proposed by Wyn Jones *et al.* (1977) whereby non-toxic compatible organic solutes (such as proline or glycine-betaine) are postulated to be accumulated in the cytoplasm to balance the osmotic potential of the vacuole where toxic non-compatible solutes (e.g. $Na^+$ and $Cl^-$) are sequestered.

Overproduction of proline in bacteria caused by altered feedback inhibition of the proline biosynthesis pathway can result in increased osmotolerance (Csonka, 1981; Smith, 1985). Hydroxyproline-resistant mutants of barley and winter wheat accumulate greater quantities of proline than wild-type (Dörffling *et al.*, 1993; Kueh and Bright, 1981). However, in the barley mutants it appears that the concentrations of proline accumulated may be an order of magnitude smaller than required to produce a significant physiological effect on osmotic stress tolerance (Lone *et al.*, 1987). In winter wheat, the hydroxyproline-resistant lines are more frost tolerant than wild-type (Dörffling *et al.*, 1993). Salt-tolerant and polyethylene glycol-resistant mutants of *Nicotiana plumbaginifolia* have been derived from protoplast culture. These exhibit higher proline levels than wild-type (Sumaryati *et al.*, 1992).

## 9.3 Pathways of synthesis of proline in bacteria

In bacteria such as *E. coli*, *Pseudomonas aeruginosa* and *Salmonella typhimurium*, proline synthesis from glutamate is catalysed by three enzymes: γ-glutamyl kinase (GK) (Equation 9.1), γ-glutamyl phosphate reductase (GPR) (also called glutamic semialdehyde dehydrogenase) (Equation 9.2) and $\Delta^1$-pyrroline-5-carboxylate reductase (P5CR) (Equation 9.4), encoded by genes *proB*, *proA* and *proC*, respectively (Adams and Frank, 1980; Baich, 1969, 1971; Csonka and Baich, 1983; Deutch *et al.*, 1982, 1984; Gamper and Moses, 1974; Hayzer and Leisinger, 1980, 1981, 1982; Hayzer and Moses, 1978a,b; Krishna and Leisinger, 1979; Krishna *et al.*, 1979; Smith *et al.*, 1984):

$$\text{glutamate} + \text{ATP} \xrightarrow[\gamma\text{-glutamyl kinase}]{Mg^{2+}} \gamma\text{-glutamyl phosphate} + \text{ADP} \quad (9.1)$$

$$\text{-glutamyl phosphate} + \text{NADPH} + H^+ \underset{\gamma\text{-glutamyl phosphate reductase}}{\longleftrightarrow} \text{glutamic-}\gamma\text{-semialdehyde} + \text{NADP}^+ + \text{Pi} \quad (9.2)$$

$$\text{glutamic-}\gamma\text{-semialdehyde} \underset{\text{spontaneous}}{\longleftrightarrow} \Delta^1\text{-pyrroline-5-carboxylate} + H_2O \quad (9.3)$$

$$\Delta^1\text{-pyrroline-5-carboxylate} + \text{NADPH} + H^+ \xrightarrow[\substack{\Delta^1\text{-pyrroline-}\\ \text{5-carboxylate}\\ \text{reductase}}]{} \text{proline} + \text{NADP}^+ \quad (9.4)$$

The step from glutamic-γ-semialdehyde (GSA) to $\Delta^1$-pyrroline-5-carboxylate (P5C) is reversible and spontaneous (Equation 9.3). The reaction catalysed by GK (Equation 9.1) is irreversible. The product of the reaction, γ-glutamyl phosphate, is extremely labile. It can be chemically trapped by reaction with hydroxylamine to yield the more stable γ-glutamyl hydroxamate which can be quantified with a ferric chloride reagent (Baich, 1969; Hayzer and Leisinger, 1980). In the absence of hydroxylamine, the γ-glutamyl phosphate product spontaneously degrades to pyroglutamate (5-oxoproline) and inorganic phosphate (Pi) (Seddon *et al.*, 1989) (Equation 9.5):

$$\gamma\text{-glutamyl phosphate} \xrightarrow[\text{spontaneous}]{} \text{pyroglutamate} + \text{Pi} + H_2O \quad (9.5)$$

The GK of *E. coli* is inhibited by proline (Baich, 1969; Rushlow *et al.*, 1984; Smith, 1985) and has a relatively low affinity for glutamate ($K_m$ = 7–10 mM) (Baich, 1969). Marked substrate activation of the enzyme is observed when the concentration of glutamate exceeds 50 mM (Baich, 1969). Proline increases the apparent $K_m$ for glutamate (Baich, 1969). Proline overproducing mutants have a GK with markedly altered feedback inhibition (Csonka, 1981; Rushlow *et al.*, 1984; Smith, 1985).

The reaction catalysed by the second enzyme (Equation 9.2) is reversible and this reversibility forms the basis of a selective assay for GPR in which P5C- ⇔ GSA-dependent, and Pi-dependent reduction of $NADP^+$ is measured. The γ-glutamyl phosphate generated in this reverse reaction spontaneously degrades to pyroglutamate and Pi (Equation 9.5) (Hayzer and Leisinger, 1981). Pyroglutamate has the following distinctive attributes: (i) unlike other amino acids it does not bind to Dowex-50-$H^+$ (Hayzer and Leisinger, 1981); (ii) it is ninhydrin negative (Hayzer and Leisinger, 1981); and (iii) it can be converted to glutamate by boiling in aqueous acid (Hayzer and Leisinger, 1981; Mazelis, 1979). The assay for GPR (P5C- ⇔ GSA- and Pi-dependent reduction of $NADP^+$) is subject to interference by an active P5CR in crude extracts (Hayzer and Leisinger, 1980). Thus, as NADPH is formed, this can be oxidized back to $NADP^+$ by the catalytic activity of P5CR (Equation 9.4). This interference by P5CR can lead to markedly non-linear $NADP^+$ reduction kinetics in bacterial crude extracts. However, GPR and P5CR can be separated easily on a diethylaminoethyl (DEAE)–cellulose column (Hayzer and Leisinger, 1980).

## 9.4 Pathways of synthesis of proline in higher plants

A number of *in vivo* labelling studies with $^{14}C$-labelled precursors (Boggess *et al.*, 1976; Morris *et al.*, 1969) or [$^{13}C$]glutamate (Heyser *et al.*, 1989a,b) or $^{15}NH_4^+/^{15}NO_3^-$ (Rhodes and Handa, 1989; Rhodes *et al.*, 1986) strongly suggest that glutamate is a major precursor of osmotic stress-induced proline accumulation in plants and that osmotic stress results in an increase of proline biosynthesis rate (Boggess *et al.*, 1976; Rhodes and Handa, 1989; Rhodes *et al.*, 1986; Stewart, 1981). The prevailing evidence (reviewed extensively by others (Delauney and Verma, 1993; Hanson and Hitz, 1982; Stewart, 1981; Thompson, 1980)) is that the increase in proline synthesis rate may, in part, involve induction or activation of enzymes of proline biosynthesis, possibly coupled with a relaxation of proline feedback inhibition control of the pathway (Boggess *et al.*, 1976; Delauney and Verma, 1993; Stewart, 1981). Interestingly, this feedback control of the proline pathway may also be modulated during development (Oaks, 1992; Oaks *et al.*, 1970). Decreased proline oxidation to glutamate (Elthon and Stewart, 1982; Huang and Cavalieri, 1979; Stewart *et al.*, 1977), decreased utilization of proline in protein synthesis (Boggess and Stewart, 1980; Stewart, 1981) and enhanced protein turnover (Fukutoku and Yamada, 1984) may all contribute to net proline accumulation. There are indications that *de novo* transcription and translation are required for stress-induced proline biosynthesis in higher plants. For example, stress-induced proline accumulation is inhibited by both cycloheximide and cordycepin in *Arabidopsis* (Verbruggen *et al.*, 1993). In contrast, because cycloheximide is inhibitory to 200 mM NaCl-induced proline accumulation in *D. spicata* cells, whereas actinomycin D is not inhibitory, Ketchum *et al.* (1991) suggest that translation, but not transcription, is necessary for production of proline in stressed cells.

In addition to the pathway of synthesis of proline from glutamate, there is also the possibility of synthesis of proline from ornithine (derived from the *N*-acetyl-glutamate pathway) in higher plants as indicated in Figure 9.1. This can occur by two routes, depending upon whether the $\alpha$- or $\delta$-$NH_2$ moiety of ornithine is transaminated (Figure 9.1) (Adams and Frank, 1980; Csonka and Baich, 1983; Delauney and Verma, 1993; Mestichelli *et al.*, 1979; Thompson, 1980). In the case of $\delta$-$NH_2$ transamination, the product is GSA which, in spontaneous equilibrium with P5C, can be converted readily to proline by P5CR (Equation 9.4). Ornithine $\delta$-transaminase has been partially purified from several plant sources, including pumpkin cotyledons (Splittstoesser and Fowden, 1973). The alternative pathway involves $\alpha$-$NH_2$ transamination of ornithine to $\alpha$-keto-$\delta$-aminovalerate $\Leftrightarrow$ $\Delta^1$-pyrroline-2-carboxylate (P2C). This product can be reduced to proline by a P2C reductase (Figure 9. 1). *In vivo* tracer studies with labelled ornithine strongly suggest that the P2C pathway of ornithine metabolism is the main route of conversion of ornithine to proline in several plant species (Mestichelli *et al.*, 1979).

There is a paucity of *in vitro* enzymological data to support the occurrence of a glutamate pathway of proline synthesis in higher plants similar to that established for bacteria (i.e. reaction sequence (Equation 9.1) – (Equation 9.4) above).

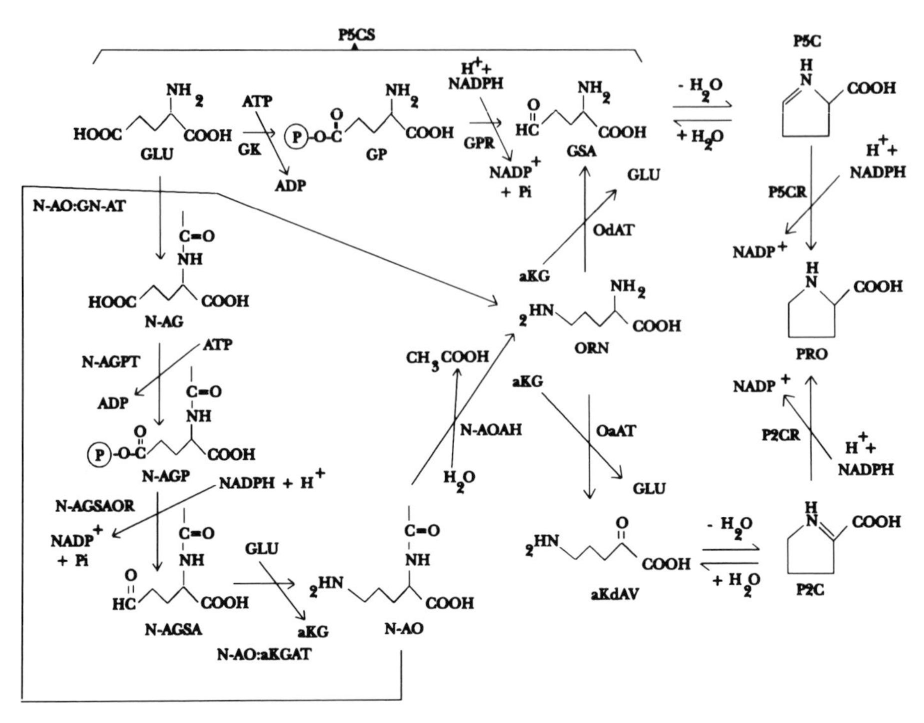


***Figure 9.1.*** *Alternative pathways of synthesis of proline in higher plants. Abbreviations used: GK, γ-glutamyl kinase; GP, γ-glutamylphosphate; GLU,* L-*glutamate; GSA, glutamic-γ-semialdehyde; GPR, γ-glutamyl phosphate reductase; aKG, α-ketoglutarate (2-oxoglutarate); aKdAV, α-keto-δ-aminovalerate; N-AG,* N-*acetyl-*L-*glutamate; N-AGP,* N-*acetylglutamyl 5-phosphate; N-AGPT, ATP:*N-*acetyl-*L-*glutamate 5-phosphotransferase; N-AGSA,* N-*acetylglutamyl-5-semialdehyde; N-AGSAOR,* N-*acetyl-*L-*glutamate-5-semialdehyde:NADP+ oxidoreductase; N-AO,* N-*acetyl-*L-*ornithine; N-AO:aKGAT,* N-*acetylornithine: α-ketoglutarate aminotransferase; N-AO:GN-AT,* $N^2$-*acetyl-*L-*ornithine:* L-*glutamate* N-*acetyltransferase; N-AOAH,* $N^2$-*acetyl-*L-*ornithine amidohydrolase; ORN,* L-*ornithine; OaAT, ornithine α-aminotransferase; OdAT, ornithine δ-aminotransferase; Pi, inorganic phosphate; P5C,* $\Delta^1$-*pyrroline-5-carboxylate; P2C,* $\Delta^1$-*pyrroline-2-carboxylate; P5CR,* $\Delta^1$-*pyrroline-5-carboxylate reductase; P5CS,* $\Delta^1$-*pyrroline-5-carboxylate synthetase; P2CR,* $\Delta^1$-*pyrroline-2-carboxylate reductase; PRO,* L-*proline.*

Although P5CR has been identified and characterized in several plant species (e.g. Krueger *et al.*, 1986; LaRosa *et al.*, 1991; Treichel, 1986), there have been no reports of GK or GPR activities in higher plants since the first report of the occurrence of *in vitro* synthesis of [$^{14}$C]GSA from [$^{14}$C]glutamate, $Mg^{2+}$, ATP and NADPH in cell-free extracts of sugar beet (Morris *et al.*, 1969). The enzyme activity described by Morris *et al.* (1969) is similar to that reported for the P5C synthase activity of mammalian cells (Lodato *et al.*, 1981; Wakabayashi and Jones, 1983).

Efforts devoted to identifying a GK activity in plant extracts by the γ-glutamyl hydroxamate assay procedure have been hindered by a large background of γ-glutamyl hydroxamate formation activity catalysed by glutamine synthetase in plant tissues (Miflin and Lea, 1977). Pyroglutamate or Pi production from glutamate, $Mg^{2+}$ and ATP cannot be employed as a specific assay for GK in crude extracts because glutamine synthetase also produces pyroglutamate and Pi when incubated in the presence of glutamate, $Mg^{2+}$ and ATP (Meister, 1983).

There does not appear to have been a systematic search for GPR in higher plants. In addition to the problem of interference from P5CR in detecting GPR activity in crude extracts alluded to above, plant tissues also have an enzyme capable of oxidizing P5C to glutamate; P5C:NAD(P)$^+$ oxidoreductase (Equation 9.7). This enzyme is the second enzyme of the two-step pathway of proline oxidation in higher plants (Elthon and Stewart, 1982; Huang and Cavalieri, 1979; Rayapati and Stewart, 1991; Stewart, 1981; Stewart and Boggess, 1978; Thompson, 1980):

$$\text{Proline} + \tfrac{1}{2}O_2 \xrightarrow[\substack{\text{Proline}\\ \text{dehydrogenase}}]{} \Delta^1\text{-pyrroline-5-carboxylate} + H_2O \qquad (9.6)$$

$$\text{GSA} \Leftrightarrow \text{P5C} + \text{NAD(P)}^+ + H_2O \xrightarrow[\substack{\text{P5C:NAD(P)}^+\\ \text{oxidoreductase}}]{} \text{glutamate} + \text{NAD(P)H} + H^+ \qquad (9.7)$$

The reaction catalysed by P5C:NAD(P)$^+$ oxidoreductase (Equation 9.7) differs from the reverse reaction catalysed by GPR (Equation 9.8) in that the latter reaction is Pi dependent, and pyroglutamate (not glutamate) is produced as product:

$$\text{P5C} \Leftrightarrow \text{GSA} + \text{NAD(P)}^+ + \text{Pi} \xrightarrow[\substack{\gamma\text{-glutamyl}\\ \text{phosphate}\\ \text{reductase}}]{} \text{pyroglutamate} + \text{Pi} + \text{NAD(P)H} + H^+ \qquad (9.8)$$

The occurrence of P5C:NAD(P)$^+$ oxidoreductase can lead to GSA- ⇔ P5C-dependent NADP$^+$ reduction which interferes with measurement of Pi- and P5C- ⇔ GSA-dependent NADP$^+$ reduction catalysed by GPR.

Given these problems associated with measurement of GPR, we have sought this enzyme activity in extracts of tobacco cells adapted to 25 g $l^{-1}$ NaCl by monitoring Pi-dependent pyroglutamate production from P5C ⇔ GSA and NADP$^+$ instead of monitoring NADP$^+$ reduction to NADPH. Tobacco cells adapted to 25 g $l^{-1}$ NaCl were employed because they accumulate high levels of proline (Binzel *et al.*, 1987) due primarily to enhanced synthesis from glutamate (Rhodes and Handa, 1989). An enzyme capable of converting P5C ⇔ GSA to pyroglutamate in the presence of Pi and NADP$^+$ was identified in partially purified tobacco cell extracts (Figure 9.2a). A Pi-dependent pyroglutamate formation activity was eluted from DEAE–Sepharose at 250 mM NaCl immediately after P5CR (Figure 9.2a). Although this assay procedure is tedious, it is less prone to interference by other enzymatic activities than is NADPH formation. As far as we are aware this is the first report of the occurrence of a GPR-like activity in extracts from higher plant cells.

## 9.5 Cloning by complementation of higher plant cDNAs encoding enzymes of proline synthesis

In contrast to the poor progress in measuring enzymes of proline synthesis in higher plant extracts, there has been rapid progress in recent years in the cloning of cDNAs encoding these enzymes from higher plants by complementation of proline-requiring mutants of *E. coli.* A cDNA encoding P5CR was cloned by direct complementation of an *E. coli proC* proline auxotroph with a soybean nodule cDNA expression library (Delauney and Verma, 1990). This cDNA then facilitated the isolation of an homologous P5CR gene from *Pisum sativum* (Williamson and Slocum, 1992). The sequence of the cDNA encoding P5CR of soybean nodules also enabled the selection of suitable primers for the isolation of a P5CR gene from *Arabidopsis thaliana* (Verbruggen *et al.*, 1993). In soybean, pea and *Arabidopsis*, P5CR transcripts increase in abundance in response to osmotic stress, indicating that P5CR gene transcription is under osmotic stress control (Delauney and Verma, 1990; Verbruggen *et al.*, 1993; Williamson and Slocum, 1992). It is still not certain, however, whether this induction of P5CR mRNA contributes to enhanced proline accumulation under stress. The P5CR cDNA from soybean has been over-expressed in tobacco (Szoke *et al.*, 1992). The transgenic plants exhibit 100-fold greater P5CR activity than wild-type. However, proline concentrations were not increased significantly in these transgenic plants (Szoke *et al.*, 1992). Thus, P5CR may not be the rate-limiting step in proline accumulation (Delauney and Verma, 1993; LaRosa *et al.*, 1991).

A cDNA encoding an ornithine-δ-aminotransferase was isolated from a mothbean (*Vigna aconitifolia*) cDNA expression library employing a *proBA* mutant of *E. coli* (CSH26) grown on nitrogen-free medium supplemented with 10 mM ornithine. This mutant cannot utilize ornithine as a proline source. Introduction of the ornithine-δ-aminotransferase cDNA from *Vigna* allowed conversion of ornithine to GSA $\Leftrightarrow$ P5C which was then readily converted to proline by P5CR encoded by the endogenous *E. coli proC* gene, facilitating growth on ornithine in the absence of exogenous proline (Delauney and Verma, 1993, Delauney *et al.*, 1993; Verma *et al.*, 1992). In striking contrast to P5CR, ornithine-δ-aminotransferase appears to be markedly down-regulated under salinity stress (Delauney *et al.*, 1993).

A cDNA encoding both GK and GPR similarly was isolated from a mothbean cDNA expression library employing *E. coli proA*, *proB* and *proBA* proline auxotrophs, and screening for cDNAs which permit growth in the absence of proline. This cDNA was found to encode a bifunctional enzyme $\Delta^1$-pyrroline-5-carboxylate synthetase (P5CS) (Hu *et al.*, 1992). The single major open reading frame of this cDNA encodes a polypeptide of 73.2 kDa which has two distinct domains exhibiting 55.3% overall similarity to *E. coli* GK and 57.9% similarity to

*E. coli* GPR, respectively (Hu *et al.*, 1992). The GK activity (as measured by the γ-glutamyl hydroxamate assay (Hayzer and Leisinger, 1980)) of this bifunctional enzyme is inhibited by proline (50% inhibition with 6 mM proline) (Hu *et al.*, 1992). This enzyme appears to be much less sensitive to feedback inhibition than the wild-type GK of *E. coli* (50% inhibition by 0.2 mM (Hu *et al.*, 1992) to 0.1 mM proline (Smith, 1985)). Northern analyses indicate that the P5CS gene is induced (particularly in leaves) by treatment of *Vigna* plants with 200 mM NaCl (Hu *et al.*, 1992).

García-Ríos *et al.* (1991, 1994) have also recently cloned a cDNA encoding both GK and GPR from a tomato fruit cDNA expression library by complementation of *proB E. coli* mutants. It was subsequently found that this cDNA (designated *PRO1*) complemented both *proA* and *proBA* mutants of *E. coli* (García-Ríos *et al.*, 1991, 1994). The nucleotide sequence of the *PRO1* cDNA revealed two protein domains; a GK domain consisting of 283 amino acids at the 5' end, and a GPR domain consisting of 429 amino acids at the 3' end (García-Ríos *et al.*, 1994). The GK domain has a 66–69% amino acid sequence similarity to the GKs of *E. coli* and yeast, and a 69% similarity to the GK domain of mothbean P5CS. The GPR domain has a 69 and 66% similarity to the corresponding enzyme from *E. coli* and the corresponding domain of mothbean P5CS, respectively (García-Ríos *et al.*, 1994). In marked contrast to mothbean P5CS, the tomato *PRO1* cDNA has an in-frame TAA (translation termination) codon between the two segments that specify GK and GPR (García-Ríos *et al.*, 1994). This codon is followed after 5 bp (i.e. at a −1 or +2 frameshift with respect to GK) by a potential translation start ATG codon (García-Ríos *et al.*, 1994). Consistent with this, when expressed in *E. coli* maxi cells, *PRO1* directs the synthesis of two separate polypeptides (30 and 45 kDa) of the approximate mass predicted from the sequence analysis (García-Ríos *et al.*, 1994).

## 9.6 Some characteristics of the *PRO1* gene products

We have recently begun to characterize the catalytic properties of the two polypeptides encoded by the *PRO1* gene by purifying these polypeptides from *E. coli proAB* mutant cells expressing *PRO1*. We first purified the *PRO1* GPR enzyme using a combination of DEAE–Sepharose, phenyl–Sepharose and ω-aminohexyl–agarose chromatography. This enzyme was assayed routinely by monitoring Pi- and P5C-dependent reduction of $NADP^+$ (Hayzer and Leisinger, 1981). The purified protein had a molecular mass (as determined by sodium dodecyl sulphate–polyacrylamide gel electrophoresis (SDS–PAGE)) of approximately 43–45 kDa (not shown), consistent with it being the larger of the two polypeptides encoded by *PRO1* (García-Ríos *et al.*, 1994). Like the bacterial enzyme (Hayzer and Leisinger, 1981), and the GPR activity detected in partially purified tobacco cell extracts (Figure 9.2a), the purified *PRO1* GPR was Pi dependent (Figure 9.2b). Pyroglutamate was produced stoichiometrically with $NADP^+$ reduction. GPR was not inhibited by 1 mM proline as determined by $NADP^+$ reduction (Figure 9.2b) and pyroglutamate formation.

Once *PRO1* GPR was purified to apparent homogeneity, we then employed this enzyme to purify GK from the *E. coli* cells expressing *PRO1*. The purified GPR was used to develop a coupled assay containing GPR, L-glutamate, $MgCl_2$, ATP and NADPH at pH 6.5, monitoring GK by its ability to initiate and sustain NADPH oxidation in this reaction mixture. Using this assay, it was found that the *PRO1* GK co-chromatographed with GPR on DEAE–Sepharose, but was subsequently separated from GPR on phenyl–Sepharose (Figure 9.3a). The purified GK had a molecular mass (as determined by SDS–PAGE) of approximately 34 kDa (not shown), consistent with it being the smaller of the two polypeptides encoded by *PRO1* (García-Ríos *et al.*, 1994).

The optimum pH for the coupled reaction of GK and GPR was found to be pH 6.0–6.5, and the optimum ratio of GK to GPR was found to about 1:5. The coupled reaction catalysed by GK + GPR was absolutely dependent on NADPH (apparent $K_m$ = 0.011 mM); no activity was observed when NADH was substituted. L-Aspartate, L-α-aminoadipic acid, L-homoserine, L-asparagine,

---

***Figure 9.2.*** *(a) Separation of P5CR and GPR activities of salt adapted tobacco cells on DEAE–Sepharose. Tobacco cells (130 g fresh weight) adapted to 25 g $l^{-1}$ NaCl were harvested and extracted, and the crude extract applied to a column of DEAE–Sepharose CL-6B as described by LaRosa* et al. *(1991). The column was eluted with a step-wise gradient of NaCl from 100 to 250 mM. NaCl concentrations are given as mM × 10 (■). Fractions of 10 ml each were collected. Total protein (▲) (μg/fraction) and P5CR (◊) nmol NADPH consumed/min/fraction) were determined as described by LaRosa* et al. *(1991). GPR was determined by incubating 100 μl aliquots of each fraction with 50 mM imidazole acetate pH 7.2, freshly prepared 0.4 mM P5C (LaRosa* et al., *1991) and 0.5 mM NADP+ in the presence (◆) or absence (□) of 5 mM $PO_4^{2-}$ for 60 min at 30°C. GPR activities are expressed in units of nmol pyroglutamate formed/h/fraction. Pyroglutamate was isolated from the enzyme reaction mixtures by collecting the aqueous wash from Dowex-50-$H^+$, subjecting this aqueous wash to acid hydrolysis (2 h at 120°C in 2 M HCl) to convert pyroglutamate to glutamate, purifying the glutamate so formed by retention on Dowex-50-$H^+$ and elution with $NH_4OH$, derivatization of the glutamate to its heptafluorobutyryl isobutyl (HFBI) derivative and quantification by gas chromatography essentially as described by Rhodes* et al. *(1986). α-Aminoadipic acid (added after the acid hydrolysis step) was used as internal standard (Rhodes, Rhodes, Bressan and Csonka, unpublished results). (b) $NADP^+$ reduction by purified* PRO1 *GPR. GPR was purified from* E. coli *cells expressing* PRO1 *by a combination of DEAE–Sepharose CL-6B, phenyl–Sepharose and ω-aminohexyl–agarose chromatography. The purified enzyme (33.8 μg/assay) was incubated in 50 mM imidazole acetate pH 7.2 in the presence of 0.4 mM P5C (LaRosa* et al. *1991) and 0.5 mM $NADP^+$ in the absence (□) or presence (◆) of 5 mM $PO_4^{2-}$ for 43 min at 30°C, monitoring absorbance at 340 nm. When 1 mM proline was added to the complete reaction mixture (◊) no inhibition of activity was observed. No $NADP^+$ reduction was observed when P5C was omitted (■) (Samaras, Csonka and Rhodes unpublished results).*

L-ornithine, L-arginine and L-citrulline were not inhibitory at 1 mM concentrations. The absence of inhibition by ornithine contrasts with regulation of the mammalian P5C synthase (Lodato *et al.*, 1981; Wakabayashi and Jones, 1983).

The coupled reaction catalysed by GK + GPR showed kinetic properties with respect to glutamate similar to those described by Baich (1969) for the GK of *E.*

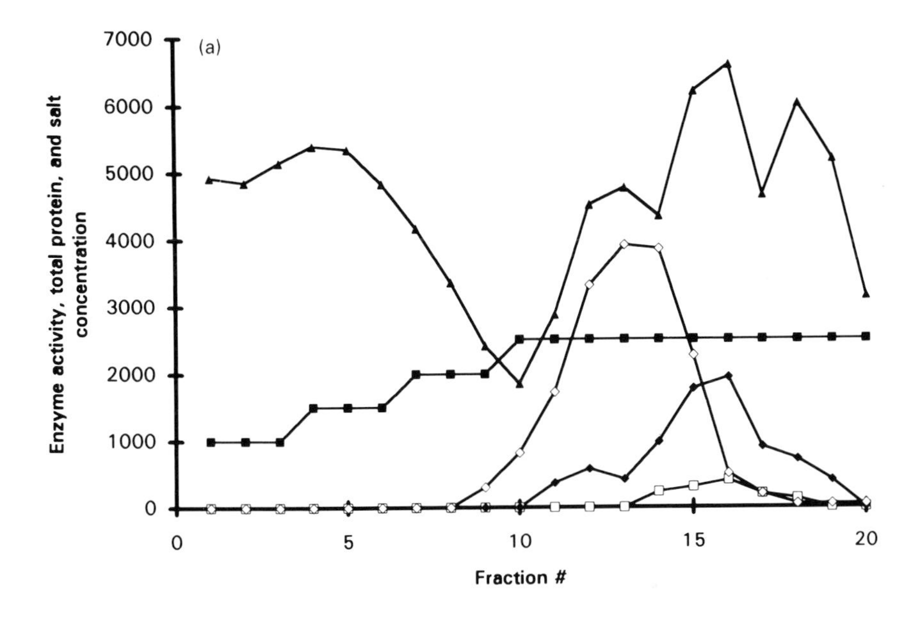


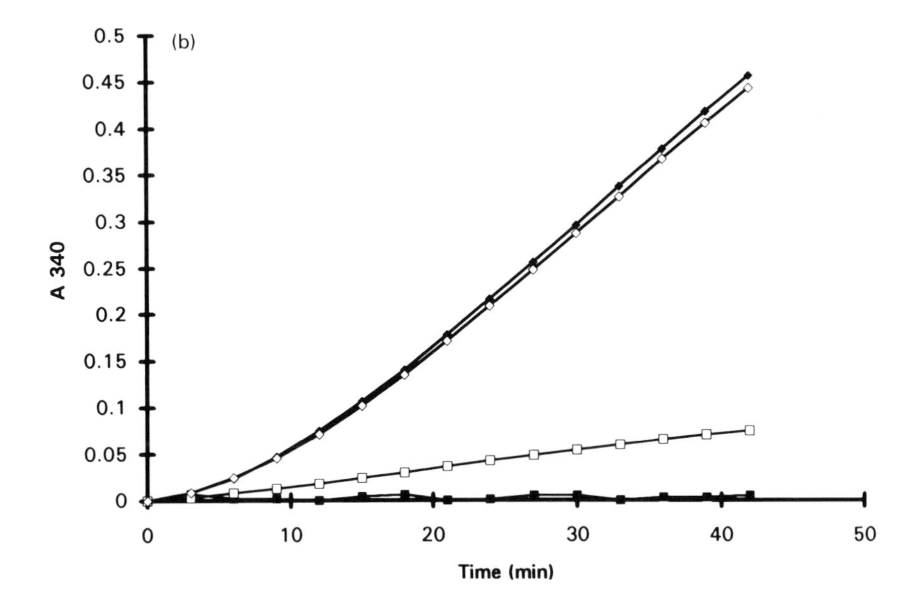

*coli*, that is a low affinity for glutamate confounded by a marked substrate activation as the concentration of glutamate is increased (not shown). The coupled reaction catalysed by GK + GPR was found to be very sensitive to inhibition by proline, but the precise inhibition kinetics were highly dependent on the concentration of glutamate employed in the asssay (Figure 9.3b). Thus, at 75 mM glutamate, 50% inhibition of the coupled activity was achieved with 0.08 mM proline whereas at 10 mM glutamate, 50% inhibition of the coupled activity was achieved with 0.02 mM proline (Figure 9.3b). As for *E. coli* GK (Baich, 1969), proline appears to increase the apparent $K_m$ for glutamate of the coupled reaction catalysed by *PRO1* GK + GPR. Although the coupled *PRO1* GK + GPR activity exhibits typical Michaelis–Menten kinetics with respect to ATP in the absence of proline (apparent $K_m$ for ATP = 2.1 mM), proline was found to be more inhibitory when the ATP concentration was reduced (< 1 mM). Thus, proline also appears to increase the apparent $K_m$ for ATP.

We are currently optimizing an assay in which the *PRO1* GK is measured in the absence of *PRO1* GPR by monitoring Pi release from ATP in the presence of $Mg^{2+}$ and glutamate, in order to ascertain whether the above kinetic properties are a function of the two enzymes acting in concert or properties manifested by the GK enzyme alone. Preliminary results (not shown) suggest that, at 75 mM glutamate, 4 mM ATP, 8 mM $MgCl_2$, and pH 6.5, 0.1 mM proline inhibits glutamate-dependent ATP hydrolysis to Pi catalysed by *PRO1* GK by 60–70%. This is very similar to the results obtained in the coupled assay, suggesting that certain of the

***Figure 9.3.*** *(a) Separation of* PRO1 *GPR and* PRO1 *GK on phenyl–Sepharose. Extracts from* E. coli *cells expressing* PRO1 *were applied to a DEAE–Sepharose CL-6B column. Peak fractions containing both GPR and GK eluting at 250 mM NaCl from DEAE–Sepharose CL-6B were brought to 350 mM NaCl and applied to a column of phenyl–Sepharose. The column was then washed with a gradient of 350 mM NaCl + 0% ethylene glycol to 0 mM NaCl + 50% ethylene glycol. Fractions were of about 5 ml each. Total protein (◆) (µg/fraction) was determined as described by LaRosa* et al. *(1991). GPR activity (■) is expressed as nmol NADPH formed/min/fraction (×10) employing the assay conditions described in Figure 9.2b legend. GK (□) was assayed by monitoring NADPH oxidation ($A_{340}$ nm) using purified GPR (11 µg), 75 mM monosodium glutamate, 4 mM ATP, 8 mM $MgCl_2$, 0.2 mM NADPH, 1 mM dithiothreitol (DTT), 50 mM MOPS–Tris, pH 6.5. GK activity is expressed as nmol NADPH consumed/min/fraction (Samaras, Csonka, García Ríos, Paino D'Urzo, Rhodes and Bressan, unpublished results). (b) Inhibition of the coupled reaction catalysed by purified* PRO1 *GK and* PRO1 *GPR by proline at different glutamate concentrations. Assays contained 2.2 µg GK and 11.4 µg GPR purifed as described in Figure 9.2(b). All assays contained 4 mM ATP, 8 mM $MgCl_2$, 0.2 mM NADPH, 1 mM DTT, 50 mM MOPS–Tris, pH 6.5, and the concentrations of proline indicated. Assays included different concentrations of monosodium glutamate: 75 mM (■); 30 mM (□); 20 mM (◆) and 10 mM (◊). Activity is expressed as the rate of NADPH oxidation as a percentage of the control (–proline) at each glutamate concentration (Samaras, Csonka and Rhodes, unpublished results).*

kinetic properties manifested in the coupled assay (Figure 9.3b) are properties of the GK enzyme alone. Pi release in the GK assay appears to be stoichiometric with NADPH oxidation in the coupled assay (not shown).

The reaction catalysed by tomato *PRO1* GK appears to be much more sensitive than the equivalent reaction catalysed by *Vigna* P5CS to feedback inhibition by

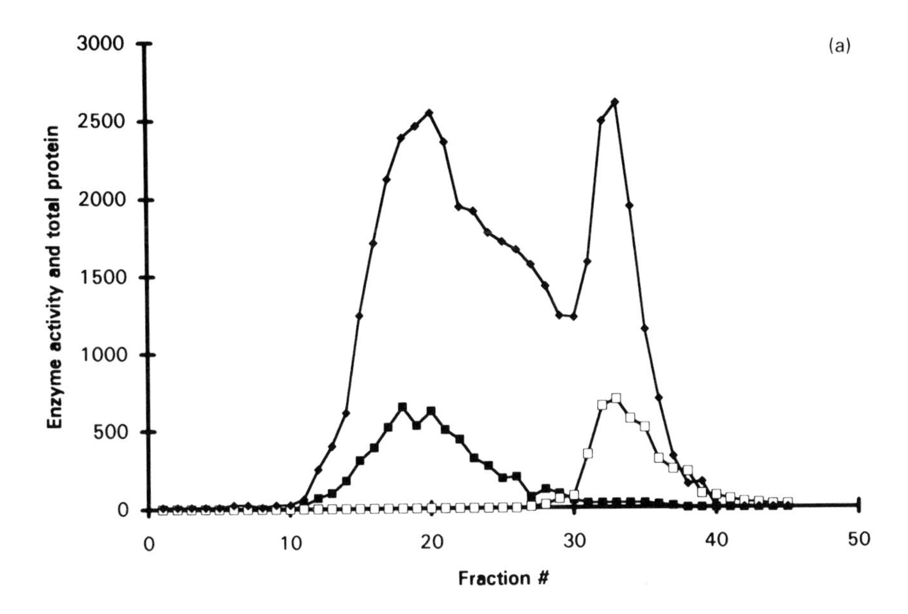


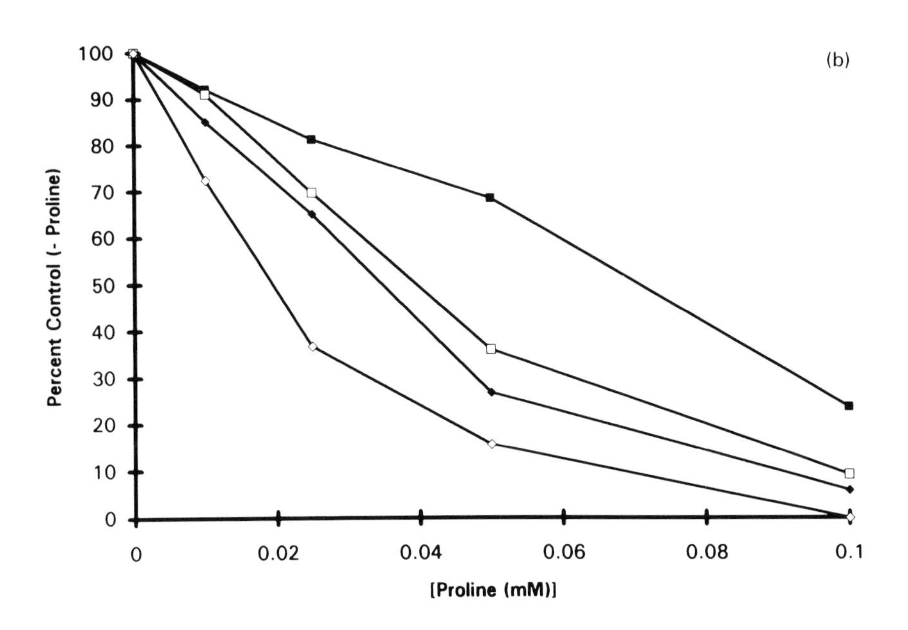

proline (50% inhibition with 0.08 mM and 6 mM, respectively) (Hu *et al.*, 1992). The reason for this is presently unclear. Delauney and Verma (1993) suggest that the reduced sensitivity of the *Vigna* P5CS enzyme to feedback inhibition by proline is consistent with *Vigna* being a proline accumulator under drought stress. However, tomato cells in culture can also produce large quantities of proline when adapted to water stress (Rhodes *et al.*, 1986).

We wish to emphasize that, while we have observed two discrete gene products from the *PRO1* gene expressed in *E. coli* maxi cells, it is not yet known whether two discrete gene products or a single hybrid enzyme (similar to the P5CS of *Vigna*) are produced by tomato plants and/or cells in culture. Antibodies against the two purified gene products currently are being prepared in order to address these questions.

Several analogues of proline were tested for their ability to inhibit the coupled reaction catalysed by the products of the *PRO1* gene. L-Azetidine-2-carboxylic acid, *N*-methyl-L-proline, 3,4-dehydro-D,L-proline, and *trans*-4-hydroxy-L-proline were severely inhibitory at 1 mM concentrations, whereas D,L-pipecolic acid, D,L-hydroxypipecolic acid, pyroglutamate, D-proline and L-prolinebetaine were not inhibitory at 1 mM. It should now be possible to select for mutant forms of the *PRO1* gene in *E. coli* which are resistant to inhibitory proline analogues (e.g. 3,4-dehydro-D,L-proline). The products of these mutant alleles could then be tested for altered feedback inhibition properties with respect to proline *in vitro*. Feedback-insensitive mutant alleles could then be transformed into higher plants to test whether they confer proline accumulation and confer reduced feedback inhibition of proline biosynthesis *in vivo*. In anticipation that such experiments may soon be feasible, we have addressed some of the problems associated with measurement of feedback inhibition of proline biosynthesis *in vivo*.

## 9.7 Measurement of feedback inhibition of proline biosynthesis *in vivo*

In plants there are several problems associated with measuring the degree of inhibition of proline synthesis by proline *in vivo*. When proline synthesis from an exogenously supplied source of labelled glutamate is monitored in the presence and absence of an exogenously supplied source of proline, results obtained can be confounded by the possibility of: (i) inhibition of labelled glutamate uptake by proline, and (ii) the possibility of rapid catabolism of proline to glutamate which could then isotopically dilute the glutamate pool serving as precursor of proline. These problems were addressed specifically in the [$^{14}$C]glutamate labelling studies described by Boggess *et al.* (1976). Similar problems apply to [$^{14}$C]acetate feeding studies (Oaks *et al.*, 1970); the extent of inhibition of [$^{14}$C]acetate uptake by proline and the degree of isotope dilution of the glutamate pool caused by proline oxidation to glutamate need to be taken into account. The approach that we have adopted to overcome some of these problems is to substitute $^{15}NH_4^+$ and $^{15}NO_3^-$ for the corresponding $^{14}$N nitrogen sources of cultured cells, monitoring the $^{15}$N-labelling of proline and several other amino acids (including alanine, glutamate, glutamine-amino-N and $\gamma$-aminobutyric acid (GABA)) using gas chromatography–mass spectroscopy (GC–MS) of N-heptafluorobutyryl isobutyl (HFBI) derivatives (Rhodes and Handa, 1989; Rhodes *et al.*, 1986). This approach has the advantage of achieving labelling of proline without supplying labelled glutamate exogenously. At the same time as the cells are transferred to $^{15}N$ medium, the cells can be supplied with deuterium-labelled $d_7$-proline. In principle, the uptake of $d_7$-proline should cause the pool of proline to expand. If the expansion of the proline pool causes specific inhibition of proline synthesis, then the $d_7$-proline should inhibit the transfer of $^{15}N$ from glutamate to proline without effect on the rate of $^{15}N$-labelling of glutamate, glutamine-amino-N, alanine and GABA. Because $d_7$-proline is 7 atomic mass units (a.m.u.) resolved from unlabelled proline and 6 a.m.u. resolved from [$^{15}N$]proline in GC–MS, the $d_7$-proline supplied does not interfere with measurement of $^{15}N$ abundance of the endogenous pool of proline derived by *de novo* synthesis from glutamate. Rapid catabolism of the supplied proline to glutamate can be monitored directly because any oxidation of the $d_7$-proline to glutamate should lead to production of $d_5$-glutamate; 5 a.m.u. resolved from endogenous glutamate and 4 a.m.u. resolved from newly synthesized [$^{15}N$]glutamate.

Figure 9.4 shows a series of electron ionization GC–MS mass spectra of a fragment ion of HFBI-proline during a 6-h time course of incubation of unadapted tobacco cells (grown on 0 g $l^{-1}$ NaCl) with $^{15}NH_4^+$ and $^{15}NO_3^-$ (Figure 9.4a) or $^{15}NH_4^+$ and $^{15}NO_3^-$ + 0.5 mM $d_7$-proline (Figure 9.4b). This particular fragment ion of proline (m/z 266 for $^{14}N,d_0$-proline) contains the four carbons of the proline ring, the single N atom of proline, but is missing the $\alpha$-carboxyl group. Since all seven deuterium atoms of $d_7$-proline are associated with the four carbons of the ring, the $d_7$-proline shows an equivalent fragment ion at m/z 273 (Figure 9.4b). Note that in the control cells supplied with $^{15}NH_4^+$ and $^{15}NO_3^-$ alone there is a gradual increase in the intensity of an ion at m/z 267 due to [$^{15}N$]proline synthesis (Figure 9.4a). In the cells supplied with $^{15}NH_4^+$ and $^{15}NO_3^-$ + 0.5 mM $d_7$-proline, the rapid uptake of $d_7$-proline leads to a diminishing signal at m/z 266 relative to the $d_7$-proline signal at m/z 273 (Figure 9.4b). We estimate that the total amount of proline increases from about 1 μmol $g^{-1}$ fresh weight to about 15 μmol $g^{-1}$ fresh weight during this 6-h treatment (not shown). Despite the dominant signal from the $d_7$-proline fragment (m/z 273), it was still possible to obtain relative ion ratios of the $^{14}N,d_0$-proline and $^{15}N,d_0$-proline at m/z 266 and 267, respectively, by selected ion monitoring over a narrow mass range (Figure 9.4c). There was no evidence for an increase in the $^{15}N,d_0$-proline fragment ion intensity at m/z 267 during the 6-h time course in the 0.5 mM $d_7$-proline-treated cells (Figure 9.4c), suggesting that the 15-fold elevated proline pool completely inhibited [$^{15}N$]proline synthesis. As will be shown below, this was not due to inhibition of $^{15}N$ uptake or glutamate synthesis.

Figure 9.5. shows equivalent electron ionization GC–MS mass spectra of a fragment ion of HFBI-glutamate during the 6-h time course (a = without proline; b = with 0.5 mM $d_7$-proline) (this particular fragment was selected because it contains the same four carbon atoms and the single N atom represented in the

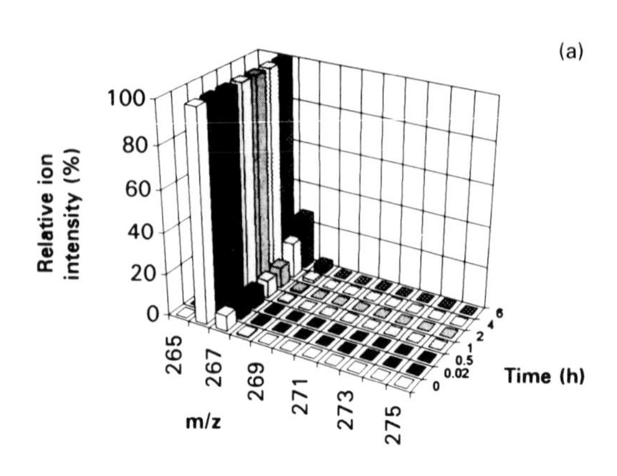


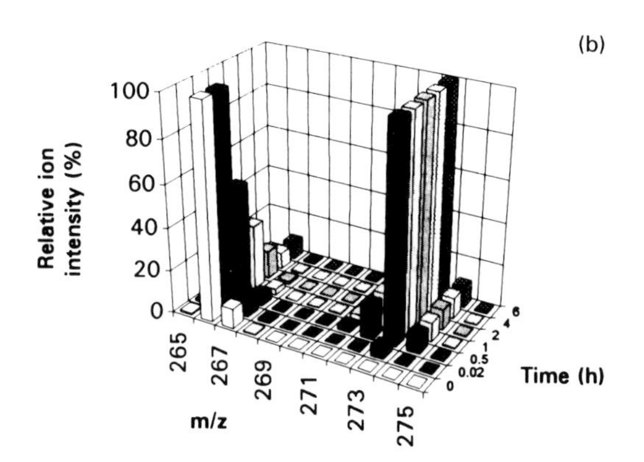


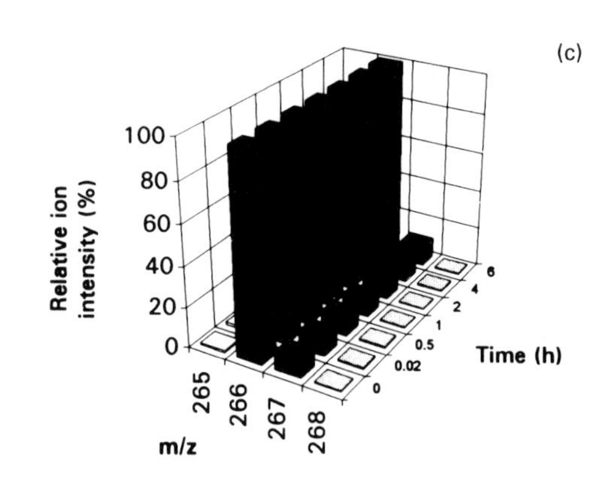


***Figure 9.4.*** *Electron ionization GC–MS mass spectra over the mass range 265–275 atomic mass units (a.m.u.) of HFBI-proline during a time course (0, 0.02, 0.5, 1, 2, 4 and 6 h) in which unadapted tobacco cells grown on 0 g $l^{-1}$ NaCl were transferred to fresh medium substituting $^{15}NH_4^+$ and $^{15}NO_3^-$ as the sole N sources at t = 0 h in the absence (a) or presence of 0.5 mM $d_7$-proline (b and c). The molecular weight (M) of the proline derivative is 367. The ion at m/z 266 corresponds to a fragment ion in which the isobutyl-derivatized α-carboxyl group is eliminated; $[M - COOC_4H_9]^+$ = M−101. The ion at m/z 267 represents natural abundance [$^{13}$C]proline and newly synthesised [$^{15}$N]proline. The ion at m/z 273 represents the equivalent fragment ion of $d_7$-proline (b). Panel c shows relative ion intensities observed at m/z 266 and 267 for the time course in the presence of $d_7$-proline using selected ion monitoring over a narrow mass range (265–268 a.m.u.) (Rhodes and Handa, unpublished results).*

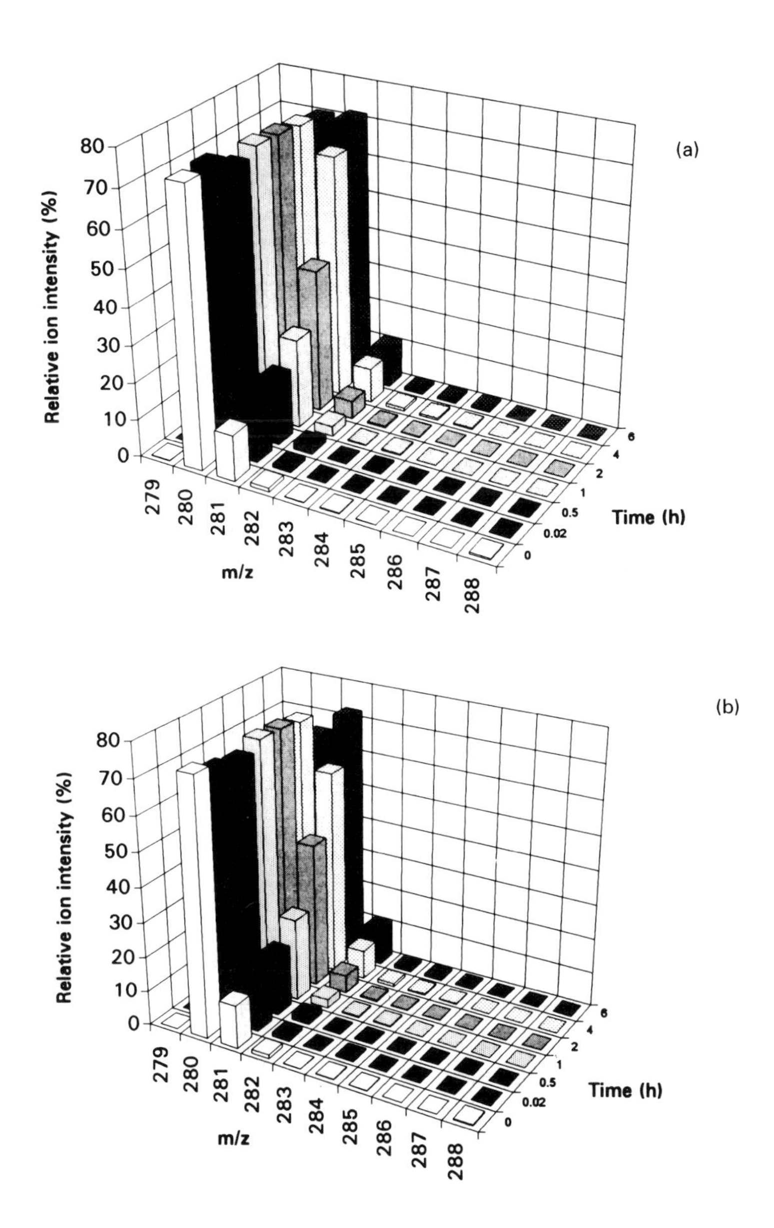


***Figure 9.5.*** *Electron ionization GC–MS mass spectra over the mass range 279–288 a.m.u. of HFBI-glutamate during the time course (0, 0.02, 0.5, 1, 2, 4 and 6 h) of incubation of tobacco cells in* $^{15}$*N-medium in the absence (a) or presence of 0.5 mM* $d_7$*-proline (b) as in Figure 9.4. The ion at m/z 280 contains the same four carbon atoms and single N atom represented in the proline fragment ion monitored in Figure 9.4 (Rhodes and Handa, unpublished results).*

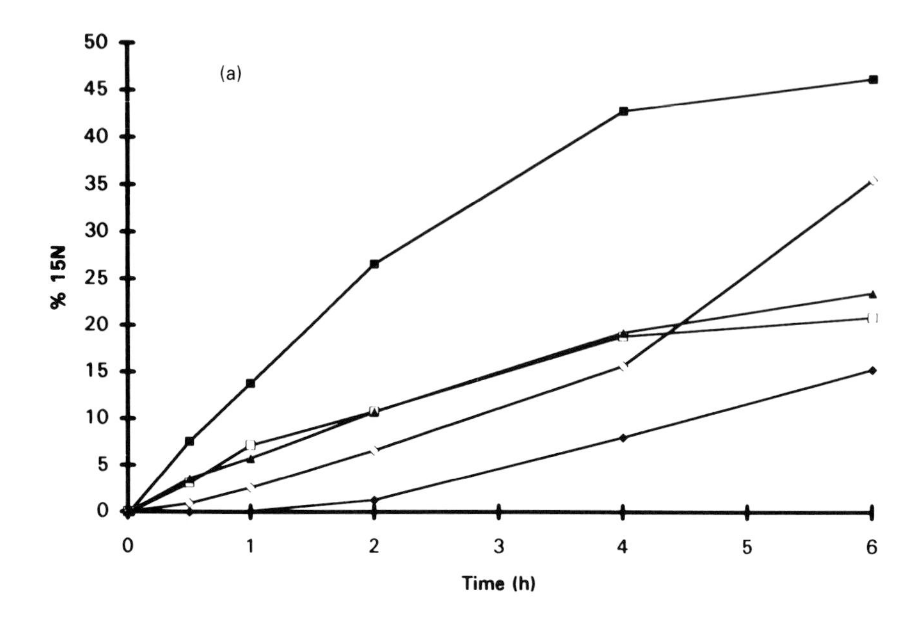


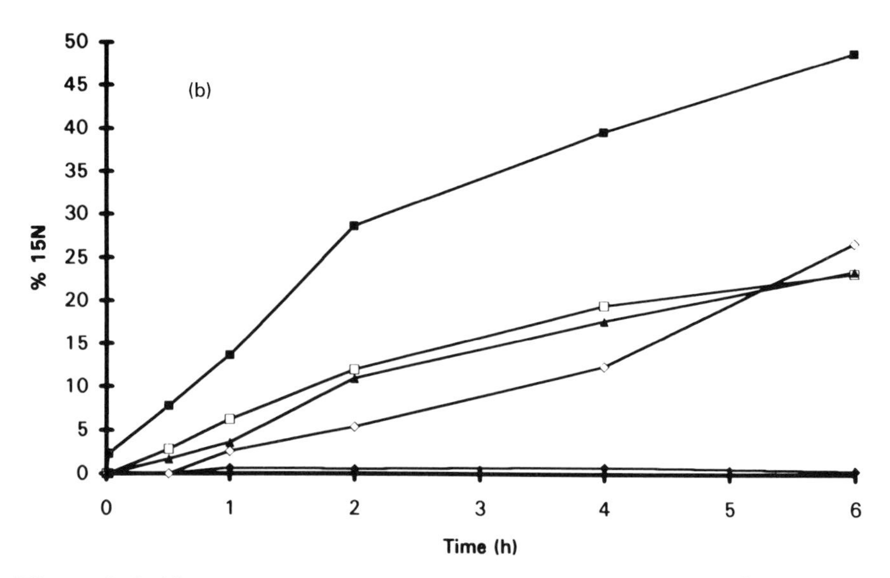


***Figure 9.6.*** *$^{15}N$ abundance of glutamate (■), glutamine-amino-N (◊), alanine (□), GABA (▲) and proline (◆) during the time course of incubation of tobacco cells in $^{15}N$-medium in the absence (a) or presence (b) of 0.5 mM $d_7$-proline as described in Figures 9.4 and 9.5. $^{15}N$ abundance was calculated from the GC–MS ion ratios of the corresponding HFBI amino acid derivatives as described by Rhodes* et al. *(1986) (Rhodes and Handa, unpublished results).*

proline fragment ion discussed above). There was no obvious effect of exogenously supplied $d_7$-proline on the rate of $^{15}N$-labelling of glutamate, as judged by the time course of changes in relative abundance of ions at m/z 280 and m/z 281. There was also no evidence for metabolism of $d_7$-proline to $d_5$-glutamate; no significant signal above background was observed at m/z 285 in the $d_7$-proline-treated cells (Figure 9.5b).

Figure 9.6 summarizes results obtained from these GC–MS analyses, calculating $^{15}N$ abundance of amino acids from the GC–MS ion ratios as described by Rhodes *et al.* (1986). The $^{15}N$-labelling patterns of glutamate, glutamine, alanine and GABA were virtually identical in cells supplied with $^{15}N$ without (Figure 9.6a) or with (Figure 9.6b) $d_7$-proline. Thus, expansion of the proline pool appears to be without effect on the rates of synthesis of glutamate, glutamine, alanine and GABA *in vivo*. The inhibition of proline biosynthesis by proline appears to be quite specific (Figure 9.6), as also concluded by Boggess *et al.* (1976).

These labelling procedures are likely to be very useful when transgenic tobacco plants and/or cell cultures expressing mutant (feedback-insensitive) alleles of *PRO1* are available. Direct measurements of proline biosynthesis rates and sensitivity to feedback inhibition by proline *in vivo* will be required to verify that these mutant alleles behave as anticipated.

## 9.8 *N*-Methylated proline derivatives in higher plants

Most reviews of proline metabolism in plants consider proline utilization in protein synthesis and oxidation of proline to glutamate as the major metabolic fates of proline (Csonka and Baich, 1983; Stewart, 1981; Thompson, 1980). This does not apply to all higher plants. Certain plant species accumulate high concentrations of *N*-methylated proline derivatives, including *N*-methyl-L-proline, *N*-methyl-*trans*-4-hydroxy-L-proline, L-prolinebetaine (*N,N*-dimethylproline; stachydrine) and *trans*-4-hydroxy-L-prolinebetaine (betonicine), some of which may be equally if not more osmoprotective than proline (Hanson, 1993; Hanson *et al.*, 1991, 1994; Naidu *et al.*, 1987, 1992; Rhodes and Hanson, 1993; Solomon *et al.*, 1994; Wyn Jones and Storey, 1981). It has long been established that *N*-methyl-L-proline (hygric acid) is an intermediate in L-prolinebetaine synthesis from L-proline in alfalfa (Essery *et al.*, 1962; Robertson and Marion, 1960), but the putative *N*-methyltransferase(s) involved have not yet been characterized:

$$\text{L-Proline} \xrightarrow{+CH_3} \text{N-methyl-L-proline} \xrightarrow{+CH_3} \text{L-prolinebetaine} \quad (9.9)$$

It is probable that proline is also a precursor of *trans*-4-hydroxy-L-prolinebetaine but the precise order of the *N*-methylation and hydroxylation reactions involved is not yet known (Hanson, 1993; Hanson *et al.*, 1994).

Prolinebetaine and *trans*-4-hydroxy-L-prolinebetaine accumulation in the Plumbaginaceae tends to be associated with species which are adapted to more chronically dry environments (Hanson *et al.*, 1994). L-Prolinebetaine and *trans*-

4-hydroxy-L-prolinebetaine are more effective osmoprotectants than L-proline in bacteria (Hanson *et al.*, 1994). Thus, the synthesis of these compounds may represent an adaptive strategy for converting proline into a compatible solute which is more effective in severe osmotic stress environments (Hanson *et al.*, 1994). Because L-prolinebetaine is not inhibitory to the reaction catalysed by the *PRO1* gene product, whereas *N*-methyl-L-proline and L-proline are both inhibitory (see Section 9.6 above), this raises the possibility that L-prolinebetaine accumulation may provide a mechanism for alleviating the proline pathway from feedback control. In the Plumbaginaceae, species which accumulate L-prolinebetaine and/or *trans*-4-hydroxy-L-prolinebetaine tend to have a lower free proline pool than species that accumulate alternative betaines (β-alaninebetaine or glycinebetaine). In principle, reduction in the proline pool size by diversion of proline into a non-inhibitory product (e.g. L-prolinebetaine) could result in enhanced flux through the proline pathway.

Because proline can serve as a chemical cue to attract phytophagous insects (Haglund, 1980), removal of proline by accumulation of L-prolinebetaine and/or *trans*-4-hydroxy-L-prolinebetaine may serve to reduce the plant's attractiveness to such predators. Moreover, because *trans*-4-hydroxy-L-prolinebetaine is a potent inhibitor of acetylcholinesterase (Friess *et al.*, 1957), the possibility exists that *trans*-4-hydroxy-L-prolinebetaine may have a dual role as an osmoprotectant and defensive agent against herbivores (Hanson and Burnet, 1994).

## 9.9 Concluding remarks

Rapid progress has been made in recent years in the cloning of cDNAs encoding key enzymes of proline biosynthesis from higher plants (Delauney and Verma, 1993; García Ríos *et al.*, 1994). Specific probes for these genes and their products will assist greatly in clarifying our understanding of how osmotic stress is perceived and transduced into an increase of proline biosynthesis rate. The mechanism(s) by which feedback control of the proline pathway is modulated by stress remains poorly understood. Metabolic engineering of this pathway is likely to play a key role in exploring this mechanism(s) and the rate-limiting step(s) of the pathway (e.g. Delauney and Verma, 1993; Szoke *et al.*, 1992).

The *N*-methyltransferase(s) and prolyl hydroxylase responsible for prolinebetaine and *trans*-4-hydroxy-L-prolinebetaine are poorly characterized (Hanson, 1993; Hanson and Burnet, 1994). The genes encoding these enzymes may prove to be as useful as the genes encoding proline biosynthetic enzymes in future efforts devoted to metabolic engineering of proline metabolism to optimize compatible osmolyte accumulation and plant resistance to osmotic stress.

## Acknowledgements

The manuscript represents Purdue University Agricultural Experiment Station Journal Article No. 14 165. This work was supported by grants from the USDA (contract # 93-37100-8871) and the McKnight Foundation.

## References


Adams, E. and Frank, L. (1980) Metabolism of proline and the hydroxyprolines. *Annu. Rev. Biochem.* **49**, 1005–1061.

Baich, A. (1969) Proline synthesis in *Escherichia coli.* A proline-inhibitable glutamic acid kinase. *Biochim. Biophys. Acta* **192**, 462–467.

Baich, A. (1971) The biosynthesis of proline in *Escherichia coli.* Phosphate-dependent glutamate γ-semialdehyde dehydrogenase (NADP), the second enzyme in the pathway. *Biochim. Biophys. Acta* **244**, 129–134.

Ballantyne, J.S. and Chamberlin, M.E. (1994) Regulation of cellular amino acid levels. In: *Cellular and Molecular Physiology of Cell Volume Regulation* (ed. K. Strange). CRC Press, Boca Raton, FL, pp.111–122.

Barnett, N.M. and Naylor, A.W. (1966) Amino acid and protein metabolism in Bermuda grass during water stress. *Plant Physiol.* **41**, 1222–1230.

Binzel, M.L., Hasegawa, P.M., Rhodes, D., Handa, S., Handa, A.K. and Bressan, R.A. (1987) Solute accumulation in tobacco cells adapted to NaCl. *Plant Physiol.* **84**, 1408–1415.

Boggess, S.F. and Stewart, C.R. (1980) The relationship between water stress induced proline accumulation and inhibition of protein synthesis in tobacco leaves. *Plant Sci. Lett.* **17**, 245–252.

Boggess, S.F., Aspinall, D. and Paleg, L.G. (1976) Stress metabolism. IX. The significance of end-product inhibition of proline biosynthesis and of compartmentation in relation to stress-induced proline accumulation. *Aust. J. Plant Physiol.* **3**, 513–525.

Brady, C.J., Gibson, T.S., Barlow, E.W.R., Spiers, J. and Wyn Jones, R.G. (1984) Salt tolerance in plants. I. Ions, compatible solutes and the stability of plant ribosomes. *Plant, Cell Environ.* **7**, 571–578.

Briens, M. and Larher, F. (1982) Osmoregulation in halophytic higher plants: a comparative study of soluble carbohydrates, polyols, betaines and free proline. *Plant, Cell Environ.* **5**, 287–292.

Cayley, S., Lewis, B.A., Guttman, H.J. and Record, M.T. Jr (1991) Characterization of the cytoplasm of *Escherichia coli* K-12 as a function of external osmolarity. Implications for protein–DNA interactions *in vivo*. *J. Mol. Biol.* **222**, 281–300.

Cayley, S., Lewis, B.A. and Record, M.T. Jr (1992) Origins of the osmoprotective properties of betaine and proline in *Escherichia coli* K-12. *J. Bacteriol.* **174**, 1586–1595.

Csonka, L.N. (1981) Proline over-production results in enhanced osmotolerance in *Salmonella typhimurium*. *Mol. Gen. Genet.* **182**, 82–86.

Csonka, L.N. (1989) Physiological and genetic responses of bacteria to osmotic stress. *Microbiol. Rev.* **53**, 121–147.

Csonka, L.N. and Baich, A. (1983) Proline biosynthesis. In: *Amino Acids, Biosynthesis and Genetic Regulation* (eds K.M. Herrmann and R.L. Somerville). Addison-Wesley, Reading, pp. 35–51.

Csonka, L.N. and Hanson, A.D. (1991) Prokaryotic osmoregulation: genetics and physiology. *Annu. Rev. Microbiol.* **45**, 569–606.

Delauney, A.J. and Verma, D.P.S. (1990) A soybean $\Delta^1$-pyrroline-5-carboxylate reductase gene was isolated by functional complementation in *Escherichia coli* and is found to be osmoregulated. *Mol. Gen. Genet.* **221**, 299–305.

Delauney, A.J. and Verma, D.P.S. (1993) Proline biosynthesis and osmoregulation in plants. *Plant J.* **4**, 215–223.

Delauney, A.J., Hu, C.-A.A., Kisher, P.B.K. and Verma, D.P.S. (1993) Cloning of ornithine δ-aminotransferase cDNA from *Vigna aconitifolia* by *trans*-complementation in *Escherichia coli* and regulation of proline synthesis. *J. Biol. Chem.* **268**, 18673–18678.

Deutch, A.H., Rushlow, K.E., Smith, C.J. and Kretschmer, P.J. (1982) *Escherichia coli* $\Delta^1$-pyrroline-5-carboxylate reductase: gene sequence, protein overproduction and purification. *Nucleic Acids Res.* **10**, 7701–7714.

Deutch, A.H., Smith, C.J. and Rushlow, K.E. (1984) Analysis of the *Escherichia coli proBA* locus by DNA and protein sequencing. *Nucleic Acids Res.* **12**, 6337–6355.

Dörffling, K., Dörffling, H. and Lesselich, G. (1993) *In vitro*-selection and regeneration of hydroxyproline-resistant lines of winter wheat with increased proline content and increased frost tolerance. *J. Plant Physiol.* **142**, 222–225.

Duncan, D.R. and Widholm, J.M. (1987) Proline accumulation and its implication in cold tolerance of regenerable maize callus. *Plant Physiol.* **83**, 703–708.

Elthon, T.E. and Stewart, C.R. (1982) Proline oxidation in corn mitochondria. *Plant Physiol.* **70**, 567–572.

Essery, J.M., McCaldin, D.J. and Marion, L. (1962) The biogenesis of stachydrine. *Phytochemistry* **1**, 209–213.

Friess, S.L., Patchett, A.A. and Witkop, B. (1957) The acetylcholinesterase surface. VII. Interference with surface binding as reflected by enzymatic response to turicine, betonicine and related heterocycles. *J. Am. Chem. Soc.* **79**, 459–462.

Fukutoku, Y. and Yamada, Y. (1984) Sources of proline-nitrogen in water-stressed soybean (*Glycine max*). II. Fate of $^{15}N$-labelled protein. *Physiol. Plant.* **61**, 622–628.

Gamper, H. and Moses, V. (1974) Enzyme organization in the proline biosynthetic pathway of *Escherichia coli*. *Biochim. Biophys. Acta* **354**, 75–87.

García-Ríos, M.G., LaRosa, P.C., Bressan, R.A., Csonka, L.N. and Hanquier, J.M. (1991) Cloning by complementation of the γ-glutamyl kinase gene from a tomato expression library. In: *Abstracts of the Third International Congress of Plant Molecular Biology, Tucson, Arizona, Oct. 6 – 11(1991)* (ed. R.B. Hallick). Abstract # 1507.

García-Ríos, M.G., Csonka, L.N., Bressan, R.A., LaRosa, P.C. and Hanquier, J. (1994) Cloning of a DNA fragment encoding γ-glutamyl kinase and γ-glutamyl phosphate reductase from a tomato cDNA library. In: *Cell Biology: Biochemical and Cellular Mechanisms of Stress Tolerance in Plants, NATO ASI Series H* (ed. J.H. Cherry). Springer, Berlin (in press).

Gibson, T.S., Spiers, J. and Brady, C.J. (1984) Salt tolerance in plants. II. *In vitro* translation of m-RNAs from salt-tolerant and salt-sensitive plants on wheat germ ribosomes. responses to ions and compatible solutes. *Plant, Cell Environ.* **7**, 579–587.

Haglund, B.M. (1980) Proline and valine – cues which stimulate grasshopper herbivory during drought stress? *Nature* **288**, 697–698.

Handa, S., Handa, A.K., Hasegawa, P.M. and Bressan, R.A. (1986) Proline accumulation and the adaptation of cultured plant cells to water stress. *Plant Physiol.* **80**, 938–945.

Hanson, A.D. (1993) Accumulation of quaternary ammonium and tertiary sulfonium compounds. In: *Plant Responses to Cellular Dehydration During Environmental Stress* (eds T.J. Close and E.A. Bray). American Society of Plant Physiologists, Rockville, MD, pp. 30–36.

Hanson, A.D. and Burnet, M. (1994) Evolution and metabolic engineering of osmoprotectant accumulation in higher plants. In: *Cell Biology: Biochemical and Cellular Mechanisms of Stress Tolerance in Plants, NATO ASI Series H* (ed. J.H. Cherry). Springer, Berlin (in press).

Hanson, A.D. and Hitz, W.D. (1982) Metabolic responses of mesophytes to plant water deficits. *Annu. Rev. Plant Physiol.* **33**, 163–203.

Hanson, A.D., Rathinasabapathi, B., Chamberlin, B. and Gage, D.A. (1991) Comparative physiological evidence that β-alanine betaine and choline-*O*-sulfate act as compatible osmolytes in halophytic *Limonium* species. *Plant Physiol.* **97**, 1199–1205

Hanson, A.D., Rathinasabapathi, B., Rivoal, J., Burnet, M., Dillon, M.O. and Gage, D.A. (1994) Osmoprotective compounds from the Plumbaginaceae: a natural experiment in metabolic engineering of stress tolerance. *Proc. Natl Acad. Sci. USA* **91**, 306–310.

Hayzer, D.J. and Leisinger, T. (1980) The gene–enzyme relationships of proline biosynthesis in *Escherichia coli. J. Gen. Microbiol.* **118**, 287–293.

Hayzer, D.J. and Leisinger, T. (1981) Proline biosynthesis in *Escherichia coli.* Stoichiometry and end-product identification of the reaction catalysed by glutamate semialdehyde dehydrogenase. *Biochem. J.* **197**, 269–274.

Hayzer, D.J. and Leisinger, T. (1982) Proline biosynthesis in *Escherichia coli.* Purification and characterization of glutamate-semialdehyde dehydrogenase. *Eur. J. Biochem.* **121**, 561–565.

Hayzer, D.J and Moses, V. (1978a) Proline biosynthesis by cell-free extracts of *Escherichia coli* and potential errors arising from the use of a bioradiological assay procedure. *Biochem. J.* **173**, 207–217.

Hayzer, D.J. and Moses, V. (1978b) The enzymes of proline biosynthesis in *Escherichia coli.* Their molecular weights and the problem of enzyme aggregation. *Biochem J.* **173**, 219–228.

Heyser, J.W., De Bruin, D., Kincaid, M.L., Johnson, R.Y., Rodriguez, M.M. and Robinson, N.J. (1989a) Inhibition of NaCl-induced proline biosynthesis by exogenous proline in halophilic *Distichlis spicata* suspension cultures. *J. Exp. Bot.* **40**, 225–232.

Heyser, J.W., Chacon, M.J. and Warren, R.C. (1989b) Characterization of L-[5-$^{13}$C]-proline biosynthesis in halophytic and nonhalophytic suspension cultures by $^{13}$C NMR. *J. Plant Physiol.* **135**, 459–466.

Hong-qi, Z., Croes, A.F. and Linskens, H.F. (1982) Protein synthesis in germinating pollen of *Petunia*: role of proline. *Planta* **154**, 199–203.

Hu, C.-A.A., Delauney, A.J. and Verma, D.P.S. (1992) A bifunctional enzyme ($\Delta^1$-pyrroline-5-carboxylate synthetase) catalyzes the first two steps in proline biosynthesis in plants. *Proc. Natl Acad. Sci. USA* **89**, 9354–9358.

Huang, A.H.C. and Cavalieri, A.J. (1979) Proline oxidase and water-stress induced proline accumulation in spinach leaves. *Plant Physiol.* **63**, 531–535.

Jones, M.M., Osmond, C.B. and Turner, N.C. (1980) Accumulation of solutes in leaves of sorghum and sunflower in response to water deficits. *Aust. J. Plant Physiol.* **7**, 193–205.

Katz, A. and Tal, M. (1980) Salt tolerance in the wild relatives of cultivated tomato: proline accumulation in callus tissue of *Lycopersicon esculentum* and *L. peruvianum. Z. Pflanzenphysiol.* **98**, 429–435.

Ketchum, R.E.B., Warren, R.C., Klima, L.J., Lopez-Gutiérrez, F. and Nabors, M.W. (1991) The mechanism and regulation of proline accumulation in suspension cultures of the halophytic grass *Distichlis spicata* L. *J. Plant Physiol.* **137**, 368–374.

Krishna, R.V. and Leisinger, T. (1979) Biosynthesis of proline in *Pseudomonas aeruginosa.* Partial purification and characterization of γ-glutamyl kinase. *Biochem. J.* **181**, 215–222.

Krishna, R.V., Beilstein, P. and Leisinger, T. (1979) Biosynthesis of proline in *Pseudomonas aeruginosa.* Properties of γ-glutamyl phosphate reductase and 1-pyrroline-5-carboxylate reductase. *Biochem. J.* **181**, 223–230.

Krueger, R., Jager, H.-J., Hintz, M. and Pahlich, E. (1986) Purification to homogeneity of pyrroline-5-carboxylate reductase of barley. *Plant Physiol.* **80**, 142–144.

Kueh, J.S.H. and Bright, S.W.J. (1981) Proline accumulation in a barley mutant resistant to *trans*-4-hydroxy-L-proline. *Planta* **153**, 166–171.

LaRosa, P.C., Rhodes, D., Rhodes, J.C., Bressan, R.A. and Csonka, L.N. (1991) Elevated accumulation of proline in NaCl-adapted tobacco cells is not due to altered $\Delta^1$-pyrroline-5-carboxylate reductase. *Plant Physiol.* **96**, 245–250.

Leigh, R.A., Ahmad, N. and Wyn Jones, R.G. (1981) Assessment of glycinebetaine and proline compartmentation by analysis of isolated beet vacuoles. *Planta* **153**, 34–41.

Lodato, R.F., Smith, R.J., Valle, D., Phang, J.M. and Aoki, T.T. (1981) Regulation of proline biosynthesis: the inhibition of pyrroline-5-carboxylate synthase activity by ornithine. *Metabolism* **30**, 908–913.

Lone, M.I., Kueh, J.S.H., Wyn Jones, R.G. and Bright, S.W.J. (1987) Influence of proline and glycinebetaine on salt tolerance of cultured barley embryos. *J. Exp. Bot.* **38**, 479–490.

Low, P.S. (1985) Molecular basis of the biological compatibility of nature's osmolytes. In: *Transport Processes, Iono- and Osmoregulation* (eds R. Gilles and M. Gilles-Baillien). Springer-Verlag, Berlin, pp. 469–477.

Manetas, Y., Petropoulou, Y. and Karabourniotis, G. (1986) Compatible solutes and their effects on phosphoenolpyruvate carboxylase of $C_4$-halophytes. *Plant, Cell Environ.* **9**, 145–151.

Mazelis, M. (1979) ATP-dependent hydrolysis of 5-oxoproline to glutamate in higher plants. In: *Nitrogen Assimilation of Plants* (eds E.J. Hewitt and C.V. Cutting). Academic Press, London, pp. 407–417.

Meister, A. (1983) Selective modification of glutathione metabolism. *Science* **220**, 472–477.

Mestichelli, L.J.J., Gupta, R.N. and Spenser. I.D. (1979) The biosynthetic route from ornithine to proline. *J. Biol. Chem.* **254**, 640–647.

Miflin, B.J. and Lea, P.J. (1977) Amino acid metabolism. *Annu. Rev. Plant Physiol.* **28**, 299–329.

Morris, C.J., Thompson, J.F. and Johnson, C.M. (1969) Metabolism of glutamic acid and N-acetylglutamic acid in leaf discs and cell-free extracts of higher plants. *Plant Physiol.* **44**, 1023–1026.

Naidu, B.P., Jones, G.P., Paleg, L.G. and Poljakoff-Mayber, A. (1987) Proline analogues in *Melaleuca* species: response of *Melaleuca lanceolata* and *M. uncinata* to water stress and salinity. *Aust. J. Plant Physiol.* **14**, 669–677.

Naidu, B.P., Paleg, L.G. and Jones, G.P. (1992) Nitrogenous compatible solutes in drought-stressed *Medicago* spp. *Phytochemistry* **31**, 1195–1197.

Oaks, A. (1992) The function of roots in the synthesis of amino acids and amides. In: *Biosynthesis and Molecular Regulation of Amino Acids in Plants* (eds B.K. Singh, H.E. Flores and J.C. Shannon). American Society of Plant Physiologists, Rockville, MD, pp. 111–120.

Oaks, A., Mitchell, D.J., Barnard, R.A. and Johnson, F.J. (1970) The regulation of proline biosynthesis in maize roots. *Can. J. Bot.* **48**, 2249–2258.

Pahlich, E., Kerres, R. and Jager, H.-J. (1983) Influence of water stress on the vacuole/extravacuole distribution of proline in protoplasts of *Nicotiana rustica*. *Plant Physiol.* **72**, 590–591.

Paleg, L.G., Douglas, T.J., van Daal, A. and Keech, D.B. (1981) Proline, betaine and other organic solutes protect enzymes against heat inactivation. *Aust. J. Plant Physiol.* **8**, 107–114.

Paleg, L.G., Stewart, G.R. and Bradbeer, J.W. (1984) Proline and glycine betaine influence

protein solvation. *Plant Physiol.* **75**, 974–978.
Pollard, A. and Wyn Jones, R.G. (1979) Enzyme activities in concentrated solutions of glycinebetaine and other solutes. *Planta* **144**, 291–298.
Rayapati, P.J. and Stewart, C.R. (1991) Solubilization of a proline dehydrogenase from maize (*Zea mays* L.) mitochondria. *Plant Physiol.* **95**, 787–791.
Rhodes, D. and Handa, S .(1989) Amino acid metabolism in relation to osmotic adjustment in plant cells. In: *Environmental Stress in Plants: Biochemical and Physiological Mechanisms, NATO ASI Series, Vol. G19* (ed. J.H. Cherry). Springer, Berlin, pp. 41–62.
Rhodes, D. and Hanson, A.D. (1993) Quaternary ammonium and tertiary sulfonium compounds in higher plants. *Annu. Rev. Plant Physiol. Plant Mol. Biol.* **44**, 357–384.
Rhodes, D., Handa, S. and Bressan, R.A. (1986) Metabolic changes associated with adaptation of plant cells to water stress. *Plant Physiol.* **82**, 890–903.
Robertson, A.V. and Marion, L. (1960) The biogenesis of alkaloids. XXV. The role of hygric acid in the biogenesis of stachydrine. *Can. J. Chem.* **38**, 396–398.
Rudolph, A.S., Crowe, J.H. and Crowe, L.M. (1986) Effects of three stabilizing agents – proline, betaine and trehalose – on membrane phospholipids. *Arch. Biochem. Biophys.* **245**, 134–143.
Rushlow, K.E., Deutch, A.H. and Smith, C.J. (1984) Identification of a mutation that relieves gamma-glutamyl kinase from allosteric feedback inhibition by proline. *Gene* **39**, 109–112.
Santarius, K.A. (1992) Freezing of isolated thylakoid membranes in complex media. VIII. Differential cryoprotection by sucrose, proline and glycerol. *Physiol. Plant.* **84**, 87–93.
Santoro, M.M., Liu, Y., Khan, S.M.A., Hou, L.-X. and Bolen, D.W. (1992) Increased thermal stability of proteins in the presence of naturally occurring osmolytes. *Biochemistry* **31**, 5278–5283.
Seddon, A.P., Zhao, K.Y. and Meister, A. (1989) Activation of glutamate by γ-glutamate kinase: formation of γ-*cis*-cycloglutamyl phosphate, an analog of γ-glutamyl phosphate. *J. Biol. Chem.* **264**, 11326–11335.
Smirnoff, N. and Cumbes, Q.J. (1989) Hydroxyl radical scavenging activity of compatible solutes. *Phytochemistry* **28**, 1057–1060.
Smirnoff, N. and Stewart, G.R. (1985) Stress metabolites and their role in coastal plants. *Vegetatio* **62**, 273–278.
Smith, C.J., Deutch, A.H. and Rushlow, K.E. (1984) Purification and characteristics of a γ-glutamyl kinase involved in *Escherichia coli* proline biosynthesis. *J. Bacteriol.* **157**, 545–551.
Smith, L.T. (1985) Characterization of a γ-glutamyl kinase from *Escherichia coli* that confers proline overproduction and osmotic tolerance. *J. Bacteriol.* **164**, 1088–1093.
Solomon, A., Beer, S., Waisel, Y., Jones, G.P. and Paleg, L.G. (1994) Effects of NaCl on the carboxylating activity of Rubisco from *Tamarix jordanis* in the presence and absence of proline-related compatible solutes. *Physiol. Plant.* **90**, 198–204.
Songstad, D.D., Duncan, D.R. and Widholm, J.M. (1990) Proline and polyamine involvement in chilling tolerance of maize suspension cultures. *J. Exp. Bot.* **41**, 289–294.
Splittstoesser, W.E. and Fowden, L. (1973) Ornithine transaminase from *Cucurbita maxima* cotyledons. *Phytochemistry* **12**, 785–790.
Stewart, C.R. (1981) Proline accumulation: biochemical aspects. In: *The Physiology and Biochemistry of Drought Resistance in Plants* (eds L.G. Paleg and D. Aspinall). Academic, Sydney, pp. 243–259.
Stewart, C.R. and Boggess, S.F. (1978) Metabolism of [5-$^3$H]proline by barley leaves and

its use in measuring the effects of water stress on proline oxidation. *Plant Physiol.* **61**, 654–657.

Stewart, C.R., Boggess, S.F., Aspinall, D. and Paleg, G. (1977) Inhibition of proline oxidation by water stress. *Plant Physiol.* **59**, 930–932.

Stewart, G.R. and Larher, F. (1980) Accumulation of amino acids and related compounds in relation to environmental stress. In: *The Biochemistry of Plants,* Volume 5 (ed. B.J. Miflin). Academic Press, New York, pp. 609–635.

Stewart, G.R. and Lee, J.A. (1974) The role of proline accumulation in halophytes. *Planta* **120**, 279–289.

Strøm, A.R., Le Rudulier, D., Jakowec, M.W., Bunnell, R.C. and Valentine, R.C. (1983) Osmoregulatory (Osm) genes and osmoprotective compounds. In: *Genetic Engineering of Plants. An Agricultural Perspective* (eds T. Kosuge, C.P. Meredith and A. Hollaender). Plenum Press, New York, pp. 39–59.

Sumaryati, S., Negrutiu, I. and Jacobs, M. (1992) Characterization and regeneration of salt- and water-stress mutants from protoplast culture of *Nicotiana plumbaginifolia* (Viviani). *Theor. Appl. Genet.* **83**, 613–619.

Szoke, A., Miao, G.-H., Hong, Z. and Verma, D.P.S. (1992) Subcellular localization of $\Delta^1$-pyrroline-5-carboxylate reductase in root/nodule and leaf of soybean. *Plant Physiol.* **99**, 1642–1649.

Tal, M. and Katz, A. (1980) Salt tolerance in the wild relatives of the cultivated tomato: the effect of proline on the growth of callus tissue of *Lycopersicon esculentum* and *L. peruvianum* under salt and water stress. *Z. Pflanzenphysiol.* **98**, 283–288.

Thompson, J.F. (1980) Arginine synthesis, proline synthesis, and related processes. In: *The Biochemistry of Plants,* Volume 5 (ed. B.J. Miflin). Academic Press, New York, pp. 375–402.

Treichel, S. (1975) The effect of NaCl on the concentration of proline in different halophytes. *Z. Pflanzenphysiol.* **76**, 56–68.

Treichel, S. (1986) The influence of NaCl on $\Delta^1$-pyrroline-5-carboxylate reductase in proline-accumulating cell suspension cultures of *Mesembryanthemum nodiflorum* and other halophytes. *Plant Physiol.* **67**, 173–181.

van Swaaji, A.C., Jacobsen, E. and Feenstra, W.J. (1985) Effect of cold hardening, wilting and exogenously applied proline on leaf proline content and frost tolerance of several genotypes of *Solanum*. *Physiol. Plant.* **64**, 230–236.

Verbruggen, N., Villarroel, R. and Van Montagu, M. (1993) Osmoregulation of a pyrroline-5-carboxylate reductase gene in *Arabidopsis thaliana*. *Plant Physiol.* **103**, 771–781.

Verma, D.P.S, Hu, C.-A.A., Delauney, A.J., Miao, G.-H. and Hong, Z. (1992) Deciphering proline biosynthesis pathways in plants by direct, trans-, and co-complementation in bacteria. In: *Biosynthesis and Molecular Regulation of Amino Acids in Plants* (eds B.K. Singh, H.E. Flores and J.C. Shannon). American Society of Plant Physiologists, Rockville, MD, pp. 128–138.

Voetberg, G.S. and Sharp, R.E. (1991) Growth of the maize primary root at low water potentials. III. Role of increased proline deposition in osmotic adjustment. *Plant Physiol.* **96**, 1125–1130.

Wakabayashi, Y. and Jones, M.E. (1983) Pyrroline-5-carboxylate synthesis from glutamate by rat intestinal mucosa. *J. Biol. Chem.* **258**, 3865–3872.

Williamson, C.L. and Slocum, R.D. (1992) Molecular cloning and evidence for osmoregulation of the $\Delta^1$-pyrroline-5-carboxylate (*proC*) gene in pea (*Pisum sativum* L.). *Plant Physiol.* **100**, 1464–1470.

Withers, L.A. and King, P.J. (1979) Proline: a novel cryoprotectant for the freeze preservation of cultured cells of *Zea mays*. *Plant Physiol.* **64**, 675–678.

Wyn Jones, R.G. and Gorham, J. (1983) Aspects of salt and drought tolerance in higher plants. In: *Genetic Engineering of Plants. An Agricultural Perspective* (eds T. Kosuge, C.P. Meredith and A. Hollaender). Plenum Press, New York, pp. 355–369.

Wyn Jones, R.G. and Storey, R. (1981) Betaines. In: *The Physiology and Biochemistry of Drought Resistance in Plants* (eds L.G. Paleg and D. Aspinall). Academic, Sydney, pp. 171–204.

Wyn Jones, R.G., Storey, R., Leigh, R.A., Ahmad, N. and Pollard, A. (1977) A hypothesis on cytoplasmic osmoregulation. In: *Regulation of Cell Membrane Activities in Plants* (eds E. Marrè and O. Cifferi). Elsevier, Amsterdam, pp. 121–136.

Yancey, P.H. (1994) Compatible and counteracting solutes. In: *Cellular and Molecular Physiology of Cell Volume Regulation* (ed. K. Strange). CRC Press, Boca Raton, FL, pp. 81–109.

Yancey, P.H., Clark, M.E., Hand, S.C., Bowlus, R.D. and Somero, G.N. (1982) Living with water stress: evolution of osmolyte systems. *Science* **217**, 1214–1222.

10

# Biosynthesis of quaternary ammonium and tertiary sulphonium compounds in response to water deficit

A.D. Hanson, J. Rivoal, M. Burnet and B. Rathinasabapathi

## 10.1 Introduction

Zwitterionic quaternary ammonium compounds (QACs) and tertiary sulphonium compounds (TSCs) are among the most effective known compatible solutes or osmoprotectants (Csonka and Hanson, 1991; Wyn Jones, 1984). Many higher plants accumulate one or more such onium compounds in response to water deficit or salinity, and there is much evidence indicating that this contributes to stress tolerance (Rhodes and Hanson, 1993; Somero, 1992). The biosynthesis and role of proline and polyols, which are the other major groups of solutes accumulating during water deficit, are discussed in Chapters 9 and 11. Recently, genetic engineering of osmoprotectant accumulation has become a promising approach to improving crop adaptation to stress (Delauney and Verma, 1993; McCue and Hanson, 1990; Tarczynski *et al.*, 1993), and this has focused attention on the metabolic pathways, enzymes and genes involved. With this prospect in mind, we first introduce the QAC and TSC osmoprotectants known in angiosperms, and then review information available on their biosynthesis and regulation. Lastly, we assess the demand for methyl group biogenesis implied by the stress-induced accumulation of onium compounds, and consider its evolutionary and engineering implications.

## 10.2 Phytochemical background

Five major QAC osmoprotectants and one TSC osmoprotectant are known to accumulate in flowering plants (Figure 10.1). Glycinebetaine appears to be the most widely distributed, occurring in at least nine diverse orders, including

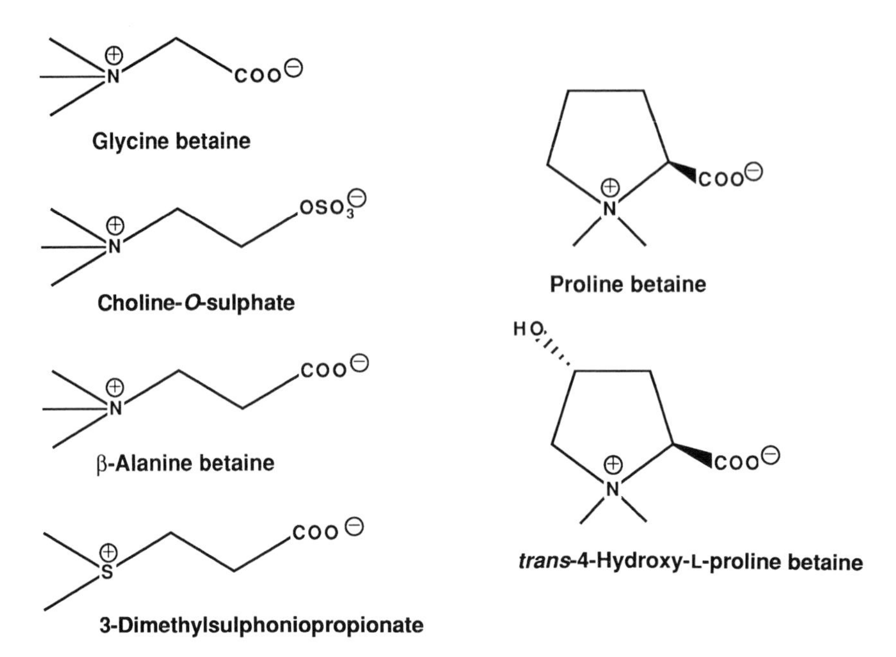


*Figure 10.1. Structures of quaternary ammonium and tertiary sulphonium compounds accumulated by flowering plants. Note that there are several possible hydroxyproline isomers, of which only one is shown.*

mono- and dicotyledons (Rhodes and Hanson, 1993; Wyn Jones, 1984; Hanson, Gage and Rhodes, unpublished). Prolinebetaine and hydroxyprolinebetaines commonly occur together, and are each known from seven dicotyledon orders (Rhodes and Hanson, 1993). Accumulation of the TSC 3-dimethylsulphoniopropionate (DMSP) has been reported only in one genus of Compositae (Storey *et al.*, 1993) and in three genera of Gramineae (Larher *et al.*, 1977; Paquet *et al.*, 1994). β-Alaninebetaine and choline-O-sulphate are known only from the Plumbaginaceae (Hanson *et al.*, 1994a).

The levels of these compounds in leaves of plants subjected to severe long-term water deficit or salinity are typically 50–250 μmol $g^{-1}$ dry weight, but can reach 400 μmol $g^{-1}$ dry weight or more. Levels in unstressed leaves are usually from two- to 10-fold lower. Maximal accumulation generally requires several days, so that QAC and TSC accumulation is a slower response to stress than proline accumulation (see Chapter 9).

## 10.3 Biosynthetic pathways and their regulation

This chapter briefly covers the three compounds for which there has been most progress. Further information on the biosynthesis of these and other compounds is summarized in recent reviews (Hanson, 1993; Rhodes and Hanson, 1993).

## 10.3.1 *Glycinebetaine*

***Pathway and enzymes.*** The biosynthetic pathway has been studied mainly in spinach and sugar beet (Chenopodiaceae), and the following account applies to these plants. In a narrow sense, glycinebetaine synthesis involves only a two-step oxidation of the primary metabolite choline via betaine aldehyde, catalysed by choline monooxygenase (CMO) and betaine aldehyde dehydrogenase (BADH). However, the choline required is synthesised *de novo*, so that, in a broader sense, the glycinebetaine biosynthesis pathway includes the steps which convert ethanolamine to choline (Summers and Weretilnyk, 1993), as shown in Figure 10.2. Enzyme activities corresponding to all the steps from ethanolamine have

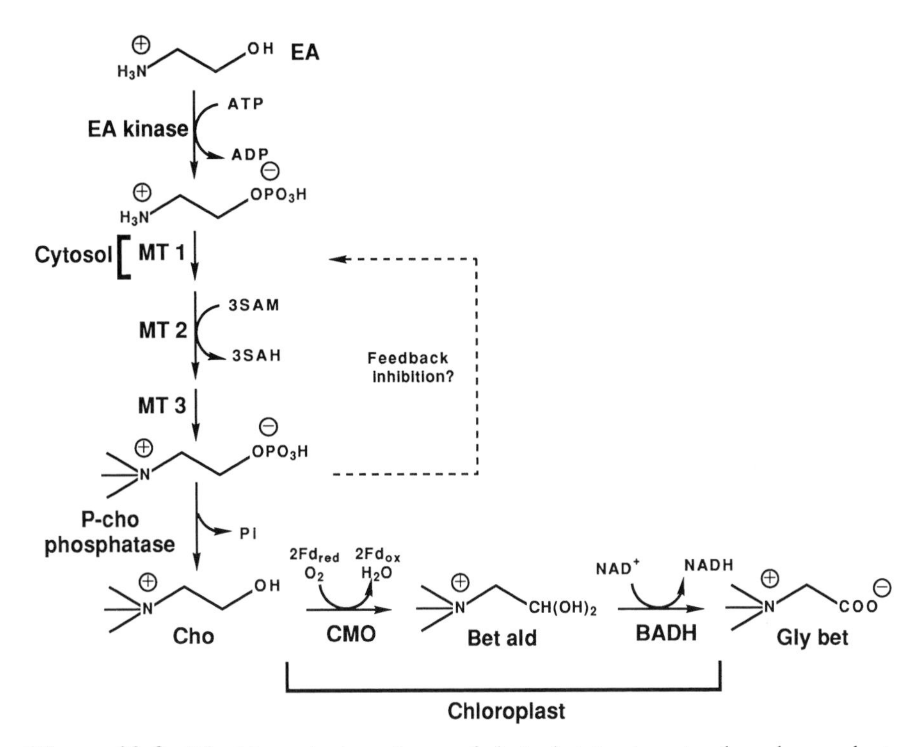


***Figure 10.2.*** *The biosynthetic pathway of glycinebetaine in spinach and sugar beet (Chenopodiaceae). Abbreviations for compounds: EA, ethanolamine; Cho, choline; Bet ald, betaine aldehyde (shown as the hydrate); Gly bet, glycinebetaine; SAM, S-adenosylmethionine; SAH, S-adenosylhomocysteine. Abbreviations for enzymes: EA kinase, ethanolamine kinase; MT 1-3;* N*-methyltransferases acting on phosphoethanolamine and its mono- and dimethyl derivatives, respectively; P-Cho phosphatase, phosphocholine phosphatase; CMO, choline monooxygenase; BADH, betaine aldehyde dehydrogenase. The subcellular localization of enzymes is indicated where known. Note that it is not yet clear whether MT 1–3 are separate enzymes, or whether the feedback loop shown acts on MT 1 itself or an earlier step. Note also that the origin of ethanolamine is unclear, but is probably from serine decarboxylation.*

been measured in extracts of unstressed and salinized spinach leaves, and are more than sufficient to account for the estimated *in vivo* flux through the pathway (Figure 10.3). Because CMO and BADH are located in the chloroplast stroma, and at least the first *N*-methyltransferase of choline synthesis is cytosolic (Wilch and Weretilnyk; cited in Rhodes and Hanson, 1993), the pathway is presumably split between these two compartments, and one or more intermediates – perhaps phosphocholine or choline – are transported across the chloroplast envelope.

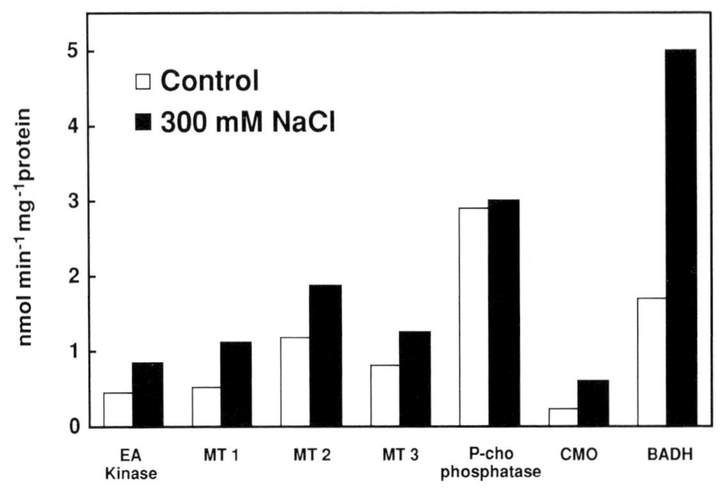


***Figure 10.3.*** *The activities of the enzymes of glycinebetaine biosynthesis in leaves of unstressed spinach plants and of plants salinized with 300 mM NaCl for at least 7 days. Data are taken from Weigel* et al. *(1986), Weretilnyk and Hanson (1989), Brouquisse* et al. *(1989) and Summers and Weretilnyk (1993). Abbreviations for enzymes are as in Figure 10.2. The endogenous rate of glycinebetaine synthesis in salinized spinach leaves can be estimated as* $\leq 0.1$ *nmol* $min^{-1}$ $mg^{-1}$ *protein, using a leaf glycinebetaine content of about 25* $\mu$*mol* $g^{-1}$ *fresh weight (Weigel* et al.*, 1986), assuming a relative growth rate of* $\leq 0.1$ *g* $g^{-1}$ $day^{-1}$*, and taking the soluble protein content of spinach leaves to be 15 mg* $g^{-1}$ *fresh weight.*

***Regulation.*** Except for phosphocholine phosphatase, the activities of all the pathway enzymes rise by about 1.5- to threefold in response to salinization (Figure 10.3). This pattern indicates that control of flux through the pathway may be shared among many steps. *In vivo* radiolabelling evidence from sugar beet provided evidence for phosphocholine hydrolysis as a metabolic control point, and for feedback inhibition of phosphocholine synthesis by phosphocholine (Hanson and Rhodes, 1983). These *in vivo* data are hard to reconcile with the high activity of phosphocholine phosphatase (Figure 10.3), but if phosphocholine or choline is indeed transported into the chloroplast (see above), the regulation observed *in vivo* may reside in the transport system.

For the last enzyme of the pathway, BADH, the salinity-induced rise in enzyme activity is due to an increase in the amount of enzyme protein (Weretilnyk and Hanson, 1989) and is associated with a higher level of BADH mRNA (McCue and Hanson, 1992a; Weretilnyk and Hanson, 1990). The signal that causes the rise in BADH mRNA level has been investigated in sugar beet; it seems not to be turgor reduction, NaCl or abscisic acid, but rather some other biochemical factor translocated from roots to leaves in salinized plants (McCue and Hanson, 1992b).

### 10.3.2 *Choline-O-sulphate*

*In vivo* radiolabelling experiments have confirmed that choline-*O*-sulphate is readily synthesised from choline in species of Plumbaginaceae (Hanson *et al.*, 1991). The pathway to choline is not yet known for the Plumbaginaceae; it may differ from that shown for spinach in Figure 10.2 because the *N*-methylation reactions involve various intermediates in different angiosperms (Rhodes and Hanson, 1993). The level of choline-*O*-sulphate rises in response to salinity alone, and rises further when $Cl^-$ is replaced isoosmotically by $SO_4^{2-}$ (Hanson *et al.*, 1991), so that the biosynthetic pathway may be regulated by both osmotic and specific ion effects.

A choline sulphotransferase (CST) catalysing the formation of choline-*O*-sulphate has been detected and partially characterized in several members of the Plumbaginaceae (Rivoal and Hanson, 1994). The activity is absent or very low in plants that do not accumulate choline-*O*-sulphate. CST is soluble and dependent on 3'-phosphoadenosine 5'-phosphosulphate as sulphate donor. The enzyme was induced (three- to fourfold) in leaves, roots and cell cultures of *Limonium perezii* treated with artificial sea water. CST was also induced in *L. perezii* cell cultures osmotically stressed with polyethylene glycol 6000.

### 10.3.3 *3-Dimethylsulphoniopropionate*

For the alga *Ulva lactuca*, Greene (1962) showed that DMSP is derived from methionine, but did not elucidate the pathway. Maw (1981) suggested a pathway in which methionine is converted to the corresponding $\alpha$-keto acid and thence to methylthiopropionate (MTP) and DMSP, as shown in Figure 10.4. *In vivo* isotope labelling studies with *Wollastonia biflora* leaf discs did not support Maw's suggestion, but showed that *S*-methylmethionine (SMM) is the first intermediate in the pathway (Hanson *et al.*, 1994b). The main evidence for this was as follows. (i) In pulse–chase experiments with [$^{14}C$]methionine, SMM had the labelling pattern expected of a pathway intermediate whereas MTP did not. (ii) [$^{14}C$]SMM was efficiently converted to DMSP but [$^{14}C$]MTP was not. (iii) Addition of unlabelled SMM, but not of MTP, reduced synthesis of [$^{14}C$]DMSP from [$^{14}C$]methionine. (iv) The dimethylsulphide group of [$^{13}CH_3$,$C^2H_3$]SMM was incorporated as a unit into DMSP. Observation (iv) also shows that the SMM cycle (Mudd and Datko, 1990) does not operate at a high level in *W. biflora* leaves.

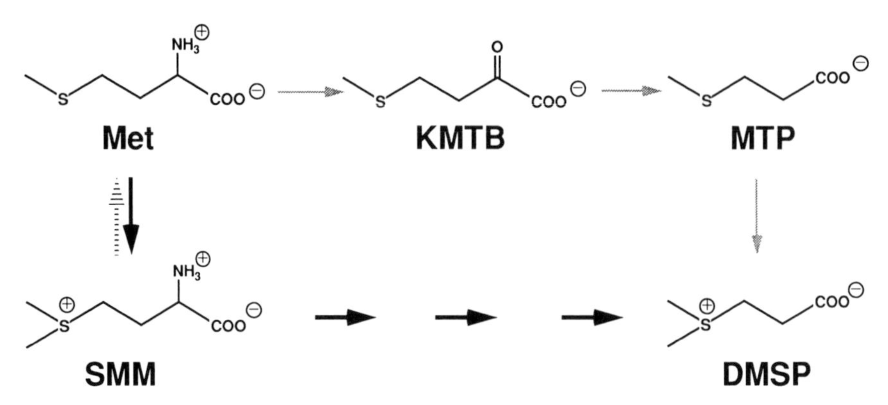


*Figure 10.4.* *Alternative proposed steps in 3-dimethylsulphoniopropionate biosynthesis from methionine. Experimental evidence supports only the route marked with heavy arrows (Hanson* et al., *1994b). Abbreviations: Met, methionine; SMM, S-methylmethionine; KMTB, 2-keto-4-methylthiobutyrate; MTP, 3-methylthiopropionate. SMM is probably widespread in flowering plants, and in some cases is interconvertible with methionine via the SMM cycle (Mudd and Datko, 1990).*

The intermediates between SMM and DMSP are not yet known; they could well be labile because none were detectable in radiolabelling experiments (Hanson *et al.*, 1994b). This would argue against either dimethylsulphoniopropylamine or the α-hydroxy derivative of SMM as intermediates, as they are stable compounds.

DMSP biosynthesis presumably is regulated both by osmotic stress and by N and S nutrition because, in salinized plants, N deficiency increases DMSP levels (Hanson *et al.*, 1994b) and S deficiency decreases them (Storey *et al.*, 1993). Stress must enhance the synthesis of SMM as well as its conversion to DMSP, because the SMM pool in unstressed plants is far too small to account for the DMSP accumulated by stressed plants (Hanson *et al.*, 1994b).

## 10.4 Demand for methyl groups

The accumulation of QACs and TSCs increases the demand for methyl groups. To set this in the context of other demands, Figure 10.5 presents a simplified methyl group budget for leaves of a glycinebetaine-accumulating plant. This budget indicates that the methyl group demand for glycinebetaine synthesis is relatively modest in unstressed leaves but becomes a major item as stress levels increase. It should be noted that the values in Figure 10.5 refer to whole leaves, and that the various demands are not all made concurrently in all leaf cells. Thus, the large methyl group demand for lignin synthesis is restricted to certain cells and developmental stages, whereas that for glycinebetaine synthesis can be incurred in stressed mesophyll cells through much of their life. Figure 10.5 thus tends to understate the impact of stress on methyl group demand in the glycinebetaine-synthesising cells themselves.

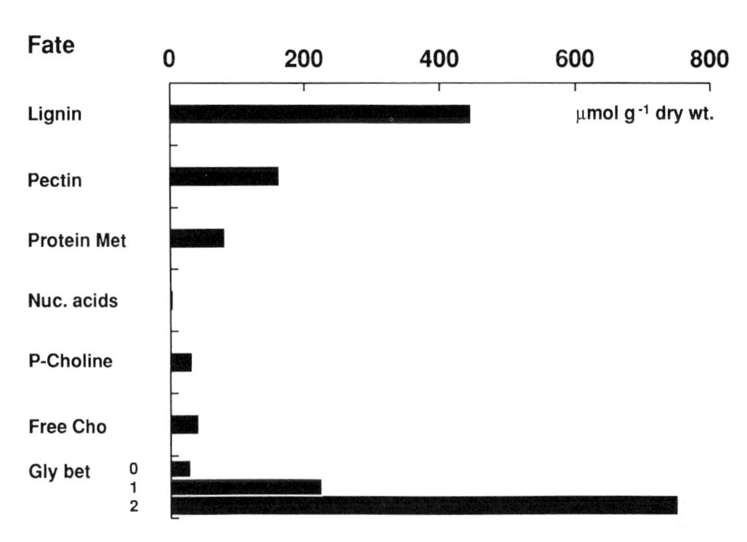


***Figure 10.5.*** *An outline methyl group budget for leaves of a glycinebetaine-accumulating $C_4$ grass. Methyl group demands were calculated from published values for biomass composition for maize (Penning de Vries* et al.*, 1974), phosphatidylcholine (P-choline) contents (Roughan and Batt, 1969), and choline and glycinebetaine levels for unstressed sorghum (0) and sorghum subjected to moderate (1) or severe (2) salt or water stress (Grieve and Maas, 1984; Monyo* et al.*, 1992). The principal assumptions made in calculating methyl demands were: that lignin is polyconiferyl alcohol (Penning de Vries* et al.*, 1974); that pectin is fully carboxymethylated polygalacturonic acid; that the mole abundance of methionine in protein is 5%; that ≤10% of nucleic acid bases are methylated. Note that the budget does not take into account turnover of the various items, which could increase methyl group demand, and that, besides promoting glycinebetaine accumulation, stress may enhance the synthesis of lignin and that of other methylated products of the phenylpropanoid pathway.*

In lignifying cells, the increased demand for methyl groups is associated with a strong increase in expression of *S*-adenosylmethionine synthetase (Kawalleck *et al.*, 1992; Peleman *et al.*, 1989). This suggests that the greater demand for methyl groups during lignification cannot be supported by normal levels of expression of the enzymes involved in one-carbon metabolism. Given that the accumulation of onium compounds may impose an even greater demand for methyl groups than lignification (Figure 10.5), it would appear that this too could require specific induction of enzymes to synthesise and transfer methyl groups. Thus, the capacity to accumulate onium compounds may depend on special adaptations in methyl group biogenesis and transfer (i.e. folate metabolism and *S*-adenosylmethionine turnover).

Within a taxon, the evolution of a high capacity for methyl group biogenesis for production of, say, glycinebetaine might predispose towards subsequent diversification of the onium compounds accumulated, as this would require only recruiting a new methyl acceptor molecule. The patterns of QAC and TSC diversity in several angiosperm families are consistent with such a process. These families include the Plumbaginaceae, Compositae, Leguminosae and Gramineae (Hanson *et al.*, 1994a,b; Rhodes and Hanson, 1993).

If the initial evolution of onium compound accumulation indeed required major adaptations in one-carbon metabolism which, once in place, facilitated diversification, there would be two implications for metabolic engineering. First, the methyl group biogenesis capacity in species which do not naturally accumulate onium compounds may not support the activity of introduced QAC or TSC biosynthesis enzymes. Progress towards testing this experimentally has been made by introducing BADH cDNAs into the non-accumulating plant tobacco (Rathinasabapathi *et al.*, 1994). The second implication is that it may be easier to change the type of onium compounds produced than to engineer the capacity for onium compound accumulation *ab initio*. Because some onium compounds may be functionally superior to others in certain types of stress environments (Hanson *et al.*, 1994a; Turner *et al.*, 1988), engineering the replacement of one by another could be a valuable option in crop improvement.

## References

Brouquisse, R., Weigel, P., Rhodes, D., Yocum, C.F. and Hanson, A.D. (1989) Evidence for a ferredoxin-dependent choline monooxygenase from spinach chloroplast stroma. *Plant Physiol.* **90**, 322–329.

Csonka, L.N. and Hanson, A.D. (1991) Prokaryotic osmoregulation: genetics and physiology. *Annu. Rev. Microbiol.* **45**, 569–606.

Delauney, A.J. and Verma, D.P.S. (1993) Proline biosynthesis and osmoregulation in plants. *Plant J.* **4**, 215–223.

Greene, R.C. (1962) Biosynthesis of dimethyl-β-propiothetin. *J. Biol. Chem.* **237**, 2251–2254.

Grieve, C.M. and Maas, E.V. (1984) Betaine accumulation in salt-stressed sorghum. *Physiol. Plant.* **61**, 167–171.

Hanson, A.D. (1993) Accumulation of quaternary ammonium and tertiary sulfonium compounds. In: *Plant Responses to Cellular Dehydration During Environmental Stress* (eds T.J. Close and E.A. Bray). American Society of Plant Physiologists, Rockville, MD, pp. 30–36.

Hanson, A.D. and Rhodes, D. (1983) $^{14}C$ Tracer evidence for synthesis of choline and betaine via phosphoryl base intermediates in salinized sugarbeet leaves. *Plant Physiol.* **71**, 692–700.

Hanson, A.D., Rathinasabapathi, B., Chamberlin, B. and Gage, D.A. (1991) Comparative physiological evidence that β-alanine betaine and choline-*O*-sulfate act as compatible osmolytes in halophytic *Limonium* species. *Plant Physiol.* **97**, 1199–1205.

Hanson, A.D., Rathinasabapathi, B., Rivoal, J., Burnet, M., Dillon, M.O. and Gage, D.A. (1994a) Osmoprotective compounds in the Plumbaginaceae: a natural experiment in metabolic engineering of stress tolerance. *Proc. Natl Acad. Sci. USA* **91**, 306–310.

Hanson, A.D., Rivoal, J., Paquet, L. and Gage, D.A. (1994b) Biosynthesis of 3-dimethylsulfoniopropionate in *Wollastonia biflora* (L.) DC.: evidence that *S*-methylmethionine is an intermediate. *Plant Physiol.* **105**, 103–110.

Kawalleck, P., Plesch, G., Hahlbrock, K. and Somssich, I. (1992) Induction by fungal elicitor of *S*-adenosyl-L-methionine synthetase and *S*-adenosyl-L-homocysteine hydrolase mRNAs in cultured cells and leaves of *Petroselinum crispum*. *Proc. Natl Acad. Sci. USA* **89**, 4713–4717.

Larher, F., Hamelin, J. and Stewart, G.R. (1977) L'acide diméthylsulfonium-3 propanoïque de *Spartina anglica*. *Phytochemistry* **16**, 2019–2020.

Maw, G.A. (1981) The biochemistry of sulphonium salts. In: *The Chemistry of the Sulphonium Group*, Part 2 (eds C.J.M. Stirling and S. Patai). John Wiley & Sons, Chichester, pp. 703–770.

McCue, K.F. and Hanson, A.D. (1990) Drought and salt tolerance: towards understanding and application. *Trends Biotechnol.* **8**, 358–362.

McCue, K.F. and Hanson, A.D. (1992a) Salt-inducible betaine aldehyde dehydrogenase from sugar beet: cDNA cloning and expression. *Plant Mol. Biol.* **18**, 1–11.

McCue, K.F. and Hanson, A.D. (1992b) Effects of soil salinity on the expression of betaine aldehyde dehydrogenase in leaves: investigation of hydraulic, ionic and biochemical signals. *Aust. J. Plant Physiol.* **19**, 555–564.

Mudd, S.H. and Datko, A.H. (1990) The *S*-methylmethionine cycle in *Lemna paucicostata*. *Plant Physiol.* **93**, 623–630.

Monyo, E.S., Ejeta, G. and Rhodes, D. (1992) Genotypic variation for glycine betaine in sorghum and its relationship to agronomic and morphological traits. *Maydica* **37**, 283–286.

Paquet, L., Rathinasabapathi, B., Saini, H., Zamir, L., Gage, D.A., Huang, Z.-H. and Hanson, A.D. (1994) Accumulation of the compatible solute 3-dimethylsulfoniopropionate in sugarcane and its relatives, but not other gramineous crops. *Aust. J. Plant Physiol.* **21**, 37–48.

Peleman, J., Boerjan, W., Engler, G., Seurinck, J., Botterman, J., Alliotte, T., Van Montagu, M. and Inze, D. (1989) Strong cellular preference in the expression of a housekeeping gene of *Arabidopsis thaliana* encoding *S*-adenosylmethionine synthetase. *Plant Cell* **1**, 81–93.

Penning de Vries, F.W.T., Brunsting, A.H.M. and Van Laar, H.H. (1974) Products, requirements and efficiency of biosynthesis: a quantitative approach. *J. Theor. Biol.* **45**, 339–377.

Rathinasabapathi, B., McCue, K.F., Gage, D.A. and Hanson, A.D. (1994) Metabolic engineering of glycine betaine synthesis: plant betaine aldehyde dehydrogenases lacking typical transit peptides are targeted to tobacco chloroplasts where they confer betaine aldehyde resistance. *Planta* **193**, 155–162.

Rhodes, D. and Hanson, A.D. (1993) Quaternary ammonium and tertiary sulfonium compounds in higher plants. *Annu. Rev. Plant Physiol. Plant Mol. Biol.* **44**, 357–384.

Rivoal, J. and Hanson, A.D. (1994) Choline-*O*-sulfate synthesis in plants: identification and partial characterization of a salinity-inducible choline sulfotransferase from *Limonium* species. *Plant Physiol.* in press.

Roughan, P.G. and Batt, R.D. (1969) The glycerolipid composition of leaves. *Phytochemistry* **8**, 363–369.

Somero, G.N. (1992) Adapting to water stress: convergence on common solutions. In: *Water and Life* (eds G.N. Somero, C.B. Osmond and C.L. Bolis). Springer, New York, pp. 3–18.

Storey, R., Gorham, J., Pitman, M.G., Hanson A.D. and Gage, D.A. (1993) Response of *Melanthera biflora* to salinity and water stress. *J. Exp. Bot.* **44**, 1551–1560.

Summers, P.S. and Weretilnyk, E.A. (1993) Choline synthesis in spinach in relation to salt stress. *Plant Physiol.* **103**, 1269–1276.

Tarczynski, M.C., Jensen, R.G. and Bohnert, H.J. (1993) Stress protection of transgenic tobacco by production of the osmolyte mannitol. *Science* **259**, 508–510.

Turner, S.M., Malin, G., Liss, P.S., Harbour, D.S. and Holligan, P.M. (1988) The seasonal variation of dimethylsulfide and dimethylsulfoniopropionate concentrations in nearshore waters. *Limnol. Oceanogr.* **33**, 364–375.

Weigel, P., Weretilnyk, E.A. and Hanson, A.D. (1986) Betaine aldehyde oxidation by spinach chloroplasts. *Plant Physiol.* **82**, 753–759.

Weretilnyk, E.A. and Hanson, A.D. (1989) Betaine aldehyde dehydrogenase from spinach leaves: purification, *in vitro* translation of the mRNA, and regulation by salinity. *Arch. Biochem. Biophys.* **271**, 56–63.

Weretilnyk, E.A. and Hanson, A.D. (1990) Molecular cloning of a plant betaine-aldehyde dehydrogenase, an enzyme implicated in adaptation to salinity and drought. *Proc. Natl Acad. Sci. USA* **87**, 2745–2749.

Wyn Jones, R.G. (1984) Phytochemical aspects of osmotic adaptation. *Recent Adv. Phytochem.* **18**, 55–78.

11

# Polyol accumulation and metabolism during water deficit

M. Popp and N. Smirnoff

## 11.1 Introduction

Water deficits in plants may be caused by abiotic factors like drought, heat, cold and salt. Under all these circumstances it is observed that certain low molecular weight organic solutes accumulate in plants, including proline, glycinebetaine (see Chapters 9 and 10), polyols such as sorbitol (Ahmad *et al.*, 1979) and cyclitols (Ford, 1982; Gorham *et al.*, 1981, 1984). Several hypotheses have been put forward to explain how this solute accumulation might ameliorate the situation of stressed organisms (Brown and Simpson, 1972; Jennings and Burke, 1990; Smirnoff and Cumbes, 1989), however, the beneficial effect for stress adaptation is still the subject of controversy (Munns, 1993).

From recent work with transgenic tobacco, there is now evidence that the capacity for mannitol production improves adaptation to salinity (Tarczynski *et al.*, 1993). Tobacco plants, which normally do not accumulate mannitol, were transformed with a bacterial gene for mannitol-1-phosphate dehydrogenase. The transgenic plants stored mannitol to an estimated cytoplasmic concentration of 100 mM. When exposed to a salt treatment of 250 mM NaCl for 30 days, the transformed plants were able to produce new roots and flowers, whereas this was not the case in the untransformed plants. The same group was able to demonstrate that, after salt and cold treatment, the mRNA for the enzyme *myo*-inositol-*O*-methyltransferase, the first step of the synthesis of the cyclitol, pinitol, was more highly expressed in the facultative halophyte *Mesembryanthemum crystallinum* (Vernon and Bohnert, 1992a,b; Vernon *et al.*, 1993). In the light of these new approaches using molecular biology techniques, it seems worthwhile to compare acyclic and cyclic polyols in their relation to water deficits.

## 11.2 Definition and description of polyols

The term 'polyol' simply refers to polyhydric alcohols. The carbon atom of these compounds can be arranged in a straight chain (acyclic polyols) or in a ring (cyclic polyols). This latter group is also called cyclitols (Micheel, 1932). Acyclic polyols are named according to the number of carbon atoms they contain, hexitols ($C_6$) being most widely distributed in higher plants (Bieleski, 1982; Lewis, 1984). The acyclic polyols are formed by reduction of sugars, for example sorbitol is related to glucose and mannitol is related to mannose. Within the cyclitols, a distinction is made according to the number of hydroxyl groups attached to the ring. The major group occurring in higher plants is the inositols ($C_6$) which have a hydroxyl group attached to each carbon atom (cyclohexanhexoles, Figure 11.1) Further groups are the quercitols (cyclohexanpentols) with five hydroxyl groups and the conduritols (cyclohexententrols) which are characterized by four hydroxyl groups and a double bond in the ring (Drew, 1984). Included in the term 'cyclitols' are their mono- and dimethylethers. This gives rise to a great variety of compounds, considering the possible isomeric configurations. Figure 11.1 illustrates the naturally occurring inositols and their monomethylethers. As with the hexitols (e.g. sorbitol from *Sorbus*), several cyclitols also have trivial names, which often indicate the plant source from which they were first isolated (e.g. pinitol, sequoyitol). This is very useful in face of the rather complicated systematic chemical names (Figure 11.1) and this method of naming was extended to the newly described 1,2-di-*O*-methyl-*muco*-inositol from *Viscum album*, which is called 'viscumitol' (Richter, 1992). Unfortunately, naming after the original plant source was not applied to D-1-*O*-methyl-*muco*-inositol and *O*-methyl-*scyllo*-inositol, therefore the abbreviations OMMI and OMSI, respectively, are suggested.

## 11.3 Distribution

As more species are examined, it is becoming clear that polyols are widely distributed and can form a substantial part of the carbohydrate pool. Acyclic polyols may be equally as important as sucrose in phloem transport, while cyclic polyols are in many cases the dominant component of the neutral fraction in leaves. There is a strong relationship between taxonomy and the type of polyol accumulated, for example sorbitol is characteristic of the Rosaceae (Lewis and Smith, 1967). It is very difficult to assess which of the two groups of polyols (acyclic and cyclic) are more widely distributed in the plant kingdom. It is important to ask such a question because water deficit will only enhance the synthesis of the polyol which is present in the well-watered plant. That means, only a plant which under sufficient water supply produces some pinitol will accumulate this compound under water shortage. According to Loescher (1987), mannitol has been reported in over 70 higher plant families. Even though pinitol does not match this number of families, it is very abundant in the Fabales alone, and thus, on a species level, is probably more widespread than the acyclic

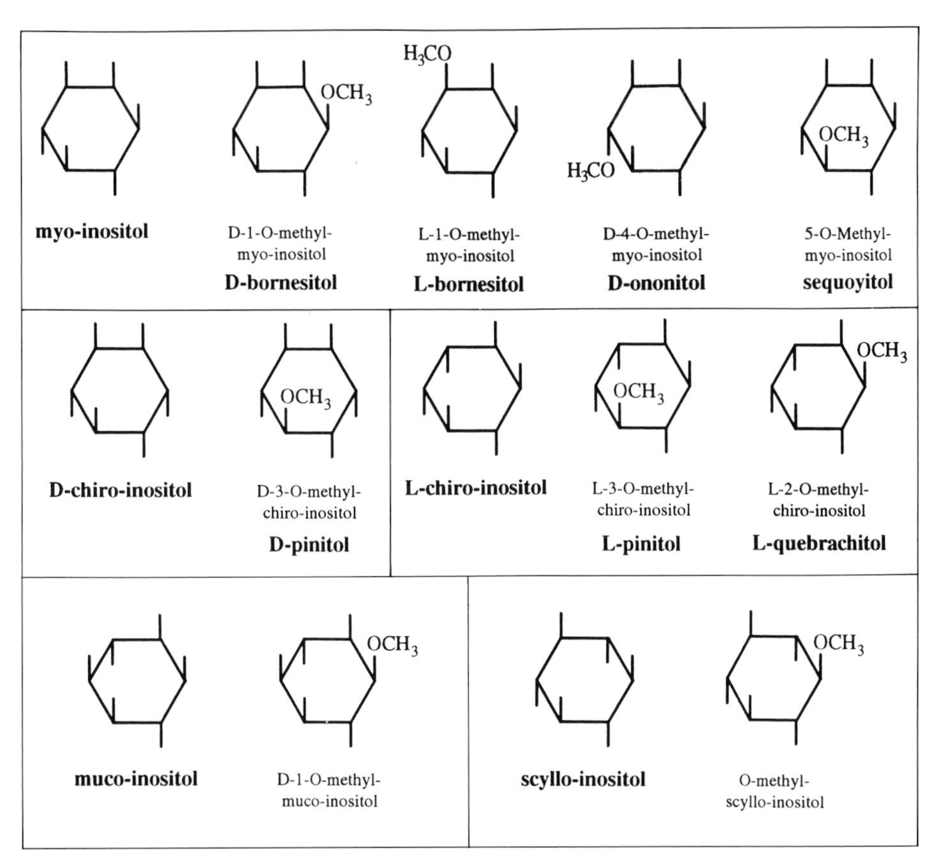


***Figure 11.1.*** *Naturally occurring inositols and their monomethylethers. The vertical bars indicate the position of hydroxyl groups. Courtesy of Dr Andreas Richter.*

polyols. As already mentioned by Drew (1984), extensive work on the occurrence of polyols has been done by Plouvier as cited in Swain (1963).

Both kinds of polyols are reported to occur in algae (Craigie, 1974), whereas so far only acyclic polyols have been detected in fungi (Jennings and Burke, 1990). This latter paper highlights the role of acyclic polyols as physiological buffering agents, whereas very little is known about the function of cyclitols.

## 11.4 Synthesis and breakdown

Focusing on those compounds in the two groups of polyols which are most abundant in higher plants reduces the number of substances to deal with to two in the case of the acyclic polyols (mannitol and sorbitol) and to a few of the inositol derivatives presented in Figure 11.1 in the case of the cyclitols. The decisive difference between the acyclic polyols, mannitol and sorbitol, and the cyclitols is that the former can be regarded as primary photosynthetic products (Loescher,

1987; Moing *et al.*, 1992). In contrast, it takes hours before radioactively labelled carbon dioxide is incorporated into cyclitols (Drew, 1984; Schilling *et al.*, 1972) even though most of them are derived from *myo*-inositol in a few steps (see below).

The biosynthetic pathway of mannitol in *Apium graveolens* (Umbelliferae) has been established. Triose phosphate, exported from the chloroplasts and converted to fructose-6-phosphate, is used for mannitol and sucrose synthesis in equal amounts. Mannitol is formed as follows (Loescher *et al.*, 1992; Rumpho *et al.*, 1983).

$$\text{Fructose-6-P} \xrightarrow{1} \text{mannose-6-P} \xrightarrow{2} \text{mannitol-1-P} \xrightarrow{3} \text{mannitol} + \text{Pi}$$

$$\text{NADPH} \curvearrowright \text{NADP}$$

where the enzymes are: 1, mannose-6-phosphate isomerase; 2, NAPDH-dependent mannose-6-phosphate reductase; 3, mannitol-1-phosphate phosphatase.

A similar pathway has been proposed for the formation of sorbitol in apple leaves from glucose-6-phosphate using an NADPH-aldose-6-phosphate reductase (Negm and Loescher, 1981) and sorbitol-6-phosphate phosphatase (Grant and ap Rees, 1981). Other aldose phosphate reductases may be responsible for the synthesis of acyclic polyols from the corresponding aldose phosphates (Negm, 1986). An alternative to the reduction of phosphorylated aldoses is the reduction of aldoses or ketoses by aldose and ketose reductases. Sorbitol in mammalian kidneys is formed from glucose by aldose reductase (Yancey, 1992). NAD(P)H-dependent aldose and ketose reductase activities have been detected in germinating soybean seeds, where they may, along with sorbitol dehydrogenase, be involved in the interconversion of glucose and fructose (Kuo *et al.*, 1990), and in *Euonymus japonica* which accumulates galactitol (Negm, 1986). The acyclic polyols are oxidized to the corresponding sugar by NAD-dependent polyol dehydrogenases (Loescher, 1987).

Pinitol in *Simmondsia chinensis* (Dittrich and Korak, 1984) and in the Leguminosae (Dittrich and Brandl, 1987) is synthesised by methylation of *myo*-inositol to form D-ononitol followed by epimerization to D-pinitol. Based on this, the following sequence has been proposed (Loewus and Loewus, 1980; Vernon and Bohnert, 1992a,b)

$$\text{Glucose-6-P} \xrightarrow{1} \textit{myo}\text{-inositol-1-P} \xrightarrow{2} \textit{myo}\text{-inositol}$$
$$\xrightarrow{3} \text{D-ononitol} \xrightarrow{4} \text{D-pinitol}$$

where the enzymes are: 1, glucose-6-phosphate cycloaldolase (*myo*-inositol-1-phosphate/synthase); 2, *myo*-inositol-1-phosphate phosphatase; 3, *myo*-inositol-O-methyltransferase with S-adenosylmethionine as the methyl donor; 4, an epimerase.

The breakdown of cyclitols is even slower than their synthesis and is often hardly detectable. During 96 h of darkness no pinitol was degraded in *M. crystallinum* (Paul and Cockburn, 1989) and similarly no decrease in the same cyclitol was observed in 44 h of darkness in clover and soybean tissue (Smith and Phillips, 1982). *Chiro*-inositol behaved in the same way in experiments with sea grasses (Drew, 1984), where the amount of cyclitol stayed constant during a 100-h dark period. Figure 11.2 compares the changes in OMMI and mannitol during a 48-h dark period in leaves from two mangrove species *Ceriops tagal* (OMMI) and *Aegiceras corniculatum* (mannitol). Excised leaves were kept in a humid chamber and polyol contents were calculated on a leaf area basis. A marked decrease in mannitol was observed in the second 24 h, while no change in OMMI could be detected. There is only one older report (Diamantoglou, 1974) in favour of a breakdown and 'mobilization' of pinitol and quercitol. In this experiment, twigs of *Pinus nigra* and *Quercus cerris* were put into water and kept in the dark for 1 week; afterwards needles and leaves, but not the stem part of twigs, were tested for their cyclitol content. The observed loss of cyclitols from needles and leaves can be also explained by transport to bark and wood, which seems likely in the light of ongoing experiments in our laboratory. In summary, while mannitol and sorbitol are easily turned over in the carbohydrate pool, the cyclitols investigated so far are metabolically rather inert.

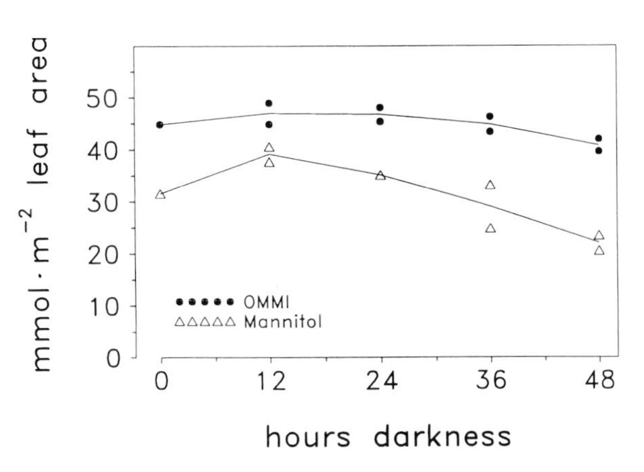


***Figure 11.2.*** *The effect of dark incubation on the OMMI content of* Ceriops tagal *leaves and the mannitol content of* Aegeiceras corniculatum *leaves. The detached leaves were incubated in a dark humid chamber and analysed for polyol content at intervals.*

## 11.5 Water deficit-induced polyol accumulation

Polyol accumulation has been reported in relation to drought, salinity and temperature extremes in higher plants, algae and fungi. Some examples will be considered in this section. Considering the metabolic inertness of cyclitols mentioned above, it is not surprising that in higher plants there are more cases of drought-induced cyclitol accumulation described in the literature than of acyclic polyols such as sorbitol and mannitol (Table 11.1).

Accumulation of soluble carbohydrates, particularly sucrose, in plants has been reported frequently as a response to water deficit even though photosynthetic carbon dioxide fixation decreases (see Chapter 8). Accumulation could ultimately be the result of reduced sink demand in the plants resulting from growth inhibi-

***Table 11.1*** *Polyol accumulation under various situations of water deficit. Where not mentioned otherwise, values refer to leaves or shoots*

| Water deficit caused by | Compound | Plants | Degree of accumulation in shoots | Reference |
|---|---|---|---|---|
| Drought | Sorbitol | *Prunus* species | Twice the values of watered controls | Ranney *et al.* (1991) |
| | OMSI<br>Ononitol | *Vigna* species | Up to 10 times the values of watered controls | Ford (1982) |
| | Pinitol<br>Ononitol<br>OMSI | 14 different legumes | Up to 10 times the values of watered controls | Ford (1984) |
| | Pinitol | *Cajanus cajan* | Fivefold increase during 26-day stress period | Keller and Ludlow (1993) |
| | | *Pinus pinaster* | Twofold increase by lowering the $\Psi$ of nutrient solution to –0.5 MPa | Nguyen and Lamont (1988) |
| Salt | Sorbitol | *Plantago maritima* | Eightfold increase in 400 mol $m^{-3}$ NaCl compared to 0 NaCl | Ahmad *et al.* (1979) |
| | | *Plantago coronopus* | 160% at 200 mol $m^{-3}$ NaCl compared to 0 | Gorham *et al.* (1981) |
| | Mannitol | *Aegiceras corniculatum* | Nearly double in 100% sea water compared to 10% sea water | Popp *et al.* (1990) |

| | | | | |
|---|---|---|---|---|
| | | *Laguncularia racemosa* | Twofold increase in 8 days up-shock (25→ 125% sea water) | Popp *et al.* (1990) |
| | Pinitol | *Honkenya peploides* | Double in 250 mol $m^{-3}$ NaCl compared to 0 NaCl | Gorham *et al.* (1981) |
| | | *Sesbania aculeata* | Threefold increase in 100 mol $m^{-3}$ NaCl compared to 0 NaCl | Gorham *et al.* (1984) |
| | | *Mesembryanthemum crystallinum* | Ninefold increase in 400 mol $m^{-3}$ NaCl compared to 0 NaCl | Paul and Cockburn (1989) |
| | OMMI | *Ceriops tagal* | 164% in 100% seawater compared to 50% sea water | Richter *et al.* (1990) |
| Cold | Mannitol | *Olea europea* | In bark tissue in winter more than double than in the warm season | Drossopoulos and Niavis (1988) |
| | Pinitol | *Robinia pseudacacia* | Bark tissue | Thonke (1990) |
| | Several cyclitols | *Viscum album* | 25% of dry matter of overwintering leaves | Richter (1989) |

tion in the same way that low temperature-induced carbohydrate accumulation could also be explained by reduced demand and decreased export of sucrose from source leaves. In the case of water deficit there is also evidence that partitioning of carbohydrate between sucrose and starch is affected. Labelling studies show that a greater proportion of carbon is incorporated into sucrose than into starch at low water potential (Bensari *et al.*, 1990; Quick *et al.*, 1992; Zrenner and Stitt, 1991). Starch breakdown might also be stimulated by water deficit (Oparka and Wright, 1988; Zrenner and Stitt, 1991). Very little is known about the regulation of polyol accumulation in higher plants during water deficit and it is difficult to assess, particularly for acyclic polyols, to what extent alterations in growth and source–sink relationships, as opposed to increased rates of synthesis, are responsible for accumulation. Sorbitol and mannitol accumulate in a number of higher plants exposed to drought and salinity (Table 11.1). The absolute concentrations may vary appreciably, depending on the plant species. While sorbitol reaches 60–70 mol m$^{-3}$ tissue water (Ahmad *et al.*, 1979), mannitol in the two mangrove species *A. corniculatum* and *Laguncularia racemosa* growing in saline water can reach 200 mol m$^{-3}$ (Table 11.1; Figure 11.3). *Plantago maritima* accumulates sorbitol as salinity increases (Ahmad *et al.*, 1979). Sorbitol also accumulates in *Plantago* species as a result of drought. This is illustrated by *P. major* leaf discs exposed to water deficit by floating them on polyethylene glycol solutions to lower the water potential. Sorbitol accumulation is much more marked than changes in other soluble sugars (Figure 11.4). Accumulation in leaf discs would suggest that the response is specific to water deficit and is not caused by alterations in source–sink relationships. The enzymology and regulation of these

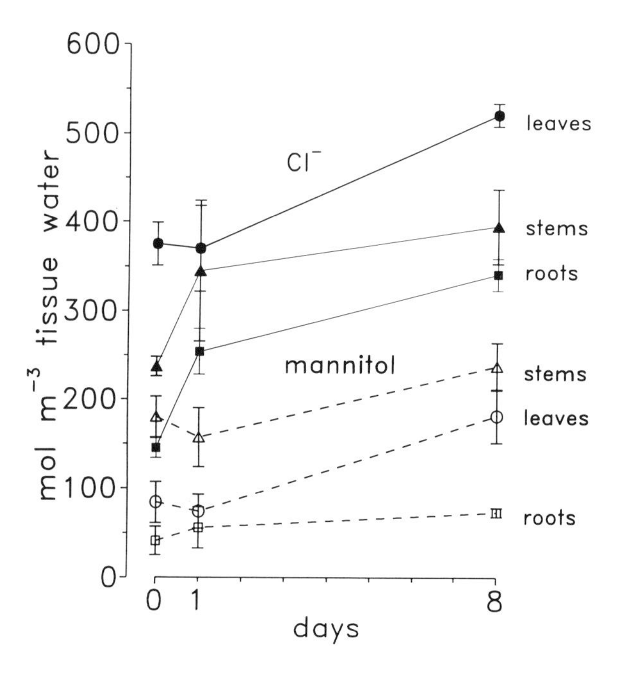


***Figure 11.3.*** *Increase in the mannitol content of different tissues of* Laguncularia racemosa *in response to increased salinity. Plants were transferred from 25% sea water to 125% sea water on day 0 (from Polania, 1990).*

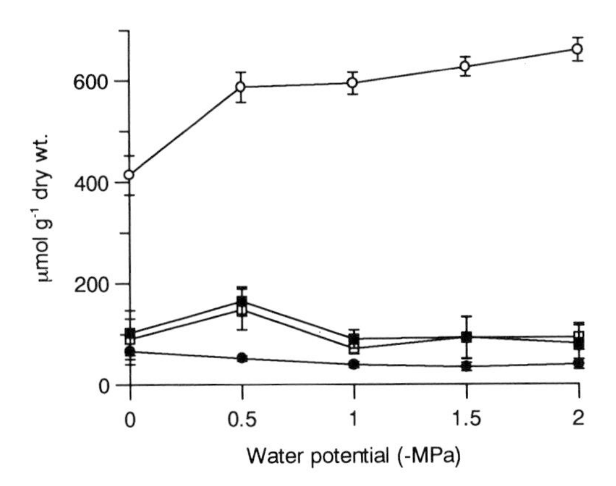


***Figure 11.4.*** *The effect of water deficit on the soluble carbohydrate pool in* Plantago major *leaf discs. Leaf discs were floated on polyethylene glycol solutions over a range of water potentials in the light for 24 h. The soluble carbohydrates were analysed by gas chromatography. The major component of the carbohydrate pool is sorbitol which preferentially accumulates at lower water potentials.* ○, *Sorbitol;* ●, *sucrose;* □, *glucose;* ■, *fructose. The bars indicate standard error (*n*=3). (Smirnoff and Raymond, unpublished data.)*

accumulations of acyclic polyols have not been investigated. Interestingly, mRNAs with homology to mammalian aldose reductases have higher abundance in some plant tissues subjected to water deficit (Bartels *et al.*, 1991). Aldose reductase activity and mRNA levels also increase in mammalian kidneys which accumulate sorbitol as an osmolyte during hyperosmotic stress (Yancey, 1992). Aldose reductase reduces glucose to sorbitol and has also been detected in some plant species (see Section 11.4).

The cyclitols pinitol, OMMI, ononitol and OMSI accumulate in various species during water deficit caused by drought and salinity (Table 11.1). In the case of pigeon pea (*Cajanus cajan*), Keller and Ludlow (1993) argued that pinitol is accumulated in the leaves because carbon flux is diverted from starch and sucrose into polyols. Pinitol concentrations in *M. crystallinum* and *Sesbania aculeata* were not higher than 10–30 mol $m^{-3}$ tissue water in the stressed plants, while in leaves of unstressed *Robinia pseudacacia* pinitol was 100 mol $m^{-3}$ (Thonke, 1990). Values for OMMI for various species in the mangrove family Rhizophoraceae ranged between 82 and 356 mol $m^{-3}$ tissue water. In the halophyte *M. crystallinum,* pinitol accumulation occurs steadily for 2 weeks after exposure to NaCl (Paul and Cockburn, 1989) but is not linked to the induction of Crassulacean acid metabolism by water deficit which is induced by NaCl (Vernon and Bohnert, 1992b). In this species, pinitol accumulation is associated with increased activity of glucose-6-phosphate cycloaldolase (Paul and Cockburn, 1989) and a *myo*-inositol-*O*-methyltransferase (Vernon and Bohnert, 1992a). The latter increase is probably transcriptionally controlled since there are higher levels of mRNA encoding the enzyme. These two enzymes are required for pinitol synthesis from glucose-6-phosphate (see Section 11.4).

A number of algae and liverworts produce acyclic polyols (including glycerol, mannitol and sorbitol) as photosynthetic products (Lewis and Smith, 1967). Levels in algae are altered by changes in external water potential usually caused by changes in salinity in marine and estuarine species (Kirst, 1990). The polyols increase and decrease in response to hyper- and hypo-osmotic conditions much more rapidly than in the higher plants considered above, and they have a clear role as compatible osmotic solutes (Kirst, 1990). High glycerol concentrations accumulate in various species of the wall-less unicellular alga *Dunaliella* which can tolerate highly saline water. Transfer of the cells from low to high salinity causes a transient shrinkage, but volume is regained within a few minutes when glycerol accumulates as a result of starch breakdown and photosynthesis. When salinity is reduced, the glycerol is rapidly metabolized to starch. The changes in glycerol are not prevented by transcription and translation inhibitors and are therefore not the result of changes in the amount of enzymes (Sadka *et al.*, 1989). The increased concentration of metabolites and regulatory molecules resulting from shrinkage could influence the activity of enzymes involved in glycerol synthesis (Bental *et al.*, 1990; Cowan *et al.*, 1992). In contrast, in the yeast *Saccharomyces cerevisiae*, in which NaCl and non-ionic osmotica cause glycerol accumulation, there is an induction of NAD-glycerol-3-phosphate dehydrogenase (André *et al.*, 1991). In the mannitol-accumulating alga *Tetraselmis subcordiformis*, mannitol-1-phosphate dehydrogenase does not change in amount but is activated by NaCl (Kirst, 1990).

## 11.6 Polyol accumulation and temperature extremes

Earlier reports on a causal interrelationship between cold hardiness and levels of sorbitol in apple trees (Ichiki and Yamaya, 1982; Raese *et al.*, 1978) have been questioned recently (Coleman *et al.*, 1992). By the induction of cold hardiness by plant growth regulators, it was shown that cold hardiness did not coincide with enhanced sorbitol and sucrose levels. However, there is further evidence for polyol accumulation at low temperatures (Table 11.2). Mannitol increased in the bark of olive trees (Drossopoulos and Niavis, 1988) and, in *R. pseudacacia*, pinitol levels in bark tissue more than doubled in late autumn, before leaf abscission (Thonke, 1990; Figure 11.5). No increase in pinitol content in *M. crystallinum* was observed by Paul and Cockburn (1989) after a 9-day period of overnight chilling. However, Vernon *et al.* (1993) showed, for the same plant species, that 78 h at 4°C induced the mRNA encoding the *myo*-inositol-*O*-methyl-transferase (*Imt*1) in a way similar to a 400 mol $m^{-3}$ NaCl treatment. Interestingly, the expression of the *Imt*1 mRNA was not influenced by exposing the roots to drought conditions in the latter set of experiments.

Regarding high temperature and polyol accumulation, there is only one report, where pinitol increased considerably in *Cicer arietinum* (Laurie, 1988 cited in Laurie and Stewart, 1990). Nevertheless, *in vitro* experiments show an ameliorating effect of polyols on heat inactivation of enzymes (see Section 11.7.4).

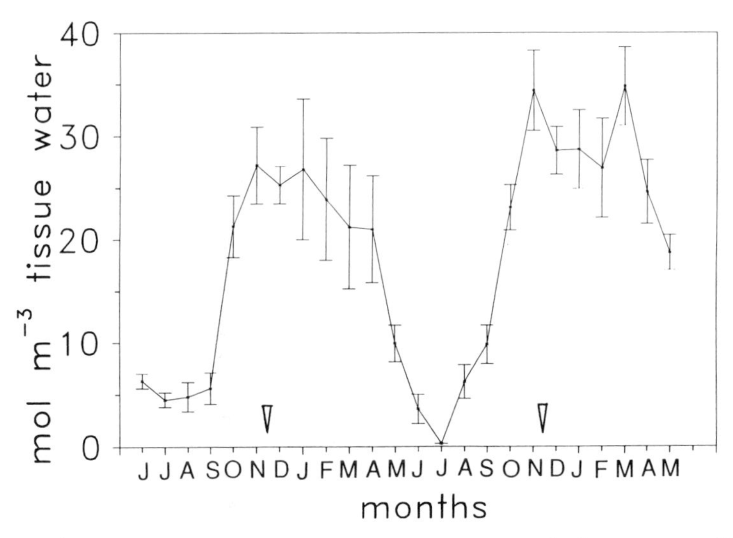


***Figure 11.5.*** *Seasonal changes in pinitol concentration in the living parts of bark tissue from* Robinia pseudacacia *over a 2-year period. The error bars are standard deviation* (n=*3–4). The arrows indicate the time of leaf abscission (from Thonke, 1990).*

## 11.7 Functions of polyols

Polyols may serve a range of functions in plants. They may serve as carbon storage and translocation compounds. Acyclic polyols are often as important as sucrose in phloem translocation from source leaves (Lewis, 1984; Lewis and Smith, 1967). Additional functions are discussed below.

### 11.7.1 *Compatible solutes and osmotic adjustment*

The term 'compatible solute' was coined originally to describe the function of polyols in sugar-tolerant yeast (Brown and Simpson, 1972) and was defined as a solute "which, at high concentration, allows an enzyme to function effectively". The role of compatible solutes is also considered in relation to proline in Chapter 10. In halophytes, it is generally accepted that compatible solutes are mainly localized in the cytoplasm where they serve for intracellular osmotic adjustment between the vacuole (rich in NaCl) and the cytoplasm (low in NaCl). Sorbitol was described as a compatible cytoplasmic solute for *P. maritima* (Ahmad *et al.*, 1979), and Gorham *et al.* (1981) assumed that pinitol may be a compatible solute in *Honkenya peploides*. Since cyclitols (with a few exceptions) are not commercially available, tests on their compatibility were not published before 1990 (Sommer *et al.*, 1990). This is also the reason why Paul and Cockburn (1989) had a question mark in their paper "Pinitol, a compatible solute in *Mesembryanthemum crystallinum*?". However, their experiments provided evidence for the cytoplasmic compartmentation of pinitol. While pinitol concentrations in the chloroplasts and in the cytosol were up to 230 and 100 mol $m^{-3}$, respectively, no pinitol was detected in vacuoles. Such a distinct localization in the cytoplasmic compartments only is probably not the case in all cyclitol-accumulating plants. On an overall leaf basis, *M. crystallinum* contained 10 mol $m^{-3}$ pinitol in the tissue sap, whereas several Fabaceae or mangroves may reach cyclitol concentrations on a tissue water basis of more than 200 mol $m^{-3}$ (Table 11.1; Figure 11.3). This is obviously too much to be completely restricted to the cytoplasm. Polyols at this concentration must make a significant contribution to osmoregulation. In species which accumulate lower levels, such as *M. crystallinum*, the polyols would have to be confined to the cytoplasm if they were to have a significant osmotic effect. Many parasitic angiosperms accumulate polyols and, although it has been proposed that they contribute to maintenance of a low water potential to drive water movement from the host plant (Richter and Popp, 1992; Stewart and Press, 1990), there is little information on the effects of water deficit on their levels. A role in turgor and volume adjustment in algae seems more likely to have been established (Kirst, 1990; Reed *et al.*, 1985).

### 11.7.2 *Hydroxyl radical scavengers*

Smirnoff and Cumbes (1989) tested various compatible solutes for hydroxyl radical scavenging activity *in vitro*. Of the compounds tested, *myo*-inositol was the most effective, closely followed by sorbitol and mannitol. Proline was less effective and glycinebetaine showed hardly any scavenging activity. In similar tests conducted with isolated cyclitols, OMMI, pinitol, ononitol, and quebrachitol proved to be even better hydroxyl radical scavengers than *myo*-inositol (Orthen *et al.*, 1994). Although there is evidence pointing to enhanced generation of hydroxyl radicals *in vivo* during water deficit and temperature extremes (see Chapter 13), the possible role of polyols in protecting cells against internally generated hydroxyl radicals awaits further investigation.

### 11.7.3 *Cryoprotectants*

Beside the cases listed in Table 11.1, we have other evidence of accumulation of cyclitols under low temperature conditions from work on trees (Lied, unpublished data). These data prompted us to examine the possible function of cyclitols as cryoprotectants in *in vitro* tests. Thylakoids (prepared from spinach) were frozen with or without added solutes and their photophosphorylation capacity was tested after thawing (Coughlan and Heber, 1982). The two cyclitols tested, OMMI and pinitol, were as effective in preventing freezing damage as was sucrose, which is a well known cryoprotectant (Terjung, 1992). Thus an interaction between cyclitols and membranes can be assumed, similar to that suggested for sugars (Crowe *et al.*, 1984).

### 11.7.4 *Heat protectants*

Both types of polyols are also able to reduce the heat inactivation of enzymes. Smirnoff and Stewart (1985) showed that mannitol and sorbitol prevent thermal denaturation of glutamine synthetase to a greater degree than proline and glycinebetaine. Laurie and Stewart (1990) demonstrated that *myo*-inositol was most effective in enhancing heat stability of glutamine synthetase isolated from chickpea. Since this plant stores large amounts of pinitol, it seems likely that methylated inositol will be similarly effective in protecting thermally sensitive enzymes. The nature of this stabilizing effect on macromolecules is still a matter for speculation; for instance it may occur via a solute–water–biopolymer interaction, as suggested by Wyn Jones (1984). There is no evidence concerning the importance of such protection *in vivo*.

## 11.8 Epilogue

Although acyclic and cyclic polyols differ substantially in their metabolic behaviour, they are very similar in their tendency to accumulate in situations of water deficit. *In vivo* evidence for their beneficial effect for stress adaptation is still scarce, but will hopefully increase by further application of molecular biology techniques. However, even now we can say that they are 'solutes for all seasons'.

## Acknowledgements

The work in M.P.'s laboratory reviewed in this article was kindly supported by grants from the Deutsche Forschungsgemeinschaft. We further wish to thank Dr Andreas Richter for valuable discussions and for providing Figure 11.1. The contributions of several co-workers (Dr J. Polania, Dr B. Thonke, Dr C. Terjung, B. Orthen and W. Lied) are gratefully acknowledged. M.P. is also grateful for the expert technical assistance of Hildegard Schwitte and the kind help of Waltraud Dörn during preparation of this manuscript.

## References

Ahmad, I., Larher, F. and Stewart, G.R. (1979) Sorbitol, a compatible osmotic solute in *Plantago maritima*. *New Phytol.* **82**, 671–678.

André, L., Hemming, A. and Adler, L. (1991) Osmoregulation in *Saccharomyces cerevisiae*. Studies on the osmotic induction of glycerol production and glycerol 3-phosphate dehydrogenase ($NAD^+$). *FEBS Lett.* **286**, 13–17.

Bartels, D., Engelhardt, K., Roncarati, R., Schneider, K., Rotter, M. and Salamini, F. (1991) An ABA and GA regulated gene expressed in barley embryo encodes an aldose reductase related protein. *EMBO J.* **10**, 1037–1043.

Bensari, M., Calmés, J. and Viala, G. (1990) Répartition du carbone fixé par photosynthèse entre l'amidon et le saccharose dans la feuille de soja. Influence d'un déficit hydrique. *Plant Physiol. Biochem.* **28**, 113–121.

Bental, M., Avron, M. and Degani, H. (1990) The role of intracellular orthophosphate in triggering osmoregulation in *Dunaliella salina*. *Eur. J. Biochem.* **188**, 117–122.

Bieleski, R.L. (1982) Sugar alcohols. In: *Encyclopedia of Plant Physiology* (eds F. Loewus and W. Tanner). Springer, Berlin, pp. 158–192.

Brown, A.D. and Simpson, J.R. (1972) Water relations of sugar-tolerant yeasts: the role of intracellular polyols. *J. Gen. Microbiol.* **72**, 589–591.

Coleman, U.K., Estabrooks, E.N., O'Hara, M., Embleton, J. and King, R.R. (1992) Seasonal changes in cold hardiness, sucrose and sorbitol in apple trees tested with plant growth regulators. *J. Hortic. Sci.* **67**, 429–435.

Coughlan, S.J., and Heber, U. (1982) The role of glycinebetaine in the protection of spinach thylakoids against freezing stress. *Planta,* **156**, 62–69.

Cowan, A.K., Rose, P.D. and Horne, L.G. (1992) *Dunaliella salina*: a model system for studying the response of plant cells to stress. *J. Exp. Bot.* **43**, 1535–1547.

Craigie, J.S. (1974) Storage products. In: *Algal Physiology and Biochemistry* (ed. W.P.D. Stewart). Blackwell Scientific Publications, London, pp. 206–235.

Crowe, L.M., Mouradian, R., Crowe, J.H., Jackson, S.A. and Womersley, C. (1984) Effects of carbohydrates on membrane stability at low water activities. *Biochim. Biophys. Acta* **769**, 141–150.

Diamantoglou, S. (1974) Über das physiologische Verhalten von Cycliten in vegetabilen Teilen höherer Pflanzen. *Biochem. Physiol. Pflanz.* **166**, 511–523.

Dittrich, P. and Brandl, A. (1987) Revision of the pathway of D-pinitol formation in Leguminosae. *Phytochemistry* **26**, 1925–1926.

Dittrich, P. and Korak, A. (1984) Novel biosynthesis of D-pinitol in *Simmondsia chinensis*. *Phytochemistry* **23**, 65–66.

Drew, E.A. (1984) Physiology and metabolism of cyclitols. In: *Storage Carbohydrates in Vascular Plants* (ed. D.H. Lewis). Cambridge University Press, Cambridge, pp. 133–155.

Drossopoulos, J.B. and Niavis, C.A. (1988) Seasonal changes of the metabolites in the leaves, bark and xylem tissues of olive tree (*Olea europaea* L.). II. Carbohydrates. *Ann. Bot.* **62**, 321–327.

Ford, C.W. (1982) Accumulation of *O*-methyl-inositol in water-stressed *Vigna* species. *Phytochemistry* **21**, 1149–1151.

Ford, C.W. (1984) Accumulation of low molecular weight solutes in water-stressed tropical legumes. *Phytochemistry* **23**, 1007–1015.

Gorham, J., Hughes, L.L. and Wyn Jones, R.G. (1981) Low-molecular-weight carbohydrates in some salt-stressed plants. *Plant Physiol.* **53**, 27–33.

Gorham, J., McDonnell, E. and Wyn Jones, R.G. (1984) Pinitol and other solutes in salt-stressed *Sesbania aculeata*. *Z. Pflanzenphysiol.* **114**, 173–178.

Grant, C.R. and ap Rees, T. (1981). Sorbitol metabolism by apple seedlings. *Phytochemistry* **20**, 1505–1511.

Ichiki, S. and Yamaya, H. (1982) Sorbitol in tracheal sap of dormant apple (*Malus domestica* Borkh.) shoots as related to cold hardiness. In: *Plant Cold Hardiness and Freezing Stress* (eds P.H. Li and A. Sakai). Academic Press, New York, pp. 181–187.

Jennings, D.H. and Burke, R.M. (1990) Compatible solutes – the mycological dimension and their role as physiological buffering agents. *New Phytol.* **116**, 277–283.

Keller, F. and Ludlow, M.M (1993) Carbohydrate metabolism in drought stressed leaves of pigeon pea (*Cajanus cajan*). *J. Exp. Bot.* **44**, 1351–1359.

Kirst, G.O. (1990) Salinity tolerance of eukaryotic marine algae. *Annu. Rev. Plant Physiol. Mol. Biol.* **41**, 21–53.

Kuo, T.M., Doehlert, D.C. and Crawford, C.G. (1990). Sugar metabolism in germinating

soybean seeds. Evidence for the sorbitol pathway in soybean axes. *Plant Physiol.* **93**, 1514–1520.
Laurie, S. and Stewart, G.R. (1990) The effects of compatible solutes on the heat stability of glutamine synthetase from chickpeas grown under different nitrogen and temperature regimes. *J. Exp. Bot.* **41**, 1451–1422.
Lewis, D.H. (1984) Physiology and metabolism of alditols. In: *Storage Carbohydrates in Vascular Plants* (ed. D.H. Lewis). Cambridge University Press, Cambridge, pp. 157–179.
Lewis, D.H. and Smith, D.C. (1967) Sugar alcohols (polyols) in fungi and green plants. I. Distribution, physiology and metabolism. *New Phytol.* **66**, 143–184.
Loescher, W.H. (1987) Physiology and metabolism of sugar alcohols in higher plants. *Physiol. Plant.* **70**, 553–557.
Loescher, W.H., Tyson, R.H., Everard, J.D., Redgewell, R.J. and Bieleski, R.L. (1992). Mannitol synthesis in higher plants. Evidence for the role and characterization of a NADPH-dependent mannose 6-phosphate reductase. *Plant Physiol.* **98**, 1396–1402.
Loewus, F.A. and Loewus, M.W. (1980) *myo*-Inositol: biosynthesis and metabolism. In: *The Biochemistry of Plants. Volume 3. Carbohydrates: Structure and Function* (ed. J. Preiss). Academic Press, New York, pp. 43–76.
Micheel, F. (1932) Übergang von der Hexosereihe in die Cyclitreihe. *Liebigs Ann. Chem.* **496**, 77–98.
Moing, A,. Carbonne, F., Rashad, M.H. and Gaudiellere, J.-P. (1992) Carbon fluxes in mature peach leaves. *Plant Physiol.* **100**, 1878–1884.
Munns, R. (1993) Physiological processes limiting plant growth in saline soils: some dogmas and hypothesis. *Plant, Cell Environ.* **16**, 15–24.
Negm, F.B. (1986) Purification and properties of an NADPH-aldose reductase (aldehyde reductase) from *Euonymus japonica* leaves. *Plant Physiol.* **80**, 972–977.
Negm, F.B. and Loescher, W.H. (1981) Characterization and partial purification of aldose 6-phosphate reductases (alditol 6-phosphate: NADP 1-oxidoreductase) from apple leaves. *Plant Physiol.* **67**, 139–142.
Nguyen, A. and Lamant, A. (1988) Pinitol and *myo*-inositol accumulation in water-stressed seedlings of maritime pine. *Phytochemistry* **27**, 3423–3427
Oparka, K.J. and Wright, K. (1988) Osmotic regulation of starch synthesis in potato tubers? *Planta* **174**, 123–126.
Orthen, B., Popp, M. and Smirnoff, N. (1994) Hydroxyl radical scavenging properties of cyclitols. *Proc. R. Soc. Edinb.* B **102**, in press.
Paul, M.J. and Cockburn, W. (1989) Pinitol a compatible solute in *Mesembryanthemum crystallinum* L.? *J. Exp. Bot.* **40**, 1093–198.
Polania, J. (1990) Anatomische und physiologische Anpassungen von Mangroven. Ph.D Thesis, University of Vienna.
Popp, M. Feldl, C., Polania, J. and Aswathappa, N. (1990) Stress metabolites in *Casuarina* and other tropical trees. In: *Fast Growing Trees and Nitrogen Fixing Trees* (eds D. Werner and P. Müller). Gustav Fischer, Stuttgart, pp. 202–207.
Quick,W.P., Chaves, M.M., Wendler, R., David, M., Rodrigues, M.L., Passaharinho, J.A., Pereira, J.S., Adcock, M.D., Leegood, R.C. and Stitt, M. (1992) The effect of water stress on photosynthetic carbon metabolism in four species grown under field conditions. *Plant, Cell Environ.* **15**, 26–35.
Raese, J.T., Williams, M.W. and Billingsley, H.D. (1978) Cold hardiness, sorbitol and sugar levels of apple shoots as influenced by controlled temperature and season. *J. Am. Soc. Hortic. Sci.* **103**, 796–801.
Ranney, T.G., Bassuk, N.L. and Whitlow, T.H. (1991) Osmotic adjustment and solute con-

stituents in leaves and roots of water-stressed cherry (*Prunus*) trees. *J. Am. Soc. Hortic. Sci.* **116**, 648–688.
Reed, R.H., Davison, I.R., Chudek, J.A. and Foster, R. (1985) The osmotic role of mannitol in the Phaeophyta: an appraisal. *Phycologia* **24**, 35–47.
Richter, A.(1989) Osmotisch wirksame Inhaltsstoffe in einheimischen Mistelarten und ihren Wirten. Ph.D Thesis, University of Vienna.
Richter, A. (1992) Viscumitol, a dimethyl-ether of *muco*-inositol from *Viscum album*. *Phytochemistry* **31**, 3925–3927.
Richter, A. and Popp, M. (1992) The physiological importance of accumulation of cyclitols in *Viscum album* L. *New Phytol.* **121**, 431–438.
Richter, A., Thonke, B. and Popp, M. (1990) 1D-1-*O*-methyl-*muco* inositol in *Viscum album* and members of the Rhizophoraceae. *Phytochemistry* **29**, 1785–1786.
Rumpho, M.E., Edward, G.E. and Loescher, W.H. (1983) A pathway for photosynthetic carbon flow to mannitol in celery leaves. Activity and localization of key enzymes. *Plant Physiol.* **73**, 869–873.
Sadka, A., Lers, A., Zamir, A. and Avron, M. (1989) A critical examination of the role of protein synthesis in the osmotic adaptation of the halotolerant alga *Dunaliella*. *FEBS Lett.* **244**, 93–98.
Schilling, N., Dittrich, P. and Kandler, O. (1972) Formation of L-quebrachitol from D-bornesitol in leaves of *Acer pseudoplatanus*. *Phytochemistry* **11**, 1401–1404.
Smirnoff, N. and Cumbes, Q.J. (1989) Hydroxyl radical scavenging of compatible solutes. *Phytochemistry* **28**, 1057–1060.
Smirnoff, N. and Stewart, R.G. (1985) Stress metabolites and their role in coastal plants. *Vegetatio* **62**, 273–278.
Smith, A.E. and Philips, D.V. (1982) Influence of sequential prolonged periods of dark and light on pinitol concentration in clover and soybean tissue. *Physiol. Plant.* **54**, 31–33.
Sommer, C., Thonke, B. and Popp, M. (1990) The compatibility of D-pinitol and 1D-1-*O*-methyl-*muco*-inositol with malate dehydrogenase activity. *Bot. Acta* **103**, 270–273.
Stewart, G.R. and Press, M.C. (1990) The physiology and biochemistry of parasitic angiosperms. *Annu. Rev. Plant Physiol. Mol. Biol.* **41**, 127–151.
Swain, T. (1963) *Chemical Plant Taxonomy*. Academic Press, London, New York.
Tarczynski, M.C., Jensen, R.G. and Bohnert, H.J. (1993) Stress protection of transgenic tobacco by production of the osmolyte mannitol. *Science* **259**, 508–510.
Terjung, Ch. (1992) Subzelluläre Funktionen der Cyclite 1D-1-*O*-methyl-*muco*-Inosit und 1D-3-*O*-Methyl-*chiro*-Inosit. Ph.D Dissertation, University of Münster.
Thonke, B. (1990) Zur Rolle der Cyclite D-Pinit und 1D-1-*O*-methyl-*muco*-Inosit in Höheren Pflanzen. Ph.D Thesis, University of Vienna.
Vernon, D.M. and Bohnert, H.J. (1992a) A novel methyl transferase induced by osmotic stress in the facultative halophyte *Mesembryanthemum crystallinum*. *EMBO J.* **11**, 2077–2085.
Vernon, D.M. and Bohnert, H.J. (1992b) Increased expression of a *myo*-inositol methyl transferase in *Mesembryanthemum crystallinum* is a part of a stress response distinct from Crassulacean acid metabolism induction. *Plant Physiol.* **99**, 1695–1698.
Vernon, D.M., Ostrem, J.-A. and Bohnert, H.J. (1993) Stress perception and response in a facultative halophyte: the regulation of salinity-induced genes in *Mesembryanthemum crystallinum*. *Plant, Cell Environ.* **16**, 437–444.
Wyn Jones, R.G. (1984) Phytochemical aspects of osmotic adaptation. In: *Phytochemical Adaptations to Stress* (eds B.N. Timmermann, C. Steelink and F.A. Loweus). Plenum Press, New York, London, pp. 55–78.

Yancey, P.H. (1992) Compatible and counteracting aspects of organic osmolytes in mammalian kidney cells *in vivo* and *in vitro*. In: *Water and Life. Comparative Analysis of Water Relationships at the Organismic, Cellular and Molecular Levels* (eds G.N. Somero, C.B. Osmond and C.L. Bolis). Springer-Verlag, Berlin, pp. 19–32.

Zrenner, R. and Stitt, M. (1991) Comparison of the effect of rapidly and gradually developing water stress on carbohydrate metabolism in spinach leaves. *Plant, Cell Environ.* **14**, 939–946.

12

# Antioxidant systems and plant response to the environment

N. Smirnoff

## 12.1 Introduction

Antioxidant systems are essential for the survival of organisms in aerobic environments. Oxygen is potentially toxic since it can be transformed by metabolic activity into more reactive forms such as superoxide, hydrogen peroxide, singlet oxygen and hydroxyl radicals (Halliwell and Gutteridge, 1989) which are collectively known as active oxygen. Elevated oxygen levels (275–300 μM) occur in chloroplasts as a result of photosynthetic oxygen evolution, while the oxygen concentration in equilibrium with air is 253 μM at 25°C (Robinson, 1988). It is difficult to assess the toxic effects of oxygen on photosynthetic organisms because photorespiration decreases productivity by the loss of previously assimilated carbon dioxide (see Chapter 4). Photorespiration also produces large amounts of hydrogen peroxide which must be removed before it causes oxidative damage. High oxygen partial pressure in the intercellular spaces of aquatic macrophytes, derived from photosynthesis, has been suggested to limit them to depths of less than 10 m because of oxygen toxicity (Peñuelas, 1987).

The aim of this chapter is to assess the role of antioxidant systems in situations where the plant is particularly prone to oxidative damage. These are: (i) photosynthetic cells exposed to normal environmental conditions as well as high light, drought and low temperature; (ii) nitrogen-fixing root nodules and (iii) the re-exposure of tissues, particularly underground parts, to oxygen after a period of oxygen shortage caused by waterlogging (post-anoxic injury). It is also well established that other environmental factors, for example UV-B radiation (see Chapter 13) and pollutants such as ozone, nitrogen oxides and sulphur dioxide are damaging, at least in part, because they generate active oxygen and other free radicals (Elstner, 1987; Mehlhorn *et al.*, 1987, 1990). These will not be covered in detail. In the first part of the chapter the formation of active oxygen and the major constituents of the antioxidant system will be described.

## 12.2 Formation and effects of active oxygen

### 12.2.1 *Formation of active oxygen in plant cells*

Molecular oxygen is not particularly reactive. Although it is a free radical with two unpaired electrons (in the $\pi^*$ antibonding orbital), they are of parallel spin. This means that it cannot accept a pair of electrons of opposite spin and, because of this spin restriction, it therefore tends to react by accepting electrons singly. It therefore reacts best with other free radicals. However, oxygen can become more reactive when transformed into more reduced or electronically excited states. These can come about by interaction of ground state (triplet) oxygen with metabolic systems and it is under these circumstances that oxygen becomes a potential hazard, as well as a necessity, of energetically efficient aerobic life.

Univalent reduction of oxygen forms superoxide. This reaction is endergonic but, after this activation, further reduction to hydrogen peroxide, hydroxyl radical and water is exergonic. Superoxide can be formed by autoxidation of various electron transport chain components in mitochondria, for example ubiquinone (Winston, 1990), and in chloroplasts (see Section 12.4). Superoxide formation can also be catalysed by microsomal membranes (Winston, 1990) and has been demonstrated in peroxisomes and glyoxysomes (Del Rio *et al.*, 1988; Sandalio *et al.*, 1988). Xanthine, aldehyde and diamine oxidases, among others, produce superoxide (Winston, 1990). Various redox-cycling xenobiotics such as the herbicide paraquat (methyl viologen) can catalyse high rates of superoxide formation by acting as electron acceptors (e.g. from reduced ferredoxin in chloroplasts) and transfering them to oxygen (Winston, 1990). Superoxide can act as an oxidizing agent (and sometimes as a reducing agent) but it is probably more damaging because of the consequent formation of hydrogen peroxide and hydroxyl radicals (Elstner, 1987; Halliwell and Gutteridge, 1989).

Hydrogen peroxide is formed by oxidases which catalyse electron transfer of two electrons to molecular oxygen, such as amino acid oxidases, glucose oxidase and glycolate oxidase. It also forms by dismutation of superoxide.

$$2O_2^- + 2H^+ \rightarrow H_2O_2 + 2O_2$$

This reaction occurs slowly at neutral pH since the protonated form of superoxide (hydroperoxyl radical, $pk_a = 4.8$) is more reactive (Halliwell and Gutteridge, 1989). However, at physiological pH (particularly in the alkaline chloroplast stroma) this reaction is speeded up by at least $10^4$-fold by the enzyme superoxide dismutase (SOD) of which various isoforms are present in all cell compartments. Hydrogen peroxide is relatively stable and is an oxidizing agent.

Further reduction of hydrogen peroxide forms the hydroxyl radical ($OH\cdot$). This species is extremely reactive and short lived, and is damaging since it reacts at the site of its formation with most organic molecules including proteins, lipids and nucleic acids. A major route for its formation *in vivo* is the iron-catalysed Haber–Weiss reaction.

$$O_2^- + Fe^{3+} \rightarrow O_2 + Fe^{2+}$$

$$Fe^{2+} + H_2O_2 \rightarrow OH\cdot + OH^- + Fe^{3+}$$

Copper and other transition metals could also catalyse the reaction (Halliwell and Gutteridge, 1989). An important feature of hydroxyl radical formation is the availability of iron or other transition metals in a suitable form, and the regulation of these to maintain low soluble levels is an important aspect of antioxidant defence.

Singlet oxygen ($^1O_2$) is formed by a change in spin state from parallel to antiparallel (Elstner, 1987; Knox and Dodge, 1985). This greatly increases its reactivity by removing the spin restriction. It can react readily with amino acids such as cysteine, methionine, tryptophan and histidine and it is a major initiator of lipid peroxidation leading to membrane damage (see Section 12.2.3). Singlet oxygen is formed by transfer of energy from photosensitizers. These are pigments which absorb light and then transfer the excitation energy to ground state oxygen, resulting in singlet oxygen formation. In plants, the most important photosensitizer is chlorophyll and singlet oxygen formation is a danger under conditions of excess excitation energy input (see Section 12.4). Haem and riboflavin and a number of secondary compounds (e.g. quinones, furanocoumarins, polyacetylenes and thiophenes) are also photosensitizers. The photodynamic secondary compounds may act as herbivore deterrents by forming singlet oxygen (Downum, 1992; Knox and Dodge, 1985). Some xenobiotics, such as the dye rose bengal, are toxic because they act as photosensitizers and the plant pathogenic fungus *Cercospora* produces a toxin cercosporin which acts as a photosensitizer in the host plant (Knox and Dodge, 1985). Some of the examples above emphasize that plants are able 'deliberately' to produce active oxygen species and functions include lignification (Olson and Varner, 1993) and pathogen attack, where active oxygen may be produced to attack pathogens or to initiate a hypersensitive response (Mehdy, 1994).

### 12.2.2 *Measurement of active oxygen*

Active oxygen formation is not easy to measure directly since most of the forms are reactive and short lived. Hydrogen peroxide is readily quantified in tissue extracts, for example using its ability to oxidize suitable substrates in the presence of peroxidase. Superoxide and hydroxyl radicals can, in principle, be measured by spin trapping (e.g. Tiron and dimethylpyrolline-N-oxide, DMPO) followed by electron paramagnetic resonance (EPR) spectroscopy to detect the resulting longer-lived organic free radical (Halliwell and Gutteridge, 1989; Miller and MacDowall, 1975). Superoxide formation in isolated chloroplasts has been measured in this way (Asada and Takahashi, 1987) but these methods are often not easy to perform *in vivo*. The involvement of hydroxyl radicals can be inferred by the action of quenchers such as dimethylsulphoxide (DMSO) which, when loaded into cells, can compete with endogenous target molecules and minimize damage. DMSO has been used to demonstrate the formation of hydroxyl radicals as a result of paraquat treatment (Babbs *et al.*, 1989). Singlet oxygen is not measured directly, but its action can be inferred by the use of quenchers (1,4-diazobicyclooctane (DABCO) and histidine). These however also react with hydroxyl radicals (Halliwell and Gutteridge, 1989). Ultraweak chemiluminescence can result from light emission as singlet oxygen returns to ground state and this increases when oxygen concentration is raised, however other chemical reactions produce light (Chapter 13). The chemiluminescence can be detected with a luminometer or liqiud scintillation counter. Because most of these measurements are equivocal and difficult to perform *in vivo*, formation of active oxygen is most often inferred from the symptoms it produces.

### 12.2.3 *Symptoms of oxidative damage*

Oxidative damage can be detected by lipid peroxidation. Polyunsaturated lipids are susceptible to peroxidation, a process in which hydroxyl radicals and singlet oxygen can react with the methylene groups forming conjugated dienes, lipid peroxy radicals and hydroperoxides. The peroxy radicals can abstract H from other unsaturated fatty acids, leading to a chain reaction of peroxidation. The peroxidation of membrane lipids leads to breakdown of structure and function. One of the products of lipid peroxidation is malondialdehyde which is often used as a measure of peroxidation. Hydrocarbons such as ethane, pentane and ethylene may also be released and these can be measured (Elstner, 1987; Halliwell and Gutteridge, 1989). Hydroxyl radicals can denature proteins and react with bases in DNA causing mutations and the aldehydes formed by lipid peroxidation can conjugate and inactivate proteins (Wolff *et al.*, 1986). Hydrogen peroxide can inactivate enzymes, particularly some of the light-activated Calvin cycle enzymes (Charles and Halliwell, 1980; Kaiser, 1979), and decarboxylate oxo-acids (Elstner, 1987). Recently, a stable free radical, probably derived from a quinone, has been detected by EPR in desiccated tissues and it is proposed as a useful marker for oxidative damage (Leprince *et al.*, 1994). Since oxidative deterioration occurs after cell damage or death, it is important to interpret measurements of oxidative damage with care.

## 12.3 The components of the antioxidant system

### 12.3.1 *Ascorbate*

Ascorbate (vitamin C) occurs in all tissues but its concentration can vary widely, usually being higher in photosynthetic cells and meristems (and some fruits). The high concentration in photosynthetic cells is a result of high concentration in chloroplasts where its concentration is around 20–30 mM (Foyer *et al.*, 1983). Measurement of ascorbate and its oxidized form, dehydroascorbate (DHA), along the length of a developing barley leaf shows that it is highest in the older segments where chloroplasts are fully developed and chlorophyll levels are highest (Figure 12.1). The permeability of the chloroplast envelope to ascorbate and DHA is low (Anderson *et al.*, 1983). It also occurs in the cell wall (Luwe *et al.*, 1993). It is not present in dry seeds and is rapidly synthesised during germination (Loewus, 1980). Usually at least 90% of the ascorbate pool is present as ascorbate (Figure 12.1). On oxidation, the monodehydroascorbate (MDHA) radical is formed. Further oxidation results in DHA. DHA is unstable, particularly at high pH and irreversibly saponifies to diketogulonic acid followed by rearrangement to 2-(*threo*-1,2,3-trihydroxypropyl)tartronic acid (Loewus, 1980) so that oxidation of the ascorbate pool can lead to its loss. It reacts non-enzymatically with superoxide, hydrogen peroxide (Elstner, 1987) and singlet oxygen (Asada and Takahashi, 1987). Ascorbate is regenerated from its oxidized forms by MDHA and DHA reductases (Section 12.4.2). Its role as an antioxidant will be discussed later (Section 12.4.2) but it also has other physiological functions which are discussed by Gander (1982). One of these is to act as a co-factor in the prolyl hydroxylase reaction which is required for hydroxyproline synthesis in the cell wall protein extensin. It can also act as a precursor for oxalate and tartrate in some plants (Loewus, 1980). It has some as yet poorly understood function in growth regulation, perhaps in the cell wall where (in some plants) ascorbate oxidase is located. Its oxidation product MDHA affects cell division, expansion and nutrient uptake (Innocenti *et al.*, 1990). MDHA may also act as an electron acceptor in the cell wall for cytochrome *b*-mediated trans-plasma membrane electron transport (Horemans *et al.*, 1994).

Despite the importance of ascorbate, very little is known about its synthesis in plants. Two possible pathways have been proposed (Loewus, 1980; Loewus *et al.*, 1990; Saito *et al.*, 1990):

1. (D-glucose) → D-galactose → D-galacturonic acid methyl ester
   → L-galactono-1,4-lactone → L-ascorbic acid
2. D-glucose → D-glucosone → L-sorbosone → L-ascorbic acid

The first scheme is known as the 'inversion' pathway. Glucose labelled in carbons 1, 2 or 6 would be predicted to produce ascorbate labelled in carbons 6, 5 and 1 respectively. This occurs in rats but not in plants (Loewus, 1980). In plants 1-, 2- and 6-labelled D-glucose and D-galactose produce ascorbate labelled in the 1, 2 and 6 carbons, respectively. The observed labelling therefore rules out the inversion of label required by pathway 1. Recent labelling and enzyme studies have provided support for pathway 2. D-[6-$^{14}$C]glucosone is a very effective precursor for ascorbate, with most of the label appearing in $C_6$ of ascorbate. L-[U-$^{14}$C]-sorbosone also acts as a precursor (Loewus *et al.*, 1990). Additionally, an NADP-dependent sorbosone dehydrogenase has been purified from bean and spinach leaves and ascorbate is the reaction product (Saito *et al.*, 1990). Despite the evidence for pathway 2, plants can rapidly convert exogenous L-galactono-1,4-lactone to ascorbate (Degara *et al.*, 1992; Hausladen and Kunert, 1990).

## 12.3.2 *Glutathione*

The tripeptide glutathione (γ-L-glutamyl-L-cysteinyl-glycine, GSH) is the major soluble thiol compound in the majority of plants which have been examined. Cysteine and methionine are present in much lower concentrations. Some species, particularly legumes, have an analogue of GSH, homoglutathione (γ-glutamyl-L-cysteinyl-β-alanine) (Zopes *et al.*, 1993). A related tripeptide γ-glutamyl-L-cysteinyl-gluatamate has been found in maize seedlings (Meuwly *et al.*, 1993) and another homologue (γ-glutamylcysteinylserine) occurs in some grasses (Klapheck *et al.*, 1992). Oxidation of the thiol group of GSH leads to the formation of oxidized glutathione (GSSG) in which two GSH molecules are joined by a disulphide bond. Reviews on GSH are given by Alscher (1989) and Smith *et al.* (1990).

GSH occurs in leaves and roots at concentrations ranging from 0.1 to 1 mM, the higher values being in leaves and meristems. Analysis of chloroplasts and vacuoles from tobacco protoplasts suggests concentrations of 0.06 mM in cytoplasm, 0.02 mM in vacuoles and 1–5 mM in chloroplasts (Foyer and Halliwell, 1976; Law *et al.*, 1983; Rennenberg, 1982). Non-aqueous isolation of chloroplasts gives concentrations of 6–13 mM, suggesting that GSH may be lost during aqueous fractionation (Klapheck *et al.*, 1987; Smith *et al.*, 1985). GSH must also occur in mitochondria but there is no information on its concentration. The GSH pool is usually 90% reduced (Law *et al.*, 1983); however, in chloroplasts isolated by non-aqueous fractionation, 30% of the pool is present as GSSG (Smith *et al.*, 1985). GSSG is reduced to GSH by GSH reductase (GR; Section 12.4.2). In desiccated tissues the GSH pool is more oxidized, being 24% GSSG in pea seeds (Kranmer and Grill, 1993), 40% GSSG in desiccated moss (Dhindsa, 1987) and 40% GSSG in severely droughted barley leaves (Smirnoff, 1993). GSH may be transported into vacuoles by a GSH-S-conjugate ATPase (Martinoia *et al.*, 1993).

The synthesis of GSH from its constituent amino acids is a two-step non-ribosomal process. γ-Glutamyl-L-cysteine is formed in an ATP-dependent reaction. The enzyme catalysing this step is γ-glutamyl-L-cysteine synthetase which has been isolated from a number of species (Rennenberg, 1982). The second step is catalysed by GSH synthetase, again using ATP, in which glycine is added to γ-glutamyl-L-cysteine to form GSH. γ-glutamyl-L-cysteine synthetase from animals is subject to feedback inhibition by GSH and this may also be the case in plants (Rennenberg, 1982). GSH can be utilized by plants as a sulphur source

---

**Figure 12.1.** *Chlorophyll concentration (a), ascorbate (Asc) and dehydroascorbate (DHA) concentration (b) and monodehydroascorbate reductase activity (c) in the primary leaves of barley (*Hordeum vulgare*) seedlings. When the primary leaves had emerged from the coleoptiles, the seedlings were transferred to low (LL: 35 μmol $m^{-2}$ $s^{-1}$) and higher (HL: 280 μmol $m^{-2}$ $s^{-1}$) light intensity conditions for 3 days. After this time, the leaves were divided into 3-cm segments and analysed (Smirnoff and Raymond, unpublished data).*

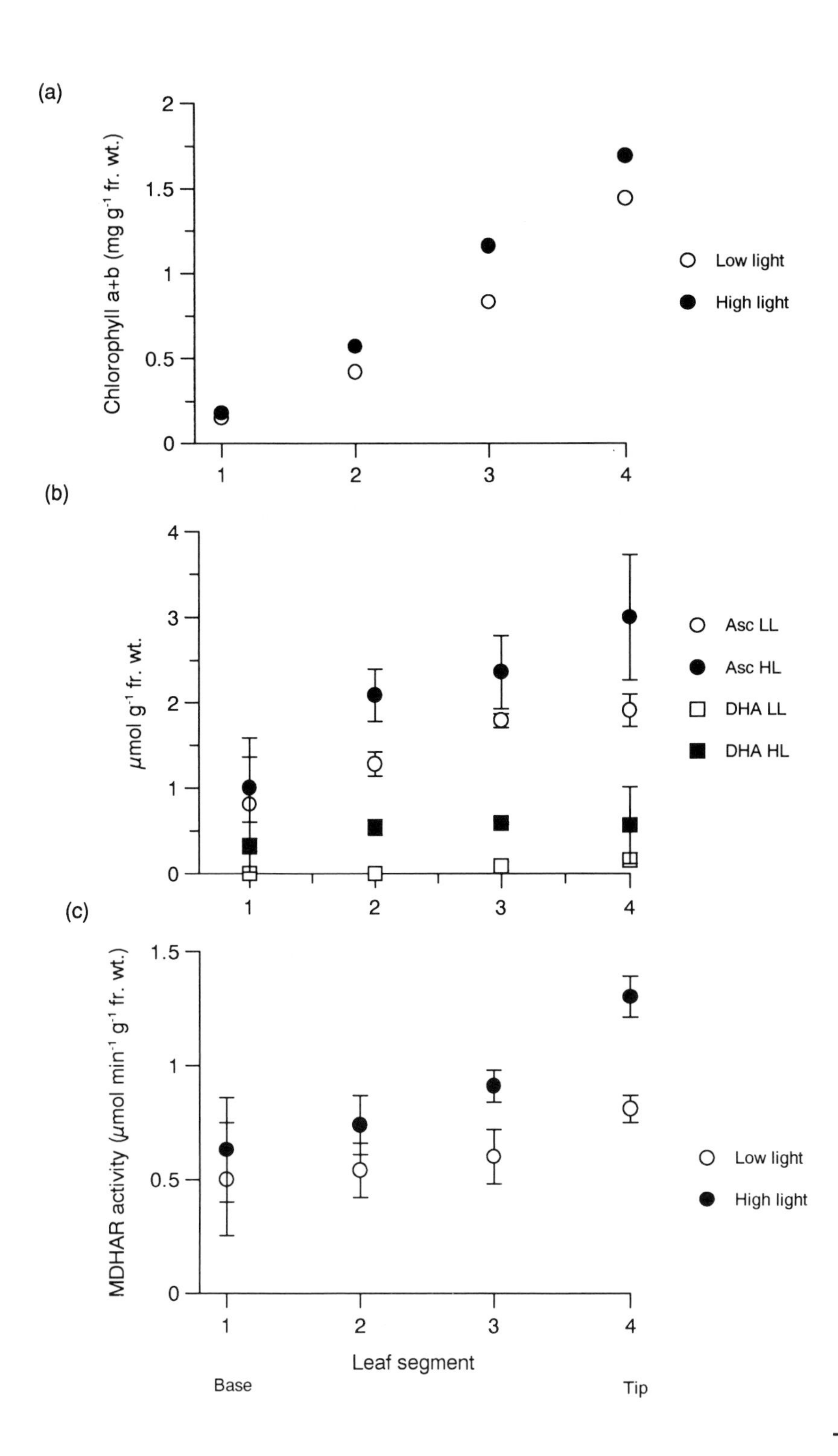
(a)
Chlorophyll a+b (mg g-1 fr. wt.)
2
1.5
1
0.5
0
1
2
3
4
Low light
High light
(b)
μmol g-1 fr. wt.
4
3
2
1
0
1
2
3
4
Asc LL
Asc HL
DHA LL
DHA HL
(c)
MDHAR activity (μmol min-1 g-1 fr. wt.)
1.5
1
0.5
0
1
2
3
4
Leaf segment
Base
Tip
Low light
High light

although the pathway for its degradation is not fully characterized. Glycine and cysteine are released and the glutamyl moiety is converted to 5-oxo-proline and then glutamate. 5-Oxo-prolinase, which converts 5-oxo-proline to glutamate in an ATP-dependent reaction, has been characterized in plants and is probably cytoplasmic (Rennenberg, 1982).

GSH synthetase occurs in the chloroplasts (25–70% of activity) and cytosol of leaves which have so far been investigated (Alscher, 1989; Klapheck *et al.*, 1987). The high GSH concentration in chloroplasts, and evidence that its synthesis may be at least partially light dependent, suggests that a large proportion may be made in chloroplasts. Concentrations increase during the day and fall to half the maximum level at night in *Picea abies* needles and tomato cotyledons (Koike and Patterson, 1988; Schupp and Rennenberg, 1988). Oxidative conditions increase GSH levels. For example, total GSH levels increase in catalase-deficient barley mutants and barley leaves treated with the catalase inhibitor aminotriazole (Smith, 1985; Smith *et al.*, 1984) under photorespiratory conditions. This treatment results in 90% of the GSH pool being oxidized to GSSG by the accumulating hydrogen peroxide. They suggest that oxidation reduces the feedback inhibition of γ-glutamyl-L-cysteine synthetase by GSH and allows more synthesis. However, since the oxidation state of the pool does not change during light/dark cycles, this cannot explain light-stimulated accumulation. More recently, May and Leaver (1993) have shown that in non-photosynthetic *Arabidopsis* cell suspension cultures aminotriazole increases GSH synthesis even though the oxidative stress is not extreme enough to increase GSSG levels. They suggest hydrogen peroxide rather than GSSG is the signal for increased GSH synthesis.

GSH, like ascorbate, has multiple functions. It is used as a sulphur transport and storage compound in plants supplied with high levels of sulphate (Rennenberg, 1984). It can be used to detoxify xenobiotics such as herbicides. This is achieved by conjugation with GSH, the reaction being catalysed by GSH-*S*-transferase enzymes (Rennenberg, 1982). The physiological role of GSH-*S*-transferases may be to conjugate lipid peroxides formed as a result of oxidative damage, and a plant enzyme able to use GSH to reduce 13-hydroperoxylinoleic and 13-hydroperoxylinolenic acids to the hydroxy-acids with production of GSSG has been cloned and expressed in *Escherichia coli* (Bartling *et al.*, 1993). These hydroperoxy acids would be major products of peroxidation in thylakoids. GSH forms the basis for synthesis of metal-binding phytochelatins ((γ-glutamyl-L-cysteinyl)$_n$-glycine, where $n$=2–7). Their synthesis is stimulated by Cu, Zn, Pb and Cd which they bind by the thiol group (Steffens, 1990). Phytochelatins are probably involved in controlling intracellular levels of free metal ions rather than in metal tolerance (De Voss *et al.*, 1992). They could be considered antioxidants to the extent that they could minimize hydroxyl radical formation by Cu-catalysed Haber–Weiss reactions (see Section 12.2.1). GSH, but not GSSG, from 0.01 to 1 mM causes induction of the expression of the same defence genes (for phytoalexin and lignin synthesis) as caused by fungal infection or elicitors when applied to *Phaseolus vulgaris* cell cultures (Wingate *et al.*, 1988). Since the effective concentrations are of the same order or less than occurs in plant cells, it is difficult to understand its significance for antifungal defence *in vivo*. GSSG is a powerful protein synthesis inhibitor (see Dhindsa, 1987; Kranmer and Grill, 1993) and its accumulation under severe oxidative stress could prevent protein synthesis.

## 12.3.3 α-*Tocopherol*

α-Tocopherol (vitamin E) is a lipophilic molecule which occurs in membranes. It is particularly concentrated in the thylakoid membranes of the chloroplasts and in leaves it is synthesised in chloroplasts (Schultz, 1990). It acts as a scavenger of active oxygen in membranes and is important in thylakoids (and mammalian eyes and skin) where light interacts with membrane-bound pigments which can potentially act as photosensitizers of singlet oxygen formation leading to lipid peroxidation (Fryer, 1992). The antioxidant and other functions of α-tocopherol can be summarized as follows (Fryer, 1992):

(i) Sacrificial scavenging of singlet oxygen (and superoxide/hydroxyl radicals). The oxidation products are tocopheryl quinones and quinone epoxides.
(ii) Physical quenching of singlet oxygen by resonance energy transfer. One molecule can deactivate up to 120 singlet oxygens before being oxidized.
(iii) Prevention of lipid peroxidation by its chain-breaking action. α-Tocopherol reduces lipid radicals and is oxidized to the α-chromanoxyl radical. α-Tocopherol is regenerated by reaction with ascorbate or GSH. Ascorbate may be the most important *in vivo* since these antioxidants together have a synergistic effect in preventing peroxidation (Figure 12.2; Kunert and Ederer, 1985).
(iv) Biophysical stabilization of membranes. α-Tocopherol decreases the permeability to ions of membranes with a high polyunsaturated fatty acid content and this may be particularly important in maintaining the trans-thylakoid pH gradient. Superoxide can de-esterify phospholipids (Senaratna and McKersie, 1986) and the free fatty acids destabilize membrane structure. α-Tocopherol binds free fatty acids and prevents membrane destabilization.

These properties make α-tocopherol an important antioxidant, particulary in photosynthetic systems, and a number of studies show that high levels of this molecule protect chloroplasts from photooxidative damage in concert with β-carotene (Fryer, 1992).

## 12.3.4 *Antioxidant enzymes*

Three enzymes in higher plants are important in reacting with and removing active oxygen species, these are SOD, catalase and ascorbate peroxidase (AP). Animals have a selenocysteine-containing glutathione peroxidase which catalyses oxidation of GSH by hydrogen peroxide and this has a major antioxidant function. GSH peroxidase activity is absent or very low in higher plants although it has been reported occasionally (Holland *et al.*, 1993). It has been detected in a moss and *Euglena* (Dhindsa, 1991; Overbauch and Fall, 1985). A cDNA clone, which is up-regulated by salinity, has been isolated from *Citrus* cell cultures and has homology to mammalian GSH peroxidase (Holland *et al.*, 1993). The importance of this enzyme in higher plants needs to be re-evaluated. The properties of SOD, catalase and AP will be described briefly below.

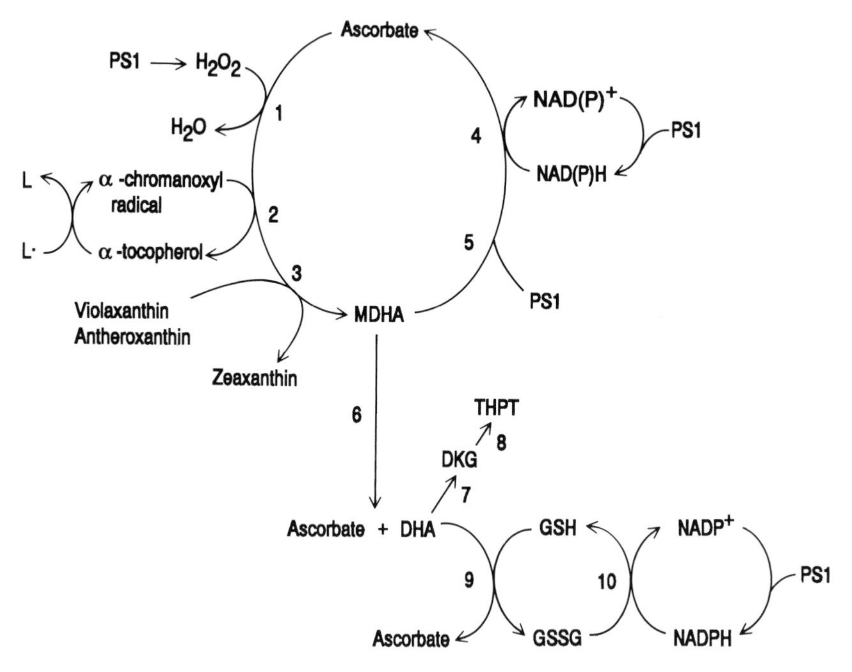


**Figure 12.2.** *The chloroplast ascorbate–glutathione cycle. (See Section 12.4 for further details). This scheme illustrates the central role of ascorbate as an antioxidant (removal of hydrogen peroxide and regeneration of α-tocopherol) and in photoprotection (formation of zeaxanthin). Hydrogen peroxide is formed from superoxide by dismutation which is catalysed by superoxide dismutase (SOD). Photoreduction of oxygen (Mehler reaction) by photosystem I (PSI) forms superoxide. The oxidation product of ascorbate, MDHA, is reduced back to ascorbate directly by an electron from PSI (reaction 5) or enzymatically (reaction 4). Any MDHA not reduced disproportionates (reaction 6) forming DHA. DHA is reduced back to ascorbate by GSH (reactions 9, 10 and 11). DHA is unstable and is irreversibly converted to other compounds (reactions 7 and 8) if it is formed more rapidly than the capacity of reactions 9 and 10. The enzymes involved are: 1, ascorbate peroxidase; 3, violaxanthin de-epoxidase; 4, monodehydroascorbate reductase (MDHAR); 9, dehydroascorbate reductase (DHAR); 10, glutathione reductase (GR). Abbreviations: L/L·, lipid/lipid radical; DKG, diketo-L-gulonic acid; THPT, 2(threo-1,2,3-trihydroxypropyl)tartronic acid.*

SOD catalyses dismutation of superoxide and maintains superoxide at lower levels than would occur if the reaction was not catalysed. The reaction products are ground state oxygen and hydrogen peroxide. Three types occur, differing in their metal co-factor and subcellular locations: Cu/ZnSOD (cytosol and stroma); MnSOD (mitochondria) and FeSOD (stroma in some species). The latter two are related proteins and the metal co-factor can be interchangeable. Cu/ZnSOD is unrelated in amino acid sequence. All the forms are nuclear encoded. The number and relative abundance of isoenzymes of each type varies between species and tissues (Bowler *et al.*, 1992; Scandalios, 1993). SOD has been one of the most widely studied enzymes in the antioxidant system and its importance is illustrated by the hypersensitivity of SOD-deficient *E. coli* and yeast mutants to oxygen (Bowler *et al.*, 1992). The genes encoding plant Cu/Zn and MnSOD have been isolated and sequenced so that their expression under different conditions can be investigated and the levels of the enzymes manipulated in transgenic plants. (Bowler *et al.*, 1992: Scandalios, 1993). These studies show that SOD expression is increased when environmental conditions cause increased active oxygen formation and are providing a powerful approach to analysing the expression and function of the antioxidant system (see Section 12.6).

Catalase breaks down hydrogen peroxide to oxygen and water. It is located in microbodies (peroxisomes, glyoxysomes and related organelles) where oxidase enzymes, such as glycolate oxidase, which produce hydrogen peroxide are located. It has a relatively low affinity for hydrogen peroxide and is often present at very high concentration in peroxisomes. Its most important function in leaves is to remove hydrogen peroxide produced during photorespiration, and catalase-deficient barley mutants show early ultrastructural damage to chloroplasts, presumably caused by lipid peroxidation, followed by death if allowed to photorespire (Smith *et al.*, 1984). Catalase activity is responsive to growth conditions. If plants are grown under low oxygen levels to eliminate photorespiration, catalase activity is decreased, returning to normal levels when plants are allowed to photorespire (Volk and Feierabend, 1989). Treatments that cause mild oxidative stress increase catalase activity (Hertwig *et al.*, 1992; Streb *et al.*, 1993). Catalase is subject to photoinactivation, partly because its haem co-factor can absorb blue light which then damages the protein, but there is also evidence that part of the inactivation is caused by light absorption in chloroplasts even though catalase is in a different organelle (Feierabend and Engel, 1986). Catalase undergoes continuous light-induced turnover involving inactivation and synthesis, Photoinactivation only becomes evident under severe photooxidative conditions (Section 12.4) or when protein synthesis inhibitors and environmental factors which inhibit protein synthesis, such as low temperature, prevent resynthesis (Feierabend and Engel, 1986; Feierabend *et al.*, 1992). This response is similar to that of the D1 protein of PSII which also turns over rapidly in the light due to photodamage. Photosynthesis suffers severe photoinhibition if its resynthesis is prevented by protein synthesis inhibitors or low temperature (Feierabend *et al.*, 1992).

Ascorbate peroxidase catalyses the oxidation of ascorbate to MDHA using hydrogen peroxide as oxidant. Two isoforms have been separated in tea and mustard leaves (Chen *et al.*, 1989; Thomsen *et al.*, 1992). The major isoform (APII) in tea leaves differs from the large number of peroxidases in plants, which can all use general substrates such as guaiacol, in its high specificity for ascorbate and its different amino acid sequence which is more similar to yeast cytochrome *c* peroxidase (Asada, 1992; Chen *et al.*, 1992). The other tea leaf isoform (API) is more similar to ascorbate peroxidase purified from soybean nodules, having high activity with guaiacol and other non-physiological substrates (Dalton *et al.*, 1986). In mustard coyledons one isoform is plasidic, the other extra-plasidic (Thomsen *et al.*, 1992). Activity of both isoforms is less in dark-grown mustard seedlings and is light-induced under phytochrome control. The enzyme could also be regulated by active oxygen because high oxygen concentration and inhibition of carotenoid synthesis (which increases singlet oxygen formation, see Section 12.4) increased the activity of both isoforms, while addition of α-tocopherol decreases activity (Thomsen *et al.*, 1992).

Three other enzymes have a role in maintaining ascorbate and glutathione in their reduced forms. These are monodehydroascorbate reductase (MDHAR), dehydroascorbate reductase (DHAR) and GR. The reactions they catalyse are shown in Figure 12.2. GR uses NADPH as reductant, MDHAR is active with NADH and NADPH, while DHAR uses GSH as its reductant. All three are localized in the chloroplasts and cytosol of leaf cells (Arrigoni *et al.*, 1981; Asada, 1992; Hossain *et al.*, 1984; Jablonski and Anderson, 1981). Different isoforms of GR are associated with different subcellular compartments (Drumm-Herrel *et al.*, 1989; Edwards *et al.*, 1990) and in pea leaves 77% of GR is chloroplastic, 3% mitochondrial and 20% cytosolic (Edwards *et al.*, 1990). All three enzymes have high affinity for their substrates and their role in maintaining ascorbate and GSH in the reduced forms is described in Section 12.4.2.

## 12.4 The ascorbate–GSH system in chloroplasts and its role in protection against excess excitation energy

### 12.4.1 *Active oxygen in chloroplasts*

Chloroplasts are particularly prone to damage from active oxygen because of elevated oxygen concentrations, presence of pigments which can act as photosensitizers and the high concentration of polyunsaturated fatty acids (linolenic and linoleic) in the thylakoid glycolipids which are targets for peroxidation. In chloroplasts, active oxygen can be formed in a number of ways (Asada and Takahashi, 1987). Superoxide and singlet oxygen formation are probably of the greatest physiological significance. Oxygen photoreduction by PSI (Mehler reaction) results in superoxide formation, which then dismutates to hydrogen peroxide, the reaction being greatly accelerated by SOD. Hydrogen peroxide can inactivate light-activated Calvin cycle enzymes (Charles and Halliwell, 1980;

Kaiser, 1979). Superoxide can be detected in isolated chloroplasts (Asada and Takahashi, 1987; Elstner, 1987; Miller and MacDowall, 1975; Robinson, 1988; Salin, 1987) and hydroxyl radicals are detected when superoxide formation is increased by methyl viologen (paraquat) treatment (Babbs *et al.*, 1989). Singlet oxygen forms by energy transfer from excited triplet state chlorophyll to oxygen (Knox and Dodge, 1985). Singlet oxygen causes peroxidation of thylakoid lipids and formation of reactive organic free radicals. α-Tocopherol (see Section 12.3.3) and β-carotene (Young and Britton, 1990), both located in thylakoids, prevent damage by singlet oxygen. β-Carotene acts as a quencher of triplet chlorophyll and singlet oxygen. Carotene-deficient mutants and plants treated with carotenoid synthesis inhibitors show rapid photooxidation of pigments and peroxidation of thylakoid lipids when exposed to light (Young and Britton, 1990). Photorespiration, originating from the oxygenase reaction of Rubisco, also leads to active oxygen formation, in this instance hydrogen peroxide formation in the peroxisomes.

Protective mechanisms which remove or prevent formation of active oxygen are required under normal circumstances since 5–10% of electrons probably flow to oxygen (~15 $\mu$mol mg chl$^{-1}$ h$^{-1}$) and result in superoxide and hydrogen peroxide formation (Hossain *et al.*, 1984). However, on exposure to excess excitation energy (i.e. more light than can be used for carbon dioxide fixation) or when carbon assimilation is inhibited, the possibility exists that more singlet oxygen (from triplet chlorophyll) and superoxide (from the Mehler reaction) will be formed. Light intensity high enough to cause photoinhibition can lead to peroxidation of thylakoid lipids and damage to thylakoid structure (Havaux *et al.*, 1991; Mishra and Singhal, 1992; Roberts *et al.*, 1991). Leaves contain photoprotective systems to minimize damage caused by excessive light. The xanthophyll cycle pigment zeaxanthin and β-carotene may aid in dissipating excess excitation energy and increase on exposure to high light (Demmig-Adams and Adams, 1992; Walker, 1992). Photorespiration and CAM could function in droughted plants to maintain high intracellular carbon dioxide by recycling (Griffiths, 1988; Chapter 8). Photorespiration may be quantitatively more important in this respect than the Mehler reaction (Wu *et al.*, 1991). There is evidence that conditions such as low temperature and drought in combination with high light increase singlet oxygen and superoxide formation in the chloroplast (Schöner and Krause, 1990; Smirnoff, 1993; Smirnoff and Colombé, 1988). In these situations, the antioxidant systems of the chloroplasts, which are described below, increase their capacity.

### 12.4.2 *The ascorbate–GSH cycle*

The high concentrations of ascorbate and GSH in chloroplasts suggest that they have an important role as antioxidant defences. A cycle involving ascorbate, GSH and the enzymes GR and DHAR, in which photosynthetically produced NADPH reduces hydrogen peroxide, was proposed by Foyer and Halliwell (1976). This was based on the occurrence of all the necessary components in chloroplasts and the rapid light-dependent oxidation of the ascorbate and GSH pools when isolated chloroplasts were exposed to hydrogen peroxide or methyl viologen (paraquat) (Law *et al.*, 1983). Earlier, Flohé and Menzel (1971) had also suggested that peroxides oxidize ascorbate in chloroplasts and the ascorbate is regenerated by reduction with GSH. The reactions of the ascorbate–GSH cycle are illustrated in Figure 12.2. Evidence for the cycle is provided by the localization of the enzymes in chloroplasts along with the light-dependent oxygen evolution which accompanies reduction of GSSG, DHA, MDHA and hydrogen peroxide in isolated and ruptured pea and spinach chloroplasts (Anderson *et al.*, 1983; Jablonski and Anderson, 1978, 1981, 1982; Nakano and Asada, 1980, 1981). The stoichiometry of the reactions and the dependence of oxygen evolution on the order of addition of the components indicates that electrons flow from NADPH to GSH, ascorbate and then to $H_2O_2$. Without this cycle all the chloroplastic ascorbate would be oxidized in 240 s of photosynthesis (Hossain *et al.*, 1984) and serious oxidative damage would ensue. Both ascorbate and GSH can also be oxidized directly and non-enzymatically by superoxide, but comparison of the concentrations of the antioxidants and rate constants of these reactions with SOD (Elstner, 1987) suggests that dismutation to hydrogen peroxide by SOD will be an order of magnitude faster in chloroplasts.

Further investigation has shown the involvement of MDHA and MDHAR in the system. The primary product of ascorbate oxidation by AP and ascorbate oxidase is MDHA rather than DHA (Hossain *et al.*, 1984). MDHAR occurs in the chloroplast and reduces MDHA to ascorbate using NAD(P)H as reductant (Hossain *et al.*, 1984). Any MDHA which is not reduced to ascorbate by MDHAR disproportionates, forming ascorbate and DHA. DHA is very unstable (particularly at the alkaline pH of the stroma, see Section 12.3.1) so that the ascorbate pool would be depleted rapidly. This may be an explanation of the use of both MDHAR and DHAR to regenerate ascorbate. The importance of maintaining ascorbate in the reduced form can be seen *in vivo* in barley leaves treated with the catalase and peroxidase inhibitor aminotriazole under photorespiratory conditions. Hydrogen peroxide accumulates in peroxisomes and then diffuses to the cytosol and chloroplasts where it oxidizes the ascorbate pool faster than MDHAR and DHAR can operate, so that after a few hours the total ascorbate pool (ascorbate + DHA) is almost completely depleted (Smirnoff, unpublished data). Loss of the ascorbate pool may therefore provide an indicator of the exposure of cells to extreme oxidative stress (Kenyon and Duke, 1985).

Some of the chloroplast AP activity is thylakoid bound and is particularly enriched in the stroma thylakoids where PSI is also concentrated (Miyake and Asada, 1992). This suggests the possibility for microcompartmentation of AP activity at the site of superoxide and hydrogen peroxide formation. Isolated thylakoids can photoreduce hydrogen peroxide in the presence of ascorbate (and methyl viologen to aid superoxide and hydrogen peroxide formation) as long as the bound AP has not been inactivated. Hydrogen peroxide and MDHA (generated by ascorbate oxidase) addition to the thylakoids induces transient photochemical quenching of fluorescence. MDHA also supports light-dependent oxygen evolution from thylakoids. These observations together show that ascorbate and bound AP reduce hydrogen peroxide near the site of its formation at PSI. The resulting MDHA radical is then reduced directly to ascorbate by the thylakoids (Miyake and Asada, 1992). Similar effects of hydrogen peroxide and MDHA on photochemical and non-photochemical fluorescence quenching have been described by Foyer and Lelandais (1993). Any MDHA which escapes this system is dealt with by the stromal ascorbate–GSH cycle (Figure 12.2).

Both the bundle sheath and mesophyll cells of $C_4$ leaves show substantial light-dependent $H_2O_2$ scavenging activity and AP activity, suggesting that hydrogen peroxide formation and detoxication by the ascorbate–GSH cycle occur in both cell types (Nakano and Edwards, 1987). Since ascorbate and GSH and the enzymes of the cycle also occur in the cytosol and mitochondria at lower levels, it seems likely that the cycle is also operative in these subcellular locations. In this case NADPH is probably derived from the oxidative pentose phosphate pathway. The lower level of activity found in non-green tissues might reflect the lower rate of hydrogen peroxide formation outside chloroplasts.

### 12.4.3 *The role of the Mehler–peroxidase reaction and xanthophyll cycle in photoprotection*

The ascorbate–GSH cycle as described above is an effective means of hydrogen peroxide photoreduction. Recent evidence from the use of isolated chloroplasts has suggested that it could also be used as a means of photoprotection when carbon dioxide supply is limited, or more generally under conditions of excess excitation energy input (Neubauer and Yamamoto, 1992, 1993; Walker, 1992). When carbon dioxide fixation is limited, some linear electron flow can be sustained by oxygen photoreduction by the Mehler reaction producing superoxide and then hydrogen peroxide. The electron flow can be enhanced further by photoreduction of the hydrogen peroxide as described above. The combined system has been termed the Mehler–peroxidase reaction. This may prevent over-reduction of the electron transport chain and allow build-up of the trans-thylakoid pH gradient. The pH gradient may be important for regulating photosynthetic electron transport in conditions of high excitation energy (Walker, 1992). It also favours zeaxanthin synthesis which is associated with increased non-photochemical fluorescence quenching (Demmig-Adams and Adams, 1992). Zeaxanthin is formed from the xanthophyll cycle carotenoids, violaxanthin and antheroxanthin, by ascorbate-dependent de-epoxidation. Excess excitation energy input leads to de-epoxidation of the xanthophyll cycle pigments and accumulation of zeaxanthin (Demmig-Adams and Adams, 1992; Walker, 1992).

Evidence for this proposal is seen in lettuce chloroplasts treated with iodoacetamide to inhibit carbon dioxide fixation. In these, illumination results in development of substantial non-photochemical quenching of chlorophyll fluorescence and zeaxanthin accumulation. This is inhibited by levels of cyanide which inhibit ascorbate peroxidase but not linear electron flow or the Mehler reaction. The Mehler reaction itself is not sufficient and is more effective when coupled to ascorbate regeneration (Neubauer and Yamamoto, 1992). The Mehler reaction on its own probably cannot generate a large pH gradient because protons may be lost from the acidic thylakoid lumen by protonation of the superoxide which then diffuses into the stroma (Walker, 1992). A high ascorbate concentration is needed for these reactions to work (Foyer and Lelandais, 1993; Neubauer and Yamamoto, 1993). The importance of these processes *in vivo* must be confirmed.

## 12.5 The effect of environmental factors on the antioxidant system

A large array of environmental factors has been suggested to affect plants, at least in part, by oxidative damage. These include, high light, low temperature, drought, desiccation, freezing, chilling in tropical species and pollutants such as ozone, nitrogen oxides and sulphur dioxide (Alscher, 1989; Doulis *et al.*, 1993; Karpinski *et al.*, 1993; Kendall and McKersie, 1989; Mehlhorn *et al.*, 1987, 1990; Smirnoff, 1993; Tanaka *et al.*, 1982; Wise and Naylor, 1987). The effects of drought and low temperature are particularly associated with high light in leaves since, as discussed above, they lead to input of excessive excitation energy (Section 12.4.1). There is much evidence to suggest that plants can acclimate to such photooxidative conditions by increasing the expression of the antioxidant system and that this often results from increased gene expression (Bowler *et al.*, 1992; Scandalios, 1993; Smirnoff, 1993). Some recent examples are described below.

Ascorbate levels are higher under conditions which favour photooxidative damage. Under high light there is turnover of ascorbate (Foyer *et al.*, 1989; Smirnoff, 1993), suggesting that high irradiance triggers ascorbate synthesis and breakdown. In longer term exposures to high irradiance, the ascorbate pool increases, particularly at low temperature, which increases the excess excitation energy (Mishra *et al.*, 1993; Smirnoff, 1993). The effect of irradiance is illustrated in Figure 12.1. Barley leaves acclimated to high light have greater concentrations of ascorbate and higher MDHAR activity than those grown at low irradiance. In this experiment there was also a small but consistent difference in the oxidation state of the pool with 90–100% as ascorbate in low light and 80% in low light, indicating greater oxidative pressure on the ascorbate pool at high irradiance. Acclimation to seasonal variation in light is also associated with seasonal fluctuations of ascorbate (Gillham and Dodge, 1987).

Cold acclimation results in higher activities of GR, AP, SOD, DHAR and MDHAR and higher GSH levels in a range of species (Doulis *et al.*, 1993; Esterbauer and Grill, 1978; Guy and Carter, 1984; Karpinski *et al.*, 1993; Mishra *et al.*, 1993; Schöner and Krause, 1990). Drought also leads to changes similar to those noted for cold acclimation (Smirnoff and Colombé, 1988) and the evidence is reviewed by Smirnoff (1993). Severe drought causes oxidation and consequent depletion of the ascorbate pool (Buckland *et al.*, 1991) and oxidation of the GSH pool (Dhindsa, 1987; Smirnoff, 1993). The effects of low temperature and drought are light dependent and appear to act by limiting the utilization of excitation energy by photosynthetic carbon dioxide fixation (Demmig-Adams and Adams, 1992; Smirnoff, 1993). Freezing and desiccation lead to oxidative and free radical damage by physical disruption of cell structure (McKersie *et al.*, 1993; Smirnoff, 1993).

The expression of a number of antioxidant enzymes, including SOD, GR and AP is increased by conditions which increase production of active oxygen. These include various pollutants, toxins and herbicides (Ansellem *et al.*, 1993; Bowler *et al.*, 1989, 1992; Mehlhorn *et al.*, 1987; Scandalios, 1993). This suggests that the photooxidative conditions described above may induce increased antioxidant defence capacity via the production of active oxygen. The increase in activity results both from increased levels of mRNA (Bowler *et al.*, 1989; Tsang *et al.*, 1991; Williamson and Scandalios, 1992) and from effects on translation (Pastori and Trippi, 1992) in different systems. Very little is known about the signals involved in the induction but, in bacteria, superoxide and hydrogen peroxide interact with transcription factors causing induction of antioxidant enzymes (Demple, 1991).

## 12.6 Manipulation of the antioxidant system in transgenic plants

The antioxidant system is a promising target for the production of transgenic plants since it is possible that an increase in one or a few enzymes could increase the tolerance of plants to conditions which cause oxidative damage. Alternatively, antisense constructs could be used to down-regulate activity and explore the function of individual components of the system. This has not yet been tried.

Various plant SOD genes have been cloned and used to produce transgenic plants. Tobacco and alfalfa expressing high levels of MnSOD targeted to chloroplasts or mitochondria are more tolerant to paraquat and freezing, respectively (Bowler *et al.*, 1991; McKersie *et al.*, 1993). Cu/ZnSOD over-expression in tobacco (Sen Gupta *et al.*, 1993) and potato (Perl *et al.*, 1993) increases tolerance to oxidative treatment. In the case of tobacco over-expressing Cu/ZnSOD in the cytoplasm and chloroplasts, the plants are markedly more resistant to loss of photosynthetic capacity (photoinhibition) when exposed to high light intensity at low temperature (Sen Gupta *et al.*, 1993). On the other hand, this approach does not always yield increases in resistance, as was found in two other cases of Cu/ZnSOD over-expression (Pitcher *et al.*, 1991; Tepperman and Dunsmuir, 1990). An *E. coli* GR gene has been expressed in the cytosol (Foyer *et al.*, 1991) and chloroplasts (Aono *et al.*, 1993) of tobacco, resulting in plants with higher GR activity. In the case of cytosolic over-expression, the plants were not more paraquat resistant although the ascorbate pool was less oxidized. Chloroplastic over-expression increased paraquat and sulphur dioxide resistance.

These experiments are promising in that they provide some direct evidence for the role of SOD and GR in protection against oxidative damage caused by paraquat, sulphur dioxide, high light/low temperature and freezing. The variation in effectiveness is interesting and might suggest that interactions with other components of the system are important. For example, Cu/ZnSOD over-expression in tobacco chloroplasts leads to increased AP activity in the transgenic plants (Sen Gupta *et al.*, 1993). Interactions between components of the antioxidant system are likely to be complex and it may be necessary to over-express a number of enzymes simultaneously to achieve increased resistance. Further discussion of the use of transgenic plants and the regulation of gene expression under oxidative conditions can be found in a recent review by Foyer *et al.* (1994).

## 12.7 Nitrogen-fixing root nodules in legumes

Nitrogen-fixing root nodules have a number of problems related to oxygen. Firstly, nitrogenase, which is located in the symbiotic bacteroids, is inactivated by oxygen. This might involve mediation of active oxygen (Becana and Rodriguez-Barrueco, 1989). Secondly, leghaemoglobin, which regulates the supply of oxygen to bacteroids, can form superoxide, hydrogen peroxide and hydroxyl radicals. Oxyleghaemoglobin can break down to form superoxide (Puppo *et al.*, 1981). This disproportionates (with the aid of SOD) to form hydrogen peroxide. Hydrogen peroxide then attacks the haem to release iron. This catalyses hydroxyl radical formation (Puppo and Halliwell, 1988) which is a potentially serious source of active oxygen since leghaemoglobin has a concentration of 3 mM in nodules (Becana and Rodriguez-Barrueco, 1989). Ferredoxin, which is an electron donor for nitrogenase, can reduce oxygen-forming superoxide. These problems are seen during senescence of nodules where Fe released from haem catalyses hydroxyl radical formation in nodule extracts (Becana and Klucas, 1992). Finally, in legumes which export ureides (allantoin and allantoic acid), the biosynthetic pathway involves oxidases which produce hydrogen peroxide in peroxisomes (Dalton *et al.*, 1986). Catalase activity in nodule peroxisomes is positively correlated with nitrogen-fixing activity and also occurs in the bacteroids (Becana and Rodriguez-Barrueco, 1989).

Nodule cytosol contains all the components of the ascorbate–GSH cycle and they are present in concentrations similar to the chloroplast (Dalton *et al.*, 1986). This suggests that the nodule cytosol has a high capacity to detoxify superoxide and hydrogen peroxide using NADPH as the reductant. The activities of the enzymes and levels of ascorbate and GSH are higher than in uninfected root tissue and increase in developing nodules in concert with nitrogen-fixing activity (Dalton *et al.*, 1986, 1991). In associations between legumes and *Bradyrhizobium* strains which form ineffective (non-fixing) nodules, expression of the system is very much lower (Dalton *et al.*, 1993). Exposure of nodulated soybean roots to high oxygen concentration causes increases in AP, GR and GSH. Much smaller effects were found in roots. Urea treatment, which inhibits nitrogen fixation, caused decreases in AP, DHAR and GR activity (Dalton *et al.*, 1991).

## 12.8 Post-anoxic injury

In waterlogged soils or in bulky tissues such as fruits, hypoxia and anoxia are potential problems since most plant tissues require oxygen for growth (see Chapter 7 for a discussion of anaerobic metabolism). Crawford and colleagues (Monk *et al.*, 1989) have suggested that part of the damaging effect of anoxia on sensitive tissues are various forms of post-anoxic injury which occur when tissues are re-exposed to air. This is analogous to the post-ischaemic injury which occurs in animals as a result of a restriction of blood flow during heart attacks or during surgical operations where tissue can suffer serious oxidative damage when oxygen supply is restored (Halliwell and Gutteridge, 1989). The mechanism for this damage is not fully characterized. One suggestion is that anoxia leads to increased levels of adenosine monophosphate. This is degraded to form hypoxanthine. The susceptible tissue may also contain xanthine dehydrogenase and this is attacked by a Ca-dependent protease (Ca homeostasis may be affected by anoxia) which converts it to a xanthine oxidase which uses oxygen instead of NADH as electron acceptor. Superoxide is a product of the reaction. The result is oxidation of hypoxanthine and consequent superoxide formation when oxygen is readmitted (Halliwell and Gutteridge, 1989). Alternatively, damage which occurs to mitochondria during anoxia may increase electron leakage to oxygen when air is readmitted. Plant organs which do not survive long periods of anoxia, for example *Iris germanica* rhizomes, show substantial lipid peroxidation when re-exposed to oxygen. In contrast, the wetland *Iris pseudacorus* survives anoxia and shows no lipid peroxidation when re-exposed to oxygen (Hunter *et al.*, 1983). The cause of this difference remains to be established and could be related either to minimal deleterious changes occurring during the anoxic period or to better antioxidant defence (Monk *et al.*, 1989). Feeding of ascorbate to chickpea seedlings improves their recovery from an anoxic treatment (Crawford and Wollenweber-Ratzner, 1992). A further possibility for oxidative damage to plants in waterlogged soil is the elevated level of soil Fe(II) formed because of low redox potential. This is more readily available for uptake than Fe(III) present in aerated soil. High iron levels in the cytosol or apoplast could result in increased hydroxyl radical formation using the oxygen diffusing from the atmosphere in aerenchyma as a source of superoxide and hydrogen peroxide, resulting in lipid peroxidation (Hendry and Brocklebank, 1985).

## 12.9 Conclusion

It is clear from the evidence reviewed in this chapter that the antioxidant system is vital for the survival of plants under normal environmental conditions. Its capacity can be increased in situations which increase the oxidative load. The exposure of leaves to conditions which limit carbon dioxide fixation or impose an excess of excitation energy, such as low temperature and drought, are particularly important, but UV-B and pollutants may be increasingly important. Good progress is being made with investigating the components of the antioxidant system and the construction of transgenic plants is beginning to contribute to a deeper understanding of its role. A number of areas require further investigation. The pathway and regulation of ascorbate synthesis must be established since it has a central role as a chloroplast antioxidant and also has functions in other aspects of cell physiology. The signals which trigger increased expression of antioxidant genes need to be understood. Finally, although it is clear that increased expression of SOD and GR can increase resistance to oxidative attack in the laboratory, it is worth considering the extent to which crops might be exposed to such conditions in the field and to what extent the antioxidant system is already optimized to deal with such situations. Perhaps, with increased pressure from pollutants and UV-B radiation, enhancement of antioxidant systems will have a beneficial effect on yield.

## References

Alscher, R.G. (1989) Biosynthesis and antioxidant function of glutathione in plants. *Physiol. Plant.* **77**, 457–464.

Anderson, J.W., Foyer, C.H. and Walker, D.A. (1983) Light-dependent reduction of dehydroascorbate and uptake of exogenous ascorbate by spinach chloroplasts. *Planta* **158**, 442–450.

Ansellem, Z., Jansen, M.A.K., Driesenaar, A.R.J. and Gressel, J. (1993) Developmental variability of photooxidative stress tolerance in paraquat-resistant *Conyza. Plant Physiol.* **103**, 1097–1106.

Aono, M., Kubo, A., Saji, H. and Kondo, N. (1993) Enhanced tolerance to photooxidative stress of transgenic *Nicotiana tabacum* with high chloroplastic glutathione reductase activity. *Plant Cell Physiol.* **34**, 129–135.

Arrigoni, O., Dipierro, S. and Borraccino, G. (1981) Ascorbate free radical reductase: a key enzyme of the ascorbic acid system. *FEBS Lett.* **125**, 242–244.

Asada, K. (1992) Ascorbate peroxidase – a hydrogen peroxide scavenging enzyme in plants. *Physiol. Plant.* **85**, 235–241.

Asada, K. and Takahashi, M. (1987) Production and scavenging of active oxygen in photosynthesis. In: *Photoinhibition* (eds D.J. Kyle, C.B. Osmond and C.J. Arntzen). Elsevier Science Publishers, Amsterdam, pp. 227–287.

Babbs, C.F., Pham, J.M. and Coolbaugh, R.C. (1989) Lethal hydroxyl radical production in paraquat-treated plants. *Plant Physiol.* **90**, 1267–1270.

Bartling, D., Radzio, R., Steiner, U. and Weiler, E.W. (1993) A glutathione S-transferase with glutathione peroxidase activity from *Arabidopsis thaliana.* Molecular cloning and functional characterization. *Eur. J. Biochem.* **216**, 579–586.

Becana, M. and Klucas, R.V. (1992) Transition metals in legume root nodules: iron-dependent free radical production increases during nodule senescence. *Proc. Natl Acad. Sci. USA* **89**, 8958–8962.

Becana, M. and Rodriguez-Barrueco, C. (1989) Protective mechanisms of nitrogenase against oxygen excess and partially-reduced oxygen intermediates. *Physiol. Plant.* **75**, 429–438.

Bowler, C., Alliote, T., De Loose, M., Van Montagu, M. and Inze, D. (1989) The induction of manganese superoxide dismutase in response to stress in *Nicotiana plumbaginifolia. EMBO J.* **8**, 31–38.

Bowler, C., Slooten, L., Vandenbranden, S., De Ryke, R., Botterman, J., Sybesma, C. and Van Montagu, M. (1991) Manganese superoxide dismutase can reduce cellular damage mediated by oxygen radicals in transgenic plants. *EMBO J.* **10**, 1723–1732.

Bowler, C., Van Montagu, M. and Inze, D. (1992) Superoxide dismutase and stress tolerance. *Annu. Rev. Plant Physiol. Mol. Biol.* **43**, 83–116.

Buckland, S.M., Price, A.H. and Hendry, G.A.F. (1991) The role of ascorbate in drought-treated *Cochlearia atlantica* Pobed. and *Armeria maritima* (Mill.) Willd. *New Phytol.* **119**, 155–160.

Charles, S.A. and Halliwell, B. (1980) Effect of hydrogen peroxide on spinach (*Spinacia oleracea*) chloroplast fructose bisphosphatase. *Biochem. J.* **189**, 373–376.

Chen, G.-X., Satoshi, S. and Asada, K. (1992) The amino acid sequence of ascorbate peroxidase from tea has a high degree of homology to that of cytochrome c peroxidase from yeast. *Plant Cell Physiol.* **33**, 109–116.

Crawford, R.M.M. and Wollenweber-Ratzer, B. (1992) Influence of ascorbate on post-anoxic growth and survival of chickpea seedlings (*Cicer arietinum* L.). *J. Exp. Bot.* **43**, 703–708.

Dalton, D.A., Russell, S.A., Hanus, F.J., Pascoe, G.A. and Evans, H.J. (1986) Enzymatic reactions of ascorbate and glutathione that prevent peroxide damage in soybean root nodules. *Proc. Natl Acad. Sci. USA* **83**, 3811–3815.

Dalton, D.A., Post, C.J. and Langeberg, L. (1991) Effects of ambient oxygen and of fixed nitrogen on concentration of glutathione, ascorbate and associated enzymes in soybean root nodules. *Plant Physiol.* **96**, 812–818.

Dalton, D.A., Langeberg, L. and Treneman, N.C. (1993) Correlations between the ascorbate-glutathione pathway and effectiveness in legume root nodules. *Physiol. Plant.* **87**, 365–370.

Degara, L., Tommasi, F., Liso, R. and Arrigoni, O. (1992) The biogenesis of galactone-γ-lactone oxidase in *Avena sativa* embryos. *Phytochemistry* **31**, 755–756.

Del Rio, L.A., Fernandez, V.M., Ruperez, F.L., Sandalio, L.M. and Palma, J.M. (1989) NADH induces the generation of superoxide radicals in leaf peroxisomes. *Plant Physiol.* **89**, 728–731.

Demmig-Adams, B. and Adams, W.W. (1992) Photoprotection and other responses to high light stress. *Annu. Rev. Plant Physiol. Mol. Biol.* **43**, 599–626.

Demple, B. (1991) Regulation of bacterial oxidative stress genes. *Annu. Rev. Genet.* **25**, 315–337.

De Voss, C.H.R., Vonk, M.J., Vooijs, R. and Schat, H. (1992) Glutathione depletion due to copper-induced phytochelatin synthesis causes oxidative stress in *Silene cucubalus*. *Plant Physiol.* **98**, 853–858.

Dhindsa, R.S. (1987) Glutathione status and protein synthesis during drought and subsequent rehydration of *Tortula ruralis*. *Plant Physiol.* **83**, 816–819.

Dhindsa, R.S. (1991) Drought stress, enzymes of glutathione metabolism, oxidation injury, and protein synthesis in *Tortula ruralis*. *Plant Physiol.* **95**, 648–651.

Doulis, A.G., Hausladen, A., Mondy, B., Alscher, R.G., Chevone, I., Hess, J.L. and Weiser, R.L. (1993) Antioxidant response and winter hardiness in red spruce (*Picea rubens* Sarg.). *New Phytol.* **123**, 365–374.

Downum, K.R. (1992) Light-activated plant defence. *New Phytol.* **122**, 401–420.

Drumm-Herrel, H., Gerhauber, U. and Mohr, H. (1989) Differential regulation by phytochrome of the appearance of plastidic and cytoplasmic isoforms of glutathione reductase in mustard (*Sinapis alba*) cotyledons. *Planta* **178**, 103–109.

Edwards, E.A., Rawsthorne, S. and Mullineaux, P.M. (1990) Subcellular distribution of multiple forms of glutathione reductase in leaves of pea (*Pisum sativum* L.). *Planta* **180**, 278–284.

Elstner, E.F. (1987) Metabolism of activated oxygen species. In: *The Biochemistry of Plants*, Volume 11 (ed. D.D. Davies). Academic Press, London, pp. 253–315.

Esterbauer, H. and Grill, D. (1978) Seasonal variation of glutathione and glutathione reductase in needles of *Picea abies*. *Plant Physiol.* **61**, 119–121.

Feierabend, J. and Engel, S. (1986) Photoinactivation of catalase *in vivo* in leaves. *Arch. Biochem. Biophys.* **251**, 567–576.

Feierabend, J., Schaan, C. and Hertwig, B. (1992) Photoinactivation of catalase occurs under both high- and low-temperature stress conditions and accompanies photoinhibition of photosystem II. *Plant Physiol.* **100**, 1554–1561.

Flohé, L. and Menzel, H. (1971) The influence of glutathione upon light-induced high-amplitude swelling and lipid peroxide formation of spinach chloroplasts. *Plant Cell Physiol.* **12**, 325–333.

Foyer, C.H. and Halliwell, B. (1976) Presence of glutathione and glutathione reductase in chloroplasts: a proposed role in ascorbic acid metabolism. *Planta* **133**, 21–25.

Foyer, C.H. and Lelandais, M. (1993) The roles of ascorbate in the regulation of photosynthesis. In: *Photosynthetic Responses to the Environment* (eds H.Y. Yamamoto and C.M. Smith). American Society of Plant Physiologists, Rockville, MD, pp. 88–101.

Foyer, C.H., Rowell, J. and Walker, D.A. (1983) Measurements of the ascorbate content of spinach leaf protoplasts and chloroplasts during illumination. *Planta* **157**, 239–244.

Foyer, C.H., Dujardyn, M. and Lemoine, Y. (1989) Responses of photosynthesis and the xanthophyll cycle and ascorbate–glutathione cycles to changes in irradiance, photoinhibition and recovery. *Plant Physiol. Biochem.* **27**, 751–760.

Foyer, C.H., Lelandais, M., Galep, C. and Kunert, K.J. (1991) Effects of elevated cytosolic glutathione reductase activity on the cellular glutathione pool and photosynthesis in leaves under normal and stress conditions. *Plant Physiol.* **97**, 863–872.

Foyer, C.H., Descourvières, P. and Kunert, K.J. (1994) Protection against oxygen radicals: an important defence mechanism studied in transgenic plants. *Plant, Cell Environ.* **17**, 507–523.

Fryer, M.J. (1992) The antioxidant effects of thylakoid vitamin E (α-tocopherol). *Plant, Cell Environ.* **15**, 381–392.

Gander, J.E. (1982) Polyhydroxy acids: Relation to hexose phosphate metabolism. In: *Encyclopedia of Plant Physiology*, Volume 13A (eds F.A. Loewus and W. Tanner). Springer, Berlin, pp. 77–102.

Gillham, D.J. and Dodge, A.D. (1987) Chloroplast superoxide and hydrogen peroxide scavenging systems from pea leaves: seasonal variations. *Plant Sci.* **50**, 105–109.

Griffiths, H. (1988) Crassulacean acid metabolism: a reappraisal of physiological plasticity in form and function. *Adv. Bot. Res.* **15**, 43–92.

Guy, C.L. and Carter, J.V. (1984) Characterization of partially purified glutathione reductase from cold-hardened and non-hardened spinach leaf tissue. *Cryobiology* **21**, 454–464.

Halliwell, B. and Gutteridge, J.M.C. (1989) *Free Radicals in Biology and Medicine*. Clarendon Press, Oxford.

Hausladen, A. and Kunert, K.J. (1990) Effects of artificially enhanced levels of ascorbate and glutathione on the enzymes monodehydroascorbate reductase, dehydroascorbate reductase and glutathione reductase in spinach *Spinacia oleracea*. *Physiol. Plant.* **79**, 384–388.

Havaux, M., Gruszecki, W.I., Dupont, I. and Leblanc, R.M. (1991) Increased heat emission and its relationship to the xanthophyll cycle in pea leaves exposed to strong light. *J. Photochem. Photobiol. B: Biol.* **8**, 361–370.

Hendry, G.A.F. and Brocklebank, K.J. (1985) Iron-induced oxygen radical metabolism in waterlogged plants. *New Phytol.* **101**, 199–206.
Hertwig, B., Streb, P. and Feierabend, J. (1992) Light dependence of catalase synthesis and degradation in leaves and the influence of interfering stress conditions. *Plant Physiol.* **100**, 1547–1553.
Holland, D., Ben-Hayyim, G., Faltin, Z., Camoin, L., Strosberg, A.D. and Eshdat, U. (1993) Molecular characterization of salt-stress associated protein in citrus: protein and cDNA homology to mammalian glutathione peroxidases. *Plant Mol. Biol.* **21**, 923–927.
Horemans, N., Asard, H. and Caubergs, R.J. (1994) The role of ascorbate free radical as an electron acceptor to cytochrome b-mediated trans-plasma membrane electron transport in higher plants. *Plant Physiol.* **104**, 1455–1458.
Hossain, M.A., Nakano, Y. and Asada, K. (1984) Monodehydroascorbate reductase in spinach chloroplasts and its participation in regeneration of ascorbate for scavenging hydrogen peroxide. *Plant Cell Physiol.* **25**, 385–395.
Hunter, M.I.S., Hetherington, A.M. and Crawford, R.M.M. (1983) Lipid peroxidation – a factor in anoxia intolerance in *Iris* species? *Phytochemistry* **22**, 1145–1147.
Innocenti, A.M., Bitonti, M.B., Arrigoni, O. and Liso, R. (1990) The size of the quiescent center in roots of *Allium cepa* L. grown with ascorbic acid. *New Phytol.* **114**, 507–509.
Jablonski, P.P. and Anderson, J.W. (1978) Light-dependent reduction of oxidised glutathione by ruptured chloroplasts. *Plant Physiol.* **61**, 221–225.
Jablonski, P.P. and Anderson, J.W. (1981) Light-dependent reduction of dehydroascorbate by ruptured pea chloroplasts. *Plant Physiol.* **67**, 1239–1244.
Jablonski, P.P. and Anderson, J.W. (1982) Light-dependent reduction of hydrogen peroxide by ruptured pea chloroplasts. *Plant Physiol.* **69**, 1407–1413.
Kaiser, W.M. (1979) Reversible inhibition of the Calvin cycle and activation of the oxidative pentose phosphate cycle in isolated intact chloroplasts by hydrogen peroxide. *Planta* **145**, 377–382.
Karpinski, S., Wingsle, G., Karpinska, B. and Hallgren, J.-E. (1993) Molecular responses to photooxidative stress in *Pinus sylvestris* (L.) II. Differential expression of CuZn-superoxide dismutases and glutathione reductase. *Plant Physiol.* **103**, 1385–1391.
Kendall, E.J. and McKersie, B.D. (1989) Free radical and freezing injury to cell membranes of winter wheat. *Physiol. Plant.* **76**, 86–94.
Kenyon, W.H. and Duke, S.O. (1985) Effect of acifluorfen on endogenous antioxidants and protective enzymes in cucumber (*Cucumis sativa* L.) cotyledons. *Plant Physiol.* **79**, 862–866.
Kioke, S. and Patterson, B.D. (1988) Diurnal variation of glutathione levels in tomato seedlings. *HortScience* **23**, 713–714.
Klapheck, S., Latus, C. and Bergmann, L. (1987) Localization of glutathione synthetase and distribution of glutathione in leaf cells of *Pisum sativum* L. *Plant Physiol.* **131**, 123–131.
Klapheck, S., Chrost, B., Starke, J. and Zimmerman, H. (1992) γ-glutamylcysteinylserine – a new homologue of glutathione in plants of the family Poaceae. *Bot. Acta* **105**, 174–179.
Knox, J.P. and Dodge, A.D. (1985) Singlet oxygen and plants. *Phytochemistry* **24**, 889–896.
Kranmer, I. and Grill, D. (1993) Content of low-molecular-weight thiols during the imbibition of pea seeds. *Physiol. Plant.* **88**, 557–562.
Kunert, K.J. and Ederer, M. (1985) Leaf aging and lipid peroxidation: The role of the antioxidants vitamin C and E. *Physiol. Plant.* **65**, 85–88.
Law, M.Y., Charles, S.A. and Halliwell, B. (1983) Glutathione and ascorbic acid in spinach

(*Spinacia oleracea*) chloroplasts. The effect of hydrogen peroxide and paraquat. *Biochem. J.* **210**, 899–903.

Leprince, O., Atherton, N.M., Deltour, R. and Hendry, G.A.F. (1994) The involvement of respiration in free radical processes during loss of desiccation tolerance in germinating *Zea mays* L. An electron paramagnetic study. *Plant Physiol.* **104**, 1333–1339.

Loewus, F.A. (1980) L-Ascorbic acid: metabolism, biosynthesis, function. In: *Biochemistry of Plants*, Volume 3 (ed. J. Preiss). Academic Press, New York, pp. 77–99.

Loewus, M.W., Bedgar, D.L., Saito, K. and Loewus, F.A. (1990) Conversion of L-sorbosone to L-ascorbate by a NADP-dependent dehydrogenase in bean and spinach leaf. *Plant Physiol.* **94**, 1492–1495.

Luwe, M.W.F., Takahama, U. and Heber, U. (1993) Role of ascorbate in detoxifying ozone in the apoplast of spinach (*Spinacia oleracea*) leaves. *Plant Physiol.* **101**, 969.

Martinoia, E., Grill, E., Tommasini, R., Kreuz, K. and Amrhein, N. (1993) ATP-dependent glutathione S-conjugate export pump in the vacuolar membrane of plants. *Nature* **364**, 247–249.

May, M.J. and Leaver, C.J. (1993) Oxidative stimulation of glutathione synthesis in *Arabidopsis thaliana* suspension cultures. *Plant Physiol.* **103**, 621–627.

McKersie, B.D., Chen, Y., de Beus, M., Bowley, S.R., Bowler, C., Inze, D., D'Halluin, K. and Botterman, J. (1993) Superoxide dismutase enhances tolerance of freezing stress in transgenic alfalfa (*Medicago sativa* L.). *Plant Physiol.* **103**, 1155–1163.

Mehdy, M.C. (1994) Active oxygen species in plant defence against pathogens. *Plant Physiol.* **105**, 467–472.

Mehlhorn, H., Cottam, D.A., Lucas, P.W. and Wellburn, A.R. (1987) Induction of ascorbate peroxidase and glutathione reductase activities by interactions of mixtures of air pollutants. *Free Rad. Res. Commun.* **3**, 193–197.

Mehlhorn, H., Tabner, B.J. and Wellburn, A.R. (1990) Electron spin resonance evidence for the formation of free radicals in plants exposed to ozone. *Physiol. Plant.* **79**, 377–383.

Meuwly, P., Thibault, P. and Rauser, W.E. (1993) γ-glutamylcysteinylglutamic acid – a new homologue of glutathione in maize. *FEBS Lett.* **336**, 472–476.

Miller, R.W. and MacDowall, F.D.H. (1975) The Tiron free radical as a sensitive indicator of chloroplastic photoautoxidation. *Biochim. Biophys. Acta* **387**, 176–187.

Mishra, R.K. and Singhal, G.S. (1992) Function of photosynthetic apparatus of intact wheat leaves under high light and heat stress and its relationship with peroxidation of thylakoid lipids. *Plant Physiol.* **98**, 1–6.

Mishra, N.P., Mishra, R.K. and Singhal, G.S. (1993) Changes in activities of anti-oxidant enzymes during exposure of intact wheat leaves to strong visible light at different temperatures in the presence of protein synthesis inhibitors. *Plant Physiol.* **102**, 903.

Miyake, C., and Asada, K. (1992) Thylakoid bound ascorbate peroxidase in spinach chloroplasts and photoreduction of its primary oxidation product, monodehydroascorbate radicals in the thylakoids. *Plant Cell Physiol.* **33**, 541–553.

Monk, L.S., Fagerstedt, K.V. and Crawford, R.M.M. (1989) Oxygen toxicity and superoxide dismutase as an antioxidant in physiological stress. *Physiol. Plant.* **76**, 738–745.

Nakano, Y. and Asada, K. (1980) Spinach chloroplasts scavenge hydrogen peroxide on illumination. *Plant Cell Physiol.* **21**, 1295–1307.

Nakano, Y. and Asada, K. (1981) Hydrogen peroxide is scavenged by ascorbate specific peroxidase in spinach chloroplasts. *Plant Cell Physiol.* **22**, 867–880.

Nakano, Y. and Edwards, G.E. (1987) Hill reaction, hydrogen peroxide scavenging, and ascorbate peroxidase activity of mesophyll and bundle sheath chloroplasts of NADP-malic enzyme type C4 species. *Plant Physiol.* **85**, 294–298.

Neubauer, C. and Yamamoto, Y.H. (1992) Mehler–peroxidase reaction mediates zeaxanthin formation and zeaxanthin-related fluorescence quenching in intact chloroplasts. *Plant Physiol.* **99**, 1354–1361.

Neubauer, C. and Yamamoto, H.Y. (1993) The role of ascorbate in the related ascorbate peroxidase, violaxanthin de-epoxidase and non-photochemical fluorescence-quenching activities. In: *Photosynthetic Responses to the Environment* (eds H.Y. Yamamoto and C.M. Smith). American Society of Plant Physiologists, Rockville, MD, pp. 166–171.

Olson, P.D. and Varner, J.E. (1993) Hydrogen peroxide and lignification. *Plant J.* **4**, 887–892.

Overbauch, J.M. and Fall, R. (1985) Characterization of a selenium-independent glutathione peroxidase from *Euglena gracilis*. *Plant Physiol.* **77**, 437–442.

Pastori, G.M. and Trippi, V.S. (1992) Oxidative stress induces a high rate of glutathione reductase synthesis in a drought-resistant maize strain. *Plant Cell Physiol.* **33**, 957–961.

Peñuelas, J. (1987) High oxygen tension inhibits vascular aquatic plant growth in deep waters. *Photosynthetica* **21**, 494–502.

Perl, A., Perl-Treves, R., Galili, G., Aviv, D., Shalgi, E., Malkin, S. and Galun, E. (1993) Enhanced oxidative stress defence in transgenic tobacco expressing tomato Cu, Zn superoxide dismutases. *Theor. Appl. Genet.* **85**, 568–576.

Pitcher, L.H., Brennan, E., Hurley, A., Dunsmuir, P., Tepperman, J.M. and Zilinskas, B.A. (1991) Overproduction of petunia chloroplastic copper zinc superoxide dismutase does not confer ozone tolerance in transgenic tobacco. *Plant Physiol.* **97**, 452–455.

Puppo, A. and Halliwell, B. (1988) Generation of hydroxyl radicals by soybean nodule leghaemoglobin. *Planta* **173**, 405–410.

Puppo, A., Rigaud, J. and Job, D. (1981) Role of superoxide anion in leghaemoglobin autoxidation. *Plant Sci. Lett.* **22**, 353–360.

Rennenberg, H. (1982) Glutathione metabolism and possible biological roles in higher plants. *Phytochemistry* **21**, 2771–2781.

Rennenberg, H. (1984) The fate of excess sulfur in higher plants. *Annu. Rev. Plant Physiol.* **35**, 121–153.

Roberts, D.R., Kristie, D.N., Thompson, J.E., Dumbroff, E.B. and Gepstein, S. (1991) *In vitro* evidence for the involvement of activated oxygen in light-induced aggregation of thylakoid proteins. *Physiol. Plant.* **82**, 389–396.

Robinson, J.M. (1988) Does oxygen photoreduction occur *in vivo*? *Physiol. Plant.* **72**, 666–680.

Saito, K., Nick, J.A. and Loewus, F.A. (1990) D-glucosone and L-sorbosone, putative intermediates of L-ascorbic acid biosynthesis in detached bean and spinach leaves. *Plant Physiol.* **94**, 1496–1500.

Salin, M.L. (1987) Toxic oxygen species and protective systems of the chloroplast. *Physiol. Plant.* **72**, 681–689.

Sandalio, L.M., Fernandez, V.M., Ruperez, F.L. and Del Rio, L.A. (1988) Superoxide radicals are produced in glyoxysomes. *Plant Physiol.* **87**, 1–4.

Scandalios, J.G. (1993) Oxygen stress and superoxide dismutases. *Plant Physiol.* **101**, 7–12.

Schöner, S. and Krause, G.H. (1990) Protective systems against active oxygen species in spinach: response to cold acclimation in excess light. *Planta* **180**, 383–389.

Schultz, G. (1990) Biosynthesis of α-tocopherol in chloroplasts of higher plants. *Fat Sci. Technol.* **92**, 86–91.

Schupp, R. and Rennenberg, H. (1988) Diurnal changes in the glutathione content of spruce needles (*Picea abies* L.). *Plant Sci.* **57**, 113–117.

Senaratna, T. and McKersie, B.D. (1986) Loss of desiccation tolerance during germination:

A free radical mechanism of injury. In: *Membranes, Metabolism and Dry Organisms* (ed. A.C. Leopold). Cornell University Press, Ithaca, pp. 85–101.

Sen Gupta, A., Webb, R.P., Holaday, A.S. and Allen, R.D. (1993) Overexpression of superoxide dismutase protects plants from oxidative stress. Induction of ascorbate peroxidase in superoxide-overproducing plants. *Plant Physiol.* **103**, 1067–1073.

Smirnoff, N. (1993) The role of active oxygen in the response of plants to water deficit and desiccation. *New Phytol.* **125**, 27–58.

Smirnoff, N. and Colombé, S.V. (1988) Drought influences the activity of the chloroplast hydrogen peroxide scavenging system. *J. Exp. Bot.* **39**, 1097–1108.

Smith, I.K. (1985) Stimulation of glutathione synthesis in photorespiring plants by catalase inhibitors. *Plant Physiol.* **79**, 1044–1047.

Smith, I.K., Kendall, A.C., Keys, A.J., Turner, J.C. and Lea, P.J. (1984) Increased levels of glutathione in a catalase-deficient mutant of barley (*Hordeum vulgare* L.). *Plant Sci. Lett.* **37**, 29–33.

Smith, I.K., Kendall, A.C., Keys, A.J., Turner, J.C. and Lea, P.J. (1985) The regulation of the biosynthesis of glutathione in leaves of barley (*Hordeum vulgare* L.). *Plant Sci.* **41**, 11–17.

Smith, I.K., Polle, A. and Rennenberg, H. (1990) Glutathione. In: *Stress Responses in Plants: Adaptation and Acclimation Mechanisms* (eds R.G. Alscher and J.R. Cumming). Wiley-Liss, New York, pp. 201–215.

Steffens, J.C. (1990) Heavy metal stress and the phytochelatin response. In: *Stress Responses in Plants: Adaptation and Acclimation Mechanisms* (eds R.G. Alscher and J.R. Cumming). Wiley-Liss, New York, pp. 377–394.

Streb, P., Michael-Knauf, A. and Feierabend, J. (1993) Preferential photoinactivation of catalase and photoinhibition of photosystem II are common early symptoms under various osmotic and chemical stress conditions. *Physiol. Plant.* **88**, 590–598.

Tanaka, K., Otsubo, T. and Kondo, N. (1982) Participation of hydrogen peroxide in the inactivation of Calvin cycle SH enzymes in $SO_2$-fumigated spinach leaves. *Plant Cell Physiol.* **23**, 1009–1018.

Tepperman, J.M. and Dunsmuir, P. (1990) Transformed plants with elevated levels of chloroplastic SOD are not more resistant to superoxide toxicity. *Plant Mol. Biol.* **14**, 501–511.

Thomsen, B., Drumm-Herrel, H. and Mohr, H. (1992) Control of the appearance of ascorbate peroxidase (EC 1.11.1.11) in mustard seedling cotyledons by phytochrome and photooxidative treatments. *Planta* **186**, 600–608.

Tsang, E.W.T., Bowler, C., Herouart, D., VanCamp, W., Villrroel, R., Genetello, C., Van Montagu, M. and Inze, D. (1991) Differential regulation of superoxide dismutases in plants exposed to environmental stress. *Plant Cell* **3**, 783–792.

Volk, S. and Feierabend, J. (1989) Photoinactivation of catalase at low temperature and its relevance to photosynthetic and peroxide metabolism in leaves. *Plant, Cell Environ.* **12**, 701–712.

Walker, D.A. (1992) Excited leaves. *New Phytol.* **121**, 325–345.

Williamson, J.D. and Scandalios, J.G. (1992) Differential response of maize catalases and superoxide dismutases to the photoactivated fungal toxin cercosporin. *Plant J.* **2**, 351–358.

Wingate, V.P.M., Lawton, M.A. and Lamb, C.J. (1988) Glutathione causes a massive and selective induction of plant defense genes. *Plant Physiol.* **87**, 206–210.

Winston, G.W. (1990) Physicochemical basis for free radical production in cells: production and defenses. In: *Stress Responses in Plants: Adaptation and Acclimation Mechanisms* (eds

R.G. Alscher and J.R. Cumming). Wiley-Liss, New York, pp. 57–86.
Wise, R.R. and Naylor, A.W. (1987) Chilling-enhanced photooxidation. Evidence for the role of singlet oxygen and superoxide in the breakdown of pigments and endogenous antioxidants. *Plant Physiol.* **83**, 278–282.
Wolff, S.P., Garner, A. and Dean, R.T. (1986) Free radicals, lipids and protein degradation. *Trends Biochem. Sci.* **11**, 27–31.
Wu, J., Neimanis, S. and Heber, U. (1991) Photorespiration is more effective than the Mehler reaction in protecting the photosynthetic apparatus against photoinhibition. *Bot. Acta* **104**, 283–291.
Young, A. and Britton, G. (1990) Carotenoids and stress. In: *Stress Responses in Plants: Adaptation and Acclimation Mechanisms* (eds R.G. Alscher and J.R. Cumming). Wiley-Liss, New York, pp. 87–112.
Zopes, H., Klapheck, S. and Bergmann, L. (1993) The function of homoglutathione and hydroxymethylglutathione for the scavenging of hydrogen-peroxide. *Plant Cell Physiol.* **34**, 515–521.

13

# Response of plants to UV-B radiation: some biochemical and physiological effects

J.F. Bornman and C. Sundby-Emanuelsson

## 13.1 Introduction

Reports of decreasing trends in stratospheric ozone concentrations (Frederick, 1993; Kerr and McElroy, 1993) during the past decade have stimulated research into the effects of different levels of elevated ultraviolet (UV) radiation on both animal and plant systems. Normally ozone absorbs a large proportion of the UV radiation before it reaches the earth's surface. Of primary concern is the increase in short wavelength UV radiation (UV-B, 290–320 nm), which is biologically effective. Although steps have been taken at various political levels to phase out most of the ozone-depleting chemicals in the very near future, this will necessarily occur at different rates in different countries across the world. Those chemicals that have been judged as critical in decreasing the amount of ozone include the chlorofluorocarbons with lifetimes of many decades. Nitrogen oxides ($NO_x$) also contribute to ozone breakdown in the stratosphere (Prinn, 1994). However, the disturbance of the ozone layer is not dependent solely on these chemicals, and therefore it is unlikely that the problem will disappear. It is almost certainly impossible to predict future scenarios accurately, given the interactive effects of other global change factors with ozone chemistry, the mechanisms of which are far from well understood.

Photosynthesis is one of the most important processes influencing plant productivity, and has therefore been the subject of quite extensive research with regard to the potential effects of an elevated UV-B radiation climate (for reviews, see Bornman, 1989; Bornman and Teramura, 1993; Jordan, 1993). This chapter will concentrate on some of the ways in which UV radiation influences plants at a physiological and biochemical level, with particular emphasis on photosynthesis.

## 13.2 Effect of UV-B radiation on plants

### 13.2.1 *Attenuation of UV radiation within leaves*

UV radiation has been shown to cause alterations in plant morphology and ultrastructure as well as in physiological and biochemical processes. Plants exposed to UV radiation commonly respond by reducing the penetration of the radiation. This screening ability is probably primarily afforded by certain phenolic compounds, in particular, flavonoids, which are water-soluble pigments absorbing in the UV region of the spectrum. Although UV-B radiation induces flavonoid production, UV-A (320–400 nm) and visible radiation also contribute. At the transcription level, UV-B radiation causes a definite increase in certain key enzymes of the flavonoid pathway (Schulze-Lefert *et al.*, 1989), for example, chalcone synthase, which must be induced for synthesis of flavonoids (Hahlbrock and Grisebach, 1979; Hahlbrock and Scheel, 1989). In the search for a UV photoreceptor with regard to increased levels of chalcone synthase and flavonoids, Ensminger and Schäfer (1992) found evidence for a flavin, although other compounds may be involved.

The attenuation of UV radiation with depth in a leaf has been demonstrated using quartz fibre optic microprobes to measure directly the gradients of radiation within leaves (Bornman and Vogelmann, 1988; Cen and Bornman, 1993; Day *et al.*, 1992; DeLucia *et al.*, 1992). Light readings are taken while a fibre optic microprobe is driven through a leaf in millimetre steps from one side towards a light source, while the opposite end of the fibre terminates in a spectroradiometer with a light detector. Using paradermal sectioning of a leaf and spectrophotometric analysis of each section to determine the gradient of UV-absorbing compounds with depth, a correlation is seen between these compounds and the attenuation of UV-B radiation through the leaf (Cen and Bornman, 1993).

### 13.2.2 *Some general effects of UV-B radiation*

Both primary and other effects of UV-B radiation induce modifications in the response of a plant, altering regulatory processes and increasing adaptive mechanisms. This may lead to changes in morphology and shifts in competitive balance between different species, ultimately affecting the yield of plants. Depending on a number of factors, such as species, cultivar and growth conditions, plant growth may be inhibited or stimulated by different levels of UV radiation (Bornman and Teramura 1993; Staxén and Bornman, 1994; Tezuka *et al.*, 1993). The seemingly different mechanisms operative in inhibition and stimulation are not well understood.

At the molecular level, some of these changes may be due to alterations in plant hormones or nucleic acids, which have been reported to be affected directly by UV-B radiation (Quaite *et al.*, 1992; Ros, 1990). Delays in the progression of cells through the cell cycle as a consequence of UV exposure have been noted especially for animal systems, although to date one report exists for *Petunia hybrida* (Staxén *et al.*, 1993).

Some of the effects of UV-B radiation on various aspects of photosynthesis have been traced back to reductions in transcript levels. For example, the reduction by UV-B radiation of the activity of ribulose-1,5-bisphosphate carboxylase, the main $CO_2$-fixing enzyme in $C_3$ plants, was found to be due to reduced RNA transcript levels for the subunits of this enzyme (Jordan *et al.*, 1992), the smaller of which is nuclear encoded, in contrast to the chloroplast-encoded larger subunit. Reduced mRNA transcript levels for the nuclear-encoded chlorophyll *a*/*b*-binding protein have also been reported, as well as for the chloroplast-encoded D1 protein (Jordan *et al.*, 1991).

### 13.2.3 *Photoinhibition by visible and UV radiation*

Under natural conditions, plants are subjected to the full solar spectrum which includes UV radiation. Apart from the more direct effect of UV radiation on specific target molecules as well as alterations to anatomical features, other environmental factors may interact to modify the response of a plant. One such example is high visible radiation which causes photoinhibition when plants absorb more light energy than they can utilize (Powles, 1984). This may lead to the situation where the rate of actual damage to photosystem II (PSII), the main target for high radiation stress, exceeds repair processes. Damage from excess absorbed energy may take the form of impairment to the electron transport chain, and in particular to the D1 polypeptide of the reaction centre. A number of adaptive and repair mechanisms are involved in order to reduce photoinhibition. Probably the most important one is the degradation and resynthesis of damaged D1 protein (Greer *et al.*, 1986). Others include quenching of excitation energy, probably due to the zeaxanthin cycle (Demmig-Adams and Adams, 1992; Chapter 12), and thermal dissipation in the antenna bed or reaction centre (Weis and Berry, 1987).

It is well known that high visible radiation can be photoinhibitory, and therefore it is of interest to investigate the response of the plant under both high visible radiation conditions and an enhanced level of UV-B radiation. Photoinhibition is usually reversible, and has been investigated largely without the simultaneous addition of UV-B radiation. The D1 polypeptide of PSII is involved in protecting the photosystem, such that it responds by rapid turnover after photoinhibition with high levels of photosynthetically active radiation (PAR). It seems that enhanced levels of UV-B radiation may compound the photoinhibitory effect, since the turnover of the D1 polypeptide is lower with visible radiation alone (in the absence of UV) than with the natural component of UV-B included (Greenberg *et al.*, 1989a). Greenberg *et al.* (1989b) postulated that the UV photoreceptor for the D1 degradation may be different from that in the visible and far-red regions. In addition, recent data show, for example, that UV-B irradiation *in vitro* causes cleavage of the D1 protein at a different site compared to that induced by visible radiation (Friso *et al.*, 1994).

Evidence for the protective role of UV-screening pigments for D1 degradation has been reported by Wilson and Greenberg (1993), who showed that the rate of D1 degradation, under short-term exposure to UV-B radiation of leaf discs from plants previously grown under UV-B radiation, was about 70% slower than in the control plants (no UV-B during growth) subsequently exposed to the same short-term UV-B radiation. Pfündel *et al.* (1992) reported that certain biochemical processes involved in protecting the plant from photoinhibition might themselves be targets of UV-B radiation. They found that the xanthophyll cycle, and more specifically the de-epoxidation of violaxanthin to zeaxanthin, was inhibited in irradiated chloroplasts and intact leaves. This cycle has been strongly implicated in the mechanisms by which excess absorbed light may be non-photochemically dissipated, since zeaxanthin appears to act as a quencher of excitation energy (Demmig-Adams and Adams, 1992; Chapter 12). Thus, the simultaneous exposure to UV radiation and high visible light may result not only in a combined stress effect, but may also disturb natural, high light protective mechanisms. The possible additive effect of UV-B radiation and photoinhibitory conditions from visible radiation will be discussed in the light of results obtained from measurement of different photosynthetic parameters in *Brassica napus*.

## 13.3 The effect of UV-B radiation on photoinhibition in *Brassica napus*

### 13.3.1 *Monitoring changes induced by UV-B radiation*

The effect of short-term, simultaneous exposure of plants to both UV-B radiation and high levels of PAR can be monitored in a number of ways. These include photosynthesis parameters, turnover of the D1 polypeptide of PSII, xanthophyll cycle turnover and estimations of the involvement of active oxygen species. Ultraweak luminescence (UL), which is very weak light emission from chemical reactions and which reflects oxidative damage (Abeles, 1986), has been used as an indicator of the ability of UV radiation to increase the production of electronically excited states, which probably involve oxygen species (Cen and Björn, 1994; Levall and Bornman, 1993; Panagopoulos *et al.*, 1989, 1990, 1992).

Plants of *B. napus* L. cv. Paroll were grown in a greenhouse chamber for 3 weeks with a photon fluence rate of 400 μmol $m^{-2}$ $s^{-1}$ PAR and a 12-h photoperiod. The addition of a small amount of UV-B (3.25 kJ $m^{-2}$ per day) was included for some of the experiments. Leaf discs were placed in Petri dishes containing either distilled water or 0.4% Tween 20, and allowed to equilibrate for 30 min under 400 μmol $m^{-2}$ $s^{-1}$. They were then exposed to either high (1600 μmol $m^{-2}$ $s^{-1}$) or low (400 μmol $m^{-2}$ $s^{-1}$) PAR (Osram projector lamps) with or without the addition of UV-B radiation (13 or 3.25 kJ $m^{-2}$ per day biologically effective radiation ($UV_{BE}$), weighted irradiance; Q-PANEL 313 UV fluorescent tubes, Largo AB, Göteborg, Sweden). Cellulose acetate removed radiation below 290 nm. Calculations for the $UV_{BE}$ radiation were based on a UV-dosage model by Björn and Murphy (1985) and the generalized plant action spectrum by Caldwell (1971), for 15 June on a cloudless day and at aerosol level 0.

Chlorophyll fluorescence measurements were carried out 0, 1, 2, 3 and 4 h after exposure to the radiation conditions and a dark adaptation period of 30 min, using a pulse-amplified modulation (PAM) fluorimeter (Walz, Effeltrich, Germany). The chlorophyll fluorescence measurements were used to determine the quantum yield (variable fluorescence/maximum fluorescence, *Fv*/*Fm*), photochemical ($q_P$) and non-photochemical quenching processes ($q_{NP}$) (Schreiber *et al.*, 1986). Turnover of D1 polypeptide in leaf discs was determined by pulse-labelling with [$^{35}$S]methionine (Sundby *et al.*, 1993). UL from the leaf discs was measured according to Panagopoulos *et al.* (1989) using a Peltier-cooled photomultiplier sensitive up to 850 nm (Hamamatsu R 928/0115/0381). Leaf discs were kept in darkness for 4 h prior to measurement in order to eliminate fluorescence from photosynthesis.

The results described below show that leaves were more photoinhibited under UV-B radiation and high PAR than under high PAR alone, and that degradation of D1 was greater. Prior exposure to UV-B radiation during growth appears to modify the response to photoinhibition, probably because of the induction of protective mechanisms. Preliminary evidence of this was seen in decreased UL in leaves of plants pre-grown with low levels of UV-B radiation. The amount of biologically effective UV-B radiation used during photoinhibitory treatment was not excessive in relation to possible future scenarios of an increased UV-B level, and yet there were clear indications of the influence of UV-B radiation in an intact system.

## 13.3.2 *Simulation of radiation conditions and response of photosynthesis*

Ozone levels vary with season, solar zenith angle, geographical location, cloudiness and aerosol content, and may change from day to day. Therefore plants may grow under cloudy conditions, yet suffer from sudden exposure to increased UV-B levels on certain days with clear skies. In this case, UV-B radiation may cause increased photoinhibitory damage to PSII. In simulation studies of the above scenario, indications of additional stress were shown by an increased turnover of the D1 polypeptide as well as by changes in chlorophyll fluorescence and UL. The induction of protective UV-screening pigments also appeared to modify plant response.

Chlorophyll fluorescence induction has been used widely to probe partial changes in the components of the photosynthetic apparatus, namely PSII. The redox state of one of the primary electron acceptors, $Q_A$, is reflected in the induction curves. The $q_P$ is related to the redox state of Q, whereas $q_{NP}$ is thought to be largely due to changes in the trans-thylakoid pH gradient or energy status of these membranes, and is correlated mainly with heat dissipation (Osmond, 1994). The $q_{NP}$ is associated with photoinhibition of PSII (Horton *et al.*, 1988; Krause and Behrend, 1986), and is considered to involve protective processes such as the xanthophyll cycle mentioned above, and/or dissipation of energy in the PSII centre itself due to inactivation of the latter (Öquist *et al.*, 1992). The fluorescence ratio *Fv/Fm* appears to be a reliable tool for monitoring photoinhibition, since linear relationships have been found between this ratio and the maximum quantum yield of oxygen evolution (Björkman and Demmig, 1987) as well as between the number of functional PSII reaction centres under photoinhibitory conditions (Öquist *et al.*, 1992).

From 10 to 50 s after the onset of fluorescence induction, evidence of photoinhibition was seen as an increase in $q_{NP}$ in leaves of *B. napus* exposed to high light (HL) of 1600 μmol $m^{-2}$ $s^{-1}$ as compared to low light (LL) conditions of 400 μmol $m^{-2}$ $s^{-1}$ (Figures 13.1 and 13.2, lower graph). Changes to $q_P$, on the other hand, can reflect photodamage (Osmond, 1994), as is suggested by the results of Figures 13.1 and 13.2 (upper graph), where a decrease in $q_P$ (~10–50 s after

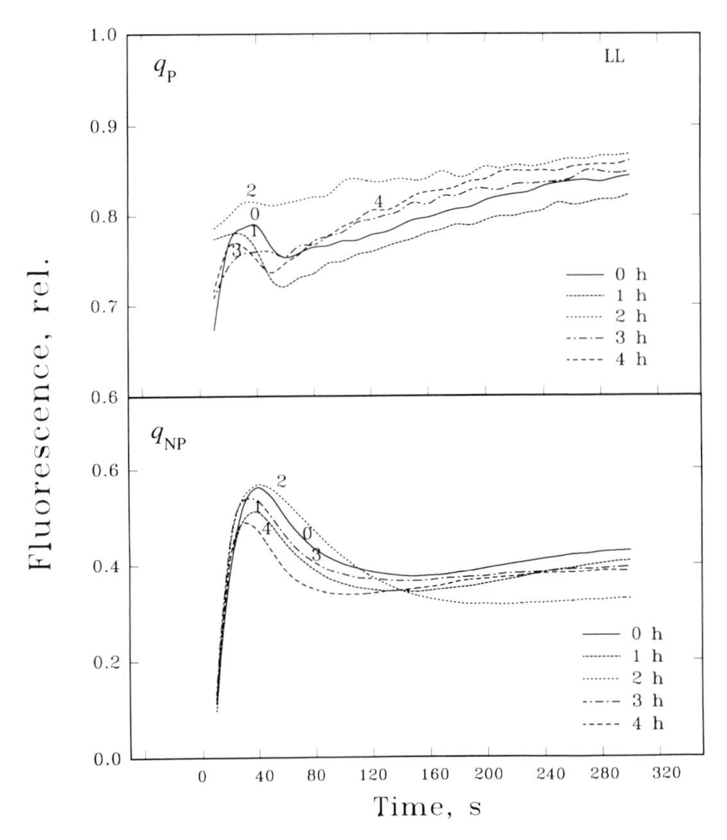


***Figure 13.1.*** *Monitoring* $q_P$ *and* $q_{NP}$ *of PSII in* B. napus *cv. Paroll with time. Leaf discs were exposed to LL conditions of 400 μmol* $m^{-2}$ $s^{-1}$ *from 0 to 4 h and sampled every hour. Excitation light for the induction curve was obtained from the LED of the PAM fluorimeter and pulses of saturating light (2000 μmol* $m^{-2}$ $s^{-1}$*) were superimposed every 10 s for 5 min.*

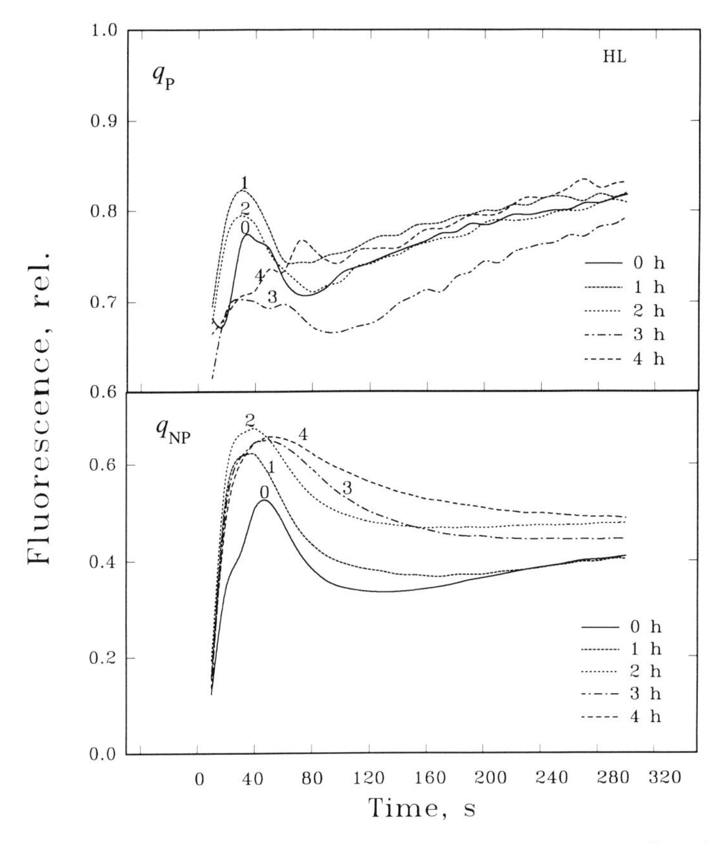


***Figure 13.2.*** *Monitoring* $q_P$ *and* $q_{NP}$ *of PSII in* B. napus *cv. Paroll with time. Leaf discs were exposed to HL conditions of 1600 μmol* $m^{-2}$ $s^{-1}$ *from 0 to 4 h and sampled every hour. Excitation light for the induction curve was obtained from the LED of the PAM fluorimeter and pulses of saturating light (2000 μmol* $m^{-2}$ $s^{-1}$*) were superimposed every 10 s for 5 min. Differences during this photoinhibitory period were most marked from 10 to 50 s after induction.*

induction) was noted after 3- and 4-h exposure to HL relative to the LL conditions (Figure 13.1). High levels of UV-B together with HL enhanced the photoinhibitory conditions of HL compared to LL conditions with and without the concomitant low levels of UV-B radiation (Figures 13.3 and 13.4). The $q_{NP}$ was markedly increased under the combination of HL and UV-B radiation compared to LL and the lower level of UV-B radiation. In contrast to the pattern seen in LL and HL conditions (Figures 13.1 and 13.2), UV-B radiation together with HL increased $q_P$ (Figures 13.3 and 13.4).

The initial fluorescence (*F*o) of dark-adapted leaves has been reported to be coupled to the dynamics of the D1 polypeptide (Bradbury and Baker, 1986; Ögren, 1991). A comparison of the non-photoinhibitory (LL) with photo-

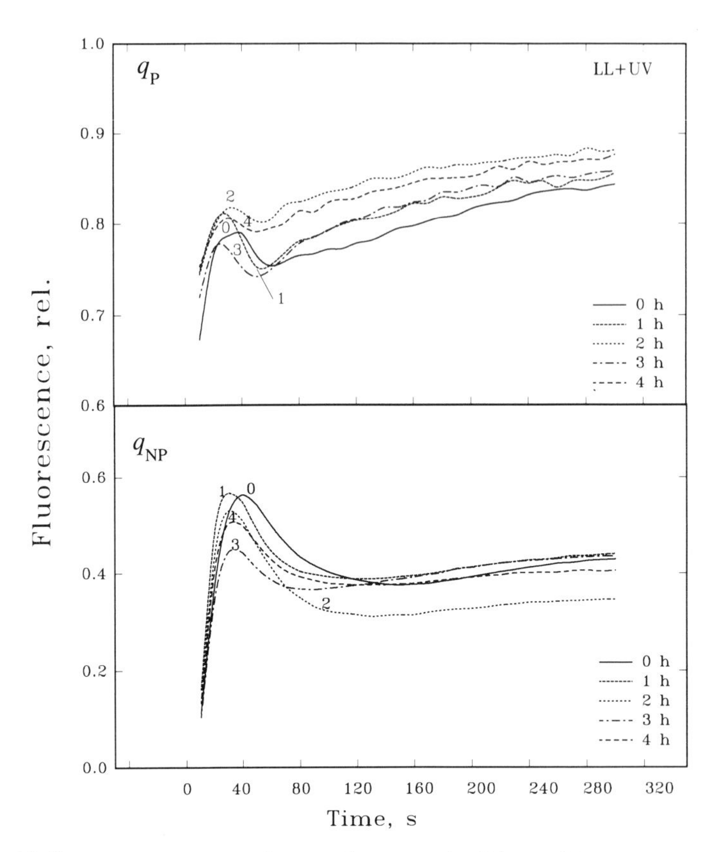


***Figure 13.3.*** *Monitoring* $q_P$ *and* $q_{NP}$ *of PSII in leaf discs of* B. napus *cv. Paroll exposed to LL conditions of 400 μmol* $m^{-2}$ $s^{-1}$ *and 3.25 kJ* $m^{-2}$ *per day of* $UV_{BE}$ *for 0–4 h. Sampling was done every hour. Excitation light for the induction curve was obtained from the LED of the PAM fluorimeter and pulses of saturating light (2000 μmol* $m^{-2}$ $s^{-1}$*) were superimposed every 10 s for 5 min.*

inhibitory conditions (HL, Figure 13.5) shows that *F*o increased under HL conditions, while the *F*m showed a decreasing trend with time. These results are indicative of photoinhibitory damage where protective mechanisms proved to be inadequate (Krause, 1988; Osmond, 1994). Under high PAR with UV-B radiation, the increase in *F*o was even more pronounced (Figure 13.6). The photosynthetic efficiency, exemplified by *F*v/*F*m, declined with HL, suggesting damage to PSII. According to Krause (1988), this decline may also signify the onset of protective mechanisms. Comparing Figure 13.5 with Figure 13.6 shows that the addition of UV-B radiation enhanced photoinhibition. Hardly any differences were found for leaves of plants exposed to LL or LL+UV radiation (Figures 13.5 and 13.6).

Effects of UV-B radiation on photoinhibition and D1 turnover have been studied mainly by exposure of plants or isolated thylakoid membranes to UV-B

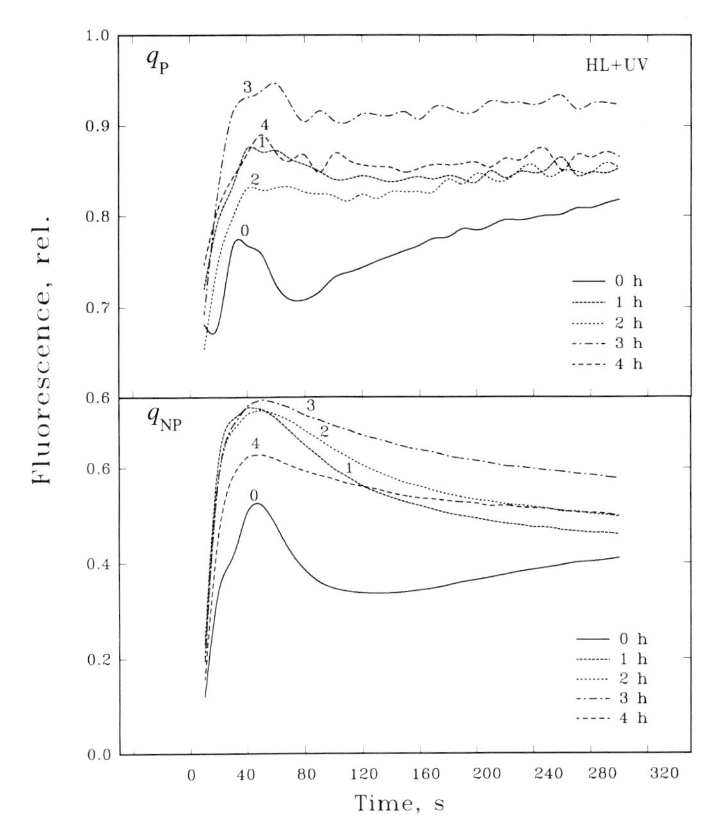


***Figure 13.4.*** *Monitoring* $q_P$ *and* $q_{NP}$ *of PSII in leaf discs of* B. napus *cv. Paroll exposed to HL conditions of 1600 µmol* $m^{-2}$ $s^{-1}$ *and 13 kJ* $m^{-2}$ *per day of* $UV_{BE}$ *for 0–4 h. Sampling was done every hour. Excitation light for the induction curve was obtained from the LED of the PAM fluorimeter and pulses of saturating light (2000 µmol* $m^{-2}$ $s^{-1}$*) were superimposed every 10 s for 5 min. Differences during this photoinhibitory period were most marked from 10 to 50 s after induction. Both* $q_P$ *and* $q_{NP}$ *increased under the combined radiation.*

irradiation alone or together with very low irradiances of visible light. Using more realistic levels of radiation, the data shown in Figure 13.6 indicate that UV-B radiation increases photoinhibition, as judged from the aggravation of the *Fv/Fm* decrease and the *Fo* increase, and the effects on quenching (Figure 13.4).

The rate of D1 degradation during photoinhibitory conditions was increased when UV-B radiation was present (Figure 13.7), as judged from chasing of the label in the D1 protein that was pre-labelled with [$^{35}$S]methionine (half-life ($t_{1/2}$) = 2.1 h at 20°C compared to $t_{1/2}$ = 3.7 h at 20°C without any UV-B irradiation). That such comparably small amounts of UV-B (compared to most other studies) gives an additive effect to the rate of D1 degradation is consistent with the observation by Greenberg *et al.* (1989a) that the rate of D1 degradation was

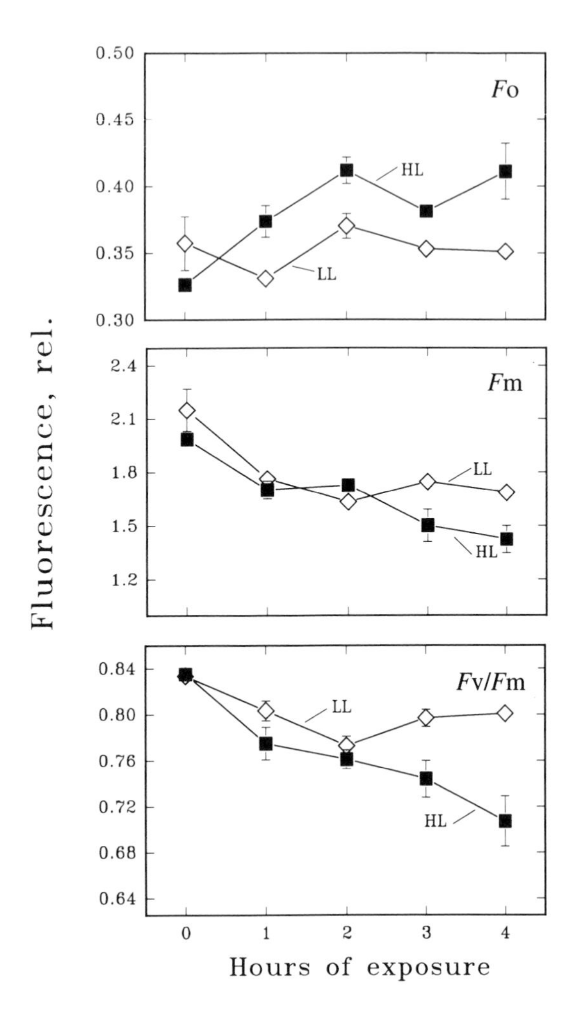


***Figure 13.5.*** *Chlorophyll fluorescence parameters* Fo, F*m* *and* F*v*/F*m* *of* B. napus *cv.* *Paroll leaf discs during 4-h exposure to LL conditions (400 μmol m*$^{-2}$ *s*$^{-1}$*) and a photoinhibitory HL treatment (1600 μmol m*$^{-2}$ *s*$^{-1}$*). Sampling was done every hour and the leaves were dark adapted for 30 min before measurement.*

decreased in *Spirodela oligorrhiza* if the UV-B present in natural sunlight at noon was filtered out.

The extent of D1 protein resynthesis after the photoinhibitory treatment with and without UV-B radiation gives different information concerning the D1 turnover. Photoinhibition caused by high levels of visible light is normally reversible and recovery occurs after a couple of hours. During this time, repair of PSII takes place with synthesis and insertion of new copies of the D1 protein (Prásil *et al.*, 1992). Therefore leaf discs were moved to low PAR and given [$^{35}$S]methionine after having been exposed to high levels of photoinhibitory light for 4 h (no [$^{35}$S]methionine present). Comparison of the amount of [$^{35}$S]methionine incorporated into D1 after 60 min in Figure 13.8 (upper graph) shows that less [$^{35}$S]methionine was incorporated into the leaf discs photoinhib-

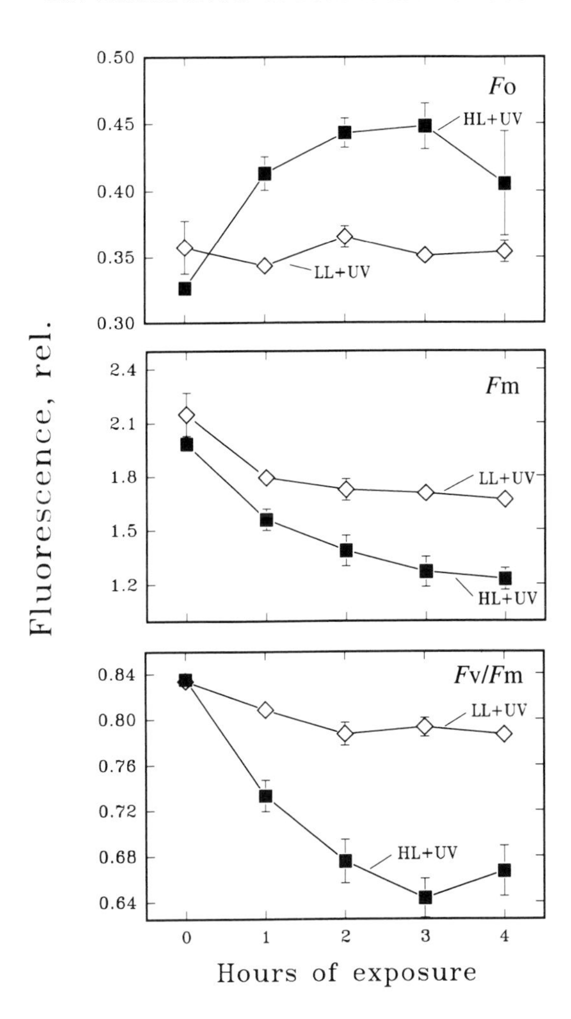


***Figure 13.6.*** *Chlorophyll fluorescence parameters* Fo, F*m* *and* F*v*/F*m* *of* B. napus *cv.* *Paroll leaf discs during 4-h exposure to LL conditions (400 µmol m*$^{-2}$ *s*$^{-1}$*) and a photoinhibitory HL treatment (1600 µmol m*$^{-2}$ *s*$^{-1}$*). During treatment conditions, the leaf discs were additionally exposed to either 3.25 or 13 kJ m*$^{-2}$ *per day of* $UV_{BE}$*. Sampling was done every hour and the leaves were dark adapted for 30 min before measurement.*

ited with HL plus UV-B radiation, than the leaf discs photoinhibited only with HL. Thus, also in this respect we can see an effect caused by a fairly realistic amount of UV-B radiation added to the photoinhibitory PAR. It is not clear at this stage how to interpret these data. One possibility is that the photoinhibitory treatment with UV-B radiation gives effects that are qualitatively different from those of 'normal' photoinhibition, and which have a negative effect on D1 repair. Thus even if the rate of D1 degradation is increased, the rate of D1 resynthesis might be hampered. This could be expected, for example, if the *psb*A transcript levels are reduced by UV-B radiation (Jordan *et al.*, 1991).

The fact that recovery from photoinhibition after 60 min lagged behind that of leaves exposed only to HL conditions during the photoinhibitory treatment was even more evident in leaves of plants pre-grown under low levels of UV-B

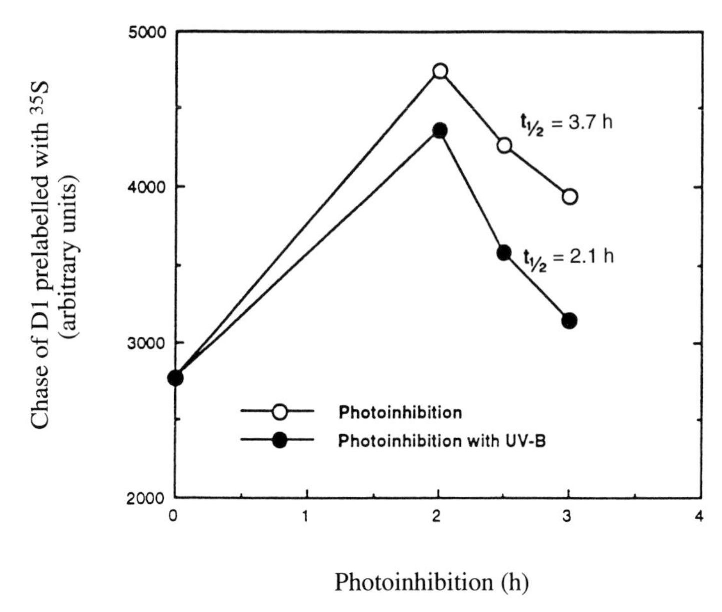


***Figure 13.7.*** *Turnover of the D1 polypeptide of* B. napus *cv. Paroll during photoinhibitory conditions (1600 µmol $m^{-2}$ $s^{-1}$) with or without the addition of UV-B radiation (13 kJ $m^{-2}$ per day of $UV_{BE}$). Leaf discs, previously labelled with [$^{35}$S]methionine, were floated on 0.4% Tween-20 and 10 mM methionine in Petri dishes and chased under 1600 µmol $m^{-2}$ $s^{-1}$ with or without UV-B radiation. Thylakoids were isolated at intervals during the chase from the leaf discs and run on SDS–PAGE as described previously (Sundby* et al., *1993). The gels were exposed for 24 h to an image plate, which was scanned on a Fuji BAS 2000 Phosphorimage, and the amount of label in the D1 polypeptide band evaluated. The values obtained from scanning of the phosphorimager screen were plotted against the chase time and fitted to an exponential curve. The rate constant* k *and the half-time* $t_{1/2}$*, which are related by* $t_{1/2} = \ln 2/k$*, were determined from a plot of ln (values from the scanning) against chase time by a straight line, where* k *represents the slope. The* $R^2$ *values for the data presented were above 0.98.*

radiation and then subjected to the same HL and enhanced UV-B regime (Figure 13.8, lower graph). It is not clear why the repair should be more impaired in the plants pre-grown with UV radiation. In this case, the lower amount of incorporation could be due to the fewer PSII centres whose D1 needed repairing as compared with plants grown without UV-B radiation.

## 13.3.3 *Reactive oxygen species*

Nucleic acids, polypeptides, carbohydrates and membrane lipids are examples of targets of free radicals and active oxygen species, the levels of which may increase upon conditions of stress, including exposure to UV-B radiation. (See Chapter

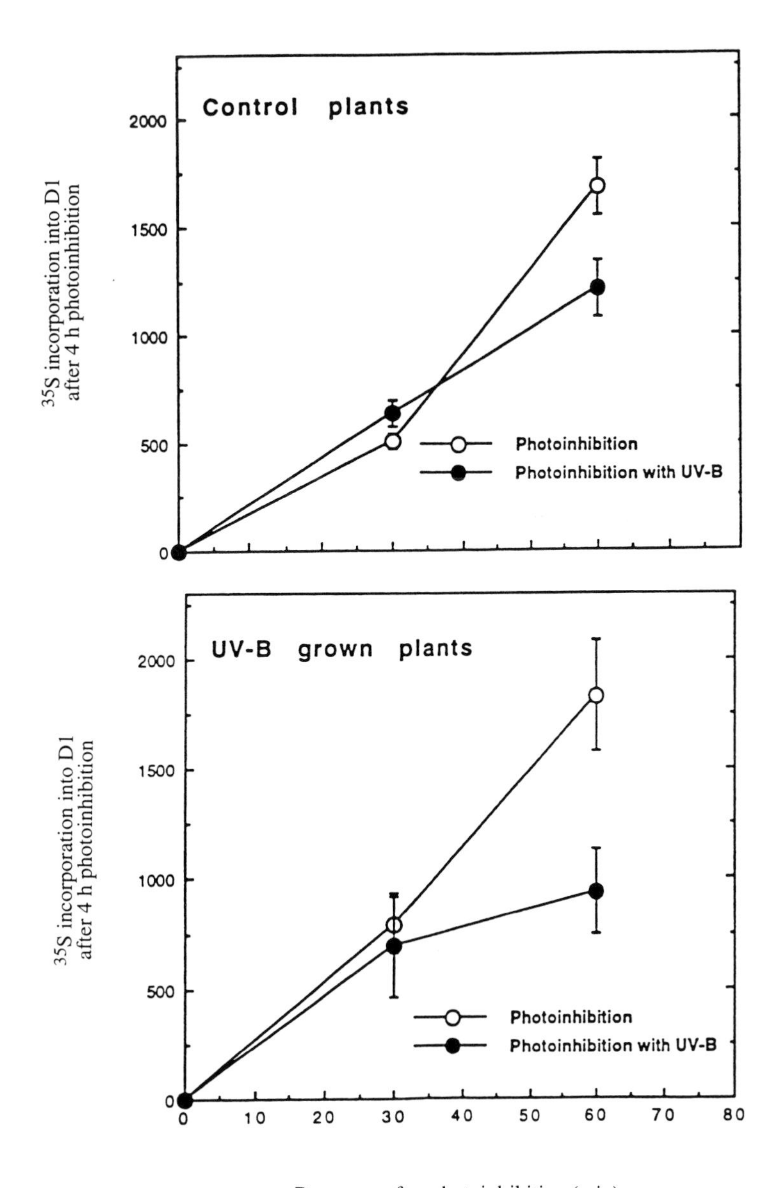


***Figure 13.8.*** *Recovery from photoinhibitory conditions and additional UV-B radiation (1600 μmol $m^{-2}$ $s^{-1}$ and 13 kJ $m^{-2}$ per day of $UV_{BE}$, respectively) in leaf discs of* B. napus *cv. Paroll. Leaf discs were exposed to LL and given [$^{35}$S]methionine after having been exposed to the photoinhibitory light for 4 h (no [$^{35}$S]methionine present). The amount of [$^{35}$S]methionine incorporated into D1 is compared (see legend to Figure 13.7 for method). Control plants are those grown previously under LL without any supplemental UV-B radiation, and UV-B grown plants are those grown previously under LL with the addition of UV-B radiation.*

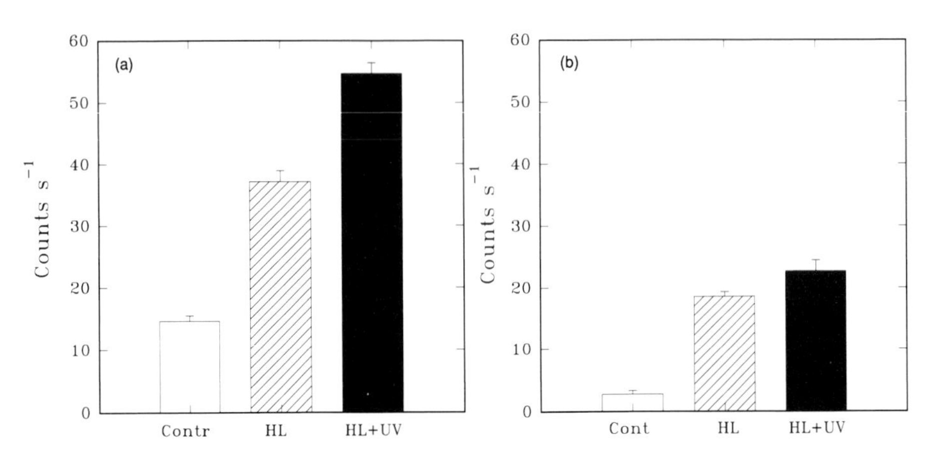


***Figure 13.9.*** *The effect of photoinhibitory light (HL) and UV-B on UL (counts $s^{-1}$) in leaf discs of* B. napus *cv. Paroll. Leaf discs were taken either from plants grown previously without UV-B radiation (a) or from plants grown under both visible and UV-B radiation (b). The leaf discs were exposed to HL and UV-B treatments for 4 h. UL was then measured after a 4-h dark period.*

12 for a discussion of free radical- and active oxygen-mediated damage.) It has been proposed that reactive oxygen intermediates are involved in photoinhibition of PSII, and more specifically in the breakdown of the D1 polypeptide (Ohad *et al.*, 1984; Schuster *et al.*, 1988). The results of Tschiersch and Ohmann (1993) indicate that singlet oxygen may not be a contributing oxygen species, whereas hydrogen peroxide, hydroxyl radicals and superoxide radicals appear to be involved in PSII photoinhibition. However, other research groups have reported that singlet oxygen is indeed involved (Jung and Kim, 1990).

By measuring the UL of certain chemical reactions, an estimate of the electronically excited states can be obtained (Abeles, 1986). Oxygen is the main motive force of this light emission from oxidation reactions (Abeles, 1986). Increases in UL have been documented after exposure of plants to UV-B radiation and fungal attack (Cen and Björn, 1994; Levall and Bornman, 1993; Panagopoulos *et al.*, 1989, 1990, 1992). Preliminary results of the present report point to an increase in UL under high PAR relative to leaves exposed to low PAR. A further increase over that found under high PAR occurred when leaf discs were exposed to supplemental UV-B radiation (Figure 13.9a). When a small amount of UV-B radiation was included during growth, the UL decreased under both high PAR and with the addition of UV radiation during treatment conditions. There was also less of an increase over high PAR when leaf discs were exposed to additional UV-B radiation during growth (Figure 13.9b), suggesting that protective mechanisms had been induced under the growth conditions with low UV-B radiation. This effect from prior exposure to UV was also seen in a lowered *F*o under photoinhibitory conditions as compared to the *F*o levels in leaves of plants not pre-grown under UV-B radiation (data not shown).

## 13.4 Concluding remarks

Clearly, UV-B radiation has an additive effect on photoinhibition caused by PAR. The mechanism for this additive effect is not clear, but it seems to be qualitatively and quantitatively different from photoinhibition induced by visible radiation, as discussed above. The influence of an increase in the UV-B portion of the solar spectrum due to a reduced ozone layer may not only manifest itself as an additional environmental factor to contend with but, through interactive events not yet fully understood, UV radiation may impair the plant's present ability to cope with the environment to which it has been adapted. Thus the use of different techniques for monitoring changes in plant response are important for furthering our understanding of a changing climate.

## Acknowledgements

We thank Ms Lena Lundh and Ms Ingun Sundén for skilled technical assistance. Financial support for the authors was provided by the Swedish Natural Science Research Council.

## References

Abeles, F.B. (1986) Plant chemiluminescence. *Annu. Rev. Plant Physiol.* **37**, 49–72.

Björkman, O. and Demmig, B. (1987) Photon yield of $O_2$ evolution and chlorophyll fluorescence characteristics at 77 K among vascular plants of diverse origin. *Planta* **170**, 489–504.

Björn, L.O. and Murphy, T.M. (1985) Computer calculation of solar ultraviolet radiation at ground level. *Physiol. Vég.* **23**, 555–561.

Bornman, J.F. (1989) Target sites of UV-B radiation in photosynthesis of higher plants. *J. Photochem. Photobiol.* **4**, 145–158.

Bornman, J.F. and Teramura, A.H. (1993) Effects of UV-B radiation on terrestrial plants. In: *Environmental UV Photobiology* (eds A.R. Young, L.O. Björn, J. Moan and W. Nultsch). Plenum Publ. Co., New York, pp. 427–471.

Bornman, J.F. and Vogelmann, T.C. (1988) Penetration of blue and UV radiation measured by fiber optics in spruce and fir needles. *Physiol. Plant.* **72**, 699–705.

Bradbury, M. and Baker, N.R. (1986) The kinetics of photoinhibition of the photosynthetic apparatus in pea chloroplasts. *Plant, Cell Environ.* **9**, 289–297.

Caldwell, M.M. (1971) Solar UV irradiation and the growth and development of higher plants. In: *Photophysiology,* Volume 6 (ed. A.C. Giese). Academic Press, New York, pp. 131–177.

Cen, Y.-P. and Björn, L.O. (1994) Action spectra for enhancement of ultraweak luminescence by ultraviolet radiation (270–340 nm) in leaves of *Brassica napus*. *J. Photochem. Photobiol.* **22**, 125–129.

Cen, Y.-P. and Bornman, J.F. (1993) The effect of exposure to enhanced UV-B radiation on the penetration of monochromatic and polychromatic UV-B radiation in leaves of *Brassica napus*. *Physiol. Plant.* **87**, 249–255.

Day, T.A., Vogelmann, T.C. and DeLucia, E.H. (1992) Are some plant life forms more effective than others in screening out ultraviolet-B radiation? *Oecologia* **92**, 513–519.

DeLucia, E.H., Day, T.A. and Vogelmann, T.C. (1992) Ultraviolet-B and visible light penetration into needles of two species of subalpine conifers during foliar development. *Plant, Cell Environ.* **15**, 921–929.

Demmig-Adams, B. and Adams, W.W. (1992) Photoprotection and other responses of plants to high light stress. *Annu. Rev. Plant Physiol. Plant Mol. Biol.* **43**, 599–626.

Ensminger, P.A. and Schäfer, E. (1992). Blue and ultraviolet-B light photoreceptors in parsley cells. *Photochem. Photobiol.* **55**, 437–447.

Frederick, J.E. (1993) Ultraviolet sunlight reaching the earth's surface: A review of recent research. *Photochem. Photobiol.* **57**, 175–178.

Friso, G., Spetea, C., Giacometti, G.M., Vass, I. and Barbato, R. (1994) Degradation of photosystem II reaction center D1-polypeptide induced by UVB radiation in isolated thylakoids. Identification and characterization of C- and N-terminal breakdown products. *Biochim. Biophys. Acta* **1184**, 78–84.

Greenberg, B.M., Gaba, V., Canaani, O., Malkin, S., Mattoo, A.K. and Edelman, M. (1989a) Separate photosensitizers mediate degradation of the 32-kDa photosystem II reaction centre polypeptide in the visible and UV spectral regions. *Proc. Natl Acad. Sci. USA* **86**, 6617–6620.

Greenberg, B.M., Gaba, V., Mattoo, A.K. and Edelman, M. (1989b) Degradation of the 32 kDa photosystem II reaction center polypeptide in UV, visible and far red light occurs through a common 23.5 kDa intermediate. *Z. Naturforsch.* **44**, 450–452.

Greer, D.H., Berry, J.A. and Björkman, O. (1986) Photoinhibition of photosynthesis in intact bean (*Phaseolus vulgaris*) leaves. Role of light and temperature and requirement for chloroplast-protein synthesis during recovery. *Planta* **168**, 253–260.

Hahlbrock, K. and Grisebach, H. (1979) Enzymic controls in the biosynthesis of lignin and flavonoids. *Annu. Rev. Plant Physiol.* **30**, 105–130.

Hahlbrock, K. and Scheel, D. (1989) Physiology and molecular biology of phenylpropanoid metabolism. *Annu. Rev. Plant Physiol. Plant Mol. Biol.* **40**, 347–369.

Horton, P., Oxborough, K., Rees, D. and Scholes, J.D. (1988) Regulation of the photochemical efficiency of photosystem II; consequences for the light response of field photosynthesis. *Plant Physiol. Biochem.* **26**, 453–460.

Jordan, B.R. (1993) The molecular biology of plants exposed to ultraviolet-B radiation and the interaction with other stresses. In: *Interacting Stresses on Plants in a Changing Climate* (eds M.B. Jackson and C.R. Black). NATO ASI Series, Volume 16, Springer-Verlag, Berlin, pp. 153–170.

Jordan, B.R., Chow, W.S., Strid, Å. and Anderson, J.M. (1991) Reduction in *cab* and *psb* A RNA transcripts in response to supplementary ultraviolet-B radiation. *FEBS Lett.* **284**, 5–8.

Jordan, B.R., He, J., Chow, W.S. and Anderson, J.M. (1992) Changes in mRNA levels and polypeptide subunits of ribulose 1,5-bisphosphate carboxylase in response to supplementary ultraviolet-B radiation. *Plant, Cell Environ.* **15**, 91–98.

Jung, J. and Kim, H.S. (1990) The chromophores as endogenous sensitizers involved in the photogeneration of singlet oxygen in spinach thylakoids. *Photochem. Photobiol.* **52**, 1003–1009.

Kerr, J.B. and McElroy, C.T. (1993) Evidence for large upward trends of ultraviolet-B radiation linked to ozone depletion. *Science* **262**, 1032–1034.

Krause, G.H. (1988) Photoinhibition of photosynthesis. An evaluation of damaging and protective mechanisms. *Physiol. Plant.* **74**, 566–574.

Krause, G.H. and Behrend, U. (1986) Delta-pH dependent chlorophyll fluorescence quenching indicating a mechanism of protection against photoinhibition of chloroplasts. *FEBS Lett.* **200**, 298–302.

Levall, M.W. and Bornman, J.F. (1993) Selection *in vitro* for UV-tolerant sugar beet (*Beta vulgaris*) somaclones. *Physiol. Plant.* **88**, 37–43.

Ögren, E. (1991) Prediction of photoinhibition of photosynthesis from measurements of fluorescence quenching components. *Planta* **184**, 538–544.

Ohad, I., Kyle, D.J. and Arntzen, C.J. (1984) Membrane polypeptide damage and repair: removal and replacement of inactivated 32-kilodalton polypeptide in chloroplast membranes. *J. Cell Biol.* **99**, 481–485.

Öquist, G., Chow, W.S. and Anderson, J.M. (1992) Photoinhibition of photosynthesis represents a mechanism for the long-term regulation of photosystem II. *Planta* **186**, 450–460.

Osmond, C.B. (1994) What is photoinhibition? Some insights from comparisons of shade and sun plants. In: *Photoinhibition of Photosynthesis: From Molecular Mechanisms to the Field* (eds N.R. Baker and J.R. Bowyer). BIOS Scientific Publishers Ltd., Oxford, pp. 1–24.

Panagopoulos, I., Bornman, J.F. and Björn, L.O. (1989) The effect of UV-B and UV-C radiation on Hibiscus leaves determined by ultraweak luminescence and fluorescence induction. *Physiol. Plant.* **76**, 461–465.

Panagopoulos, I., Bornman, J.F. and Björn, L.O. (1990) Effects of ultraviolet radiation and visible light on growth, fluorescence induction, ultraweak luminescence and peroxidase activity in sugar beet plants. *J. Photobiochem. Photobiol.* **8**, 73–87.

Panagopoulos, I., Bornman, J.F. and Björn, L.O. (1992) Response of sugar beet plants to ultraviolet-B (280–320 nm) radiation and *Cercospora* leaf spot disease. *Physiol. Plant.* **84**, 140–145.

Pfündel, E., Pan, R.-S. and Dilley, R.A. (1992) Inhibition of violaxanthin deepoxidation by ultraviolet-B radiation in isolated chloroplasts and intact leaves. *Plant Physiol.* **98**, 1372–1380.

Powles, S.B. (1984) Photoinhibition of photosynthesis induced by visible light. *Annu. Rev. Plant Physiol.* **35**, 15–44.

Prásil, O., Adir, N. and Ohad, I. (1992) Dynamics of photosystem II: mechanism of photoinhibition and recovery processes. In: *The Photosystems: Structure, Function and Molecular Biology* (ed. J. Barber). Elsevier Science Publishers, New York, pp. 295–348.

Prinn, R.G. (1994) The interactive atmosphere: Global atmospheric–biospheric chemistry. *Ambio* **23**, 50–61.

Quaite, F.E., Sutherland, B.M. and Sutherland, J.C. (1992) Quantitation of pyrimidine dimers in DNA from UV-B irradiated alfalfa (*Medicago sativa* L.) seedlings. *Appl. Theor. Electrophor.* **2**, 171–176.

Ros, J. (1990) On the effect of UV-radiation on elongation growth of sunflower seedlings (*Helianthus annuus* L.). *Resc. Karlsruher Beitr. Entw. Ökophysiol.* **8**, 1–157.

Schreiber, U., Schliwa, U. and Bilger, W. (1986) Continuous recording of photochemical and non-photochemical chlorophyll fluorescence quenching with a new type of modulation fluorometer. *Photosynth. Res.* **10**, 51–62.

Schulze-Lefert, P., Dangl, J.L., Becker-André, M., Hahlbrock, K. and Schulz, W. (1989) Inducible *in vivo* DNA footprints define sequences necessary for UV light activation of the parsley chalcone synthase gene. *EMBO J.* **8**, 651–656.

Schuster, G., Timberg, R. and Ohad, I. (1988) Turnover of thylakoid photosystem II proteins during photoinhibition of *Chlamydomonas reinhardtii*. *Eur. J. Biochem.* **177**, 403–410.

Staxén, I. and Bornman, J.F. (1994) A morphological and cytological study of *Petunia hybrida* exposed to UV-B radiation. *Physiol. Plant.* **91**, 735–740.

Staxén, I., Bergounioux, C. and Bornman, J.F. (1993) Effect of ultraviolet radiation on cell division and microtubule organisation in *Petunia hybrida* protoplasts. *Protoplasma* **173**, 70–76.

Sundby, C., McCaffery, S. and Anderson, J.M. (1993) Turnover of the photosystem 2 D1 protein in higher plants under photoinhibitory and non-photoinhibitory irradiance. *J. Biol. Chem.* **268**, 25476–25482.

Tezuka, T., Hotta, T. and Watanabe, I. (1993) Growth promotion of tomato and radish plants by solar UV radiation reaching the earth's surface. *J. Photochem. Photobiol.* **19**, 61–66.

Tschiersch, H. and Ohmann, E. (1993) Photoinhibition in *Euglena gracilis*: Involvement of reactive oxygen species. *Planta* **191**, 316–323.

Weis, E. and Berry, J.A. (1987) Quantum efficiency of photosystem II in relation to "energy"-dependent quenching of chlorophyll fluorescence. *Biochim. Biophys. Acta* **894**, 198–208.

Wilson, M.I. and Greenberg, B. M. (1993) Protection of the D1 photosystem II reaction center protein from degradation in ultraviolet radiation following adaptation of *Brassica napus* L. to growth in ultraviolet-B. *Photochem. Photobiol.* **57**, 556–563.

# Index

ALSO AVAILABLE FROM BIOS SCIENTIFIC PUBLISHERS LTD

# Photoinhibition of Photosynthesis

## From molecular mechanisms to the field

**N.R. Baker & J.R. Bowyer (Eds)**
respectively University of Essex, UK; and Royal Holloway and Bedford New College, London, UK

A comprehensive treatise on photoinhibition which provides an authoritative, up-to-date review of the important molecular, environmental and physiological issues.

### Contents

What is photoinhibition? Some insights from comparisons of shade and sun plants, *C.B.Osmond*; Elucidating the molecular mechanisms of photoinhibition by studying isolated photosystem II reaction centres, *A.Telfer & J.Barber*; Light-induced reactions impairing electron transfer through photosystem II, *S.Styring & C.Jegerschöld*; Photoprotection in photosystem II - the role of cytochrome b559, *J.Whitmarsh et al*; Excitation energy transfer between chlorophylls and carotenoids. A proposed molecular mechanism for non-photochemical quenching, *T.G.Owens*; The role of light-harvesting compex II in energy quenching, *P.Horton & A.Ruban*; Mechanisms for scavenging reactive molecules generated in chloroplasts under light stress, *K.Asada*; Light-induced proteolysis of photosystem II reaction centre and light-harvesting complex II proteins in isolated preparations, *B.Andersson*; Light-induced degradation of the photosystem II reaction centre D1 protein *in vivo*: an integrative approach, *I.Ohad et al*; Regulation of the synthesis of D1 and D2 proteins of photosystem II, *J.Nickelsen & J.-D.Rochaix*; D1 protein turnover: response to photodamage or regulatory mechanism? *C.Critchley*; The role of early light-induced proteins (ELIPs) during light stress, *I.Adamska & K.Kloppstech*; Photoinhibition and the light environment within leaves, *J.N.Nishio et al*; Factors determining the nature of the light dosage response curve of leaves, *J.W.Leverenz*; Photosynthetic response to sunflecks and light gaps: mechanisms and constraints, *R.W.Pearcy*; The responses of thylakoid electron transport and light utilization efficiency to sink limitation of photosynthesis, *J.Harbinson*; Drought stress and high light effects on leaf photosynthesis, *G.Cornic*; Depressions of photosynthesis in crops with water deficits, *D.R.Ort et al*; Photoinhibition induced by low temperatures, *G.H.Krause*; Photoinhibition of crop photosynthesis in the field at low temperatures, *N.R.Baker et al*; The role of photoinhibition during tree seedling establishment at low temperatures, *M.C.Ball*; Depressions of photosynthesis in mangrove canopies, *J.M.Cheeseman*; Light regimes and energy balance in plant canopies, *V.P.Gutschick*; Light utilization and photoinhibition of photosynthesis in marine phytoplankton, P.G.Falkowski et al.; The signifance of photoinhibition for photosynthetic productivity, *E.Ögren*; The cost of photoinhibition to plant communities, *J.A.Raven*.

Hardback; 496 pages; 1-872748-03-1; 1994

# ORDERING DETAILS

**Main address for orders**

**BIOS Scientific Publishers Ltd**
**St Thomas House, Becket Street,**
**Oxford OX1 1SJ, UK**
**Tel: +44 1865 726286**
**Fax: +44 1865 246823**

**Australia and New Zealand**
DA Information Services
648 Whitehorse Road, Mitcham, Victoria 3132, Australia
Tel: (03) 873 4411
Fax: (03) 873 5679

**India**
Viva Books Private Ltd
4346/4C Ansari Road, New Delhi 110 002, India
Tel: 11 3283121
Fax: 11 3267224

**Singapore and South East Asia**
(Brunei, Hong Kong, Indonesia, Korea, Malaysia, the Philippines, Singapore, Taiwan, and Thailand)
Toppan Company (S) PTE Ltd
38 Liu Fang Road, Jurong, Singapore 2262
Tel: (265) 6666
Fax: (261) 7875

**USA and Canada**
Books International Inc
PO Box 605, Herndon, VA 22070, USA
Tel: (703) 435 7064
Fax: (703) 689 0660

Payment can be made by cheque or credit card (Visa/Mastercard, quoting number and expiry date). Alternatively, a *pro forma* invoice can be sent.

Prepaid orders must include £2.50/US$5.00 to cover postage and packing for one item and £1.25/US$2.50 for each additional item.